Knowledge-Based Bioinformatics

Knowledge-Based Bioinformatics

From Analysis to Interpretation

Edited by

Gil Alterovitz • Marco Ramoni

*Harvard Medical School and Massachusetts Institute of Technology,
Boston, USA*

A John Wiley and Sons, Ltd., Publication

Library of Congress Cataloging-in-Publication Data

Knowledge based bioinformatics : from analysis to interpretation / edited by Gil Alterovitz, Marco Ramoni.
       p. ; cm.
   Includes bibliographical references and index.
   ISBN 978-0-470-74831-2 (cloth)
   1. Bioinformatics. 2. Expert systems (Computer science) I. Alterovitz, Gil. II. Ramoni, Marco F.
   [DNLM: 1. Computational Biology. 2. Expert Systems. 3. Medical Informatics.
   4. Molecular Biology. QU 26.5 K725 2010]
   QH324.25.K66 2010
   572.80285 – dc22

                                                                                    2010010927

A catalogue record for this book is available from the British Library.

ISBN: 978-0-470-74831-2

Typeset in 10/12pt Times-Roman by Laserwords Private Limited, Chennai, India

Contents

Preface

The information generated by progressive biomedical research is increasing rapidly, resulting in a tremendous increase in the biological data resource, including protein and gene databases, model organism databases, annotation databases, biomolecular interaction databases, microarray data, scientific literature data, and much more. The challenge is in representation, integration, analysis and interpretation of the available knowledge and data. The book, *Knowledge-Based Bioinformatics: From Analysis to Interpretation*, is an endeavor to address the above challenges. The driving force is the need for more background information and broader coverage of recent developments in the field of knowledge-based systems and data-analysis approaches, and their applications to deal with issues that arise from the current increase of biological data in genomic and proteomic research. Also, opportunity exists in utilizing these vast amounts of valuable information for benefit in fitness and disease conditions.

Knowledge-Based Bioinformatics: From Analysis to Interpretation, introduces knowledge-driven approaches, methods, and implementation techniques for bioinformatics. The book includes coverage from data-driven Bayesian networks to ontology-based analysis with applications in the field of bioinformatics. It is divided into four sections. The first section provides an overview of knowledge-driven approaches. Chapter 1, *Knowledge-based bioinformatics*, presents the current status of biomedical research and significance of knowledge-driven approaches in analyzing the data generated. The focus is on current utilization of the approaches and further enhancement required for advancing the biomedical knowledge. Chapter 2, *Knowledge-driven approaches to genome-scale analysis*, further explains the concept and covers various systems used for supporting biomedical discovery in genome-scale data. It emphasizes the importance of the knowledge-driven approaches for utilizing the existing knowledge, and challenges to overcome in their development and application. Chapter 3, *Technologies and best practices for building bio-ontologies*, reviews the process of building bio-ontologies, analyzing the benefits and problems of modeling biological knowledge axiomatically, especially with regards to automated reasoning. It also focuses on various knowledge representation languages, tools and community-level best practices to help the reader to make informed decisions when building bio-ontologies. In Chapter 4, *Design, implementation and updating of knowledge bases*, the focus is on architecture of knowledge bases. It describes various bioinformatics knowledge bases and the approach taken to meet the challenges of acquisition, maintenance, and interpretation of large amounts of data, and the methodology to efficiently mine the data.

In the second section, the focus shifts from knowledge-driven approaches to data-analysis approaches. Chapter 5, *Classical statistical learning in bioinformatics*, reviews various statistical methods and recent advances in analysis and interpretation of the data. Also in this chapter, classical concerns with multiple testing with focus on the empirical Bayes method, practical issues to be considered in treatments for genomics, various investigative analysis procedures, and traditional and modern classification procedures are reviewed. Chapter 6, *Bayesian methods in genomics and proteomics studies*, provides further insight into the Bayesian methods. The chapter focuses on concepts in Bayesian methods, computational methods for statistical inference of Bayesian models, and their applications in genomics and proteomics. Chapter 7, *Automatic text analysis for bioinformatics knowledge discovery*, introduces the basic concepts and current methodologies applied in biomedical text mining. The chapter provides an outlook on recent advances in automatic literature analysis and the contribution to knowledge discovery in the biomedical domain as well as integration of bioinformatics knowledge bases and the results from automatic literature analysis.

The third section covers gene and protein information. Chapter 8, *Fundamentals of gene ontology functional annotation*, reviews the current approach to functional annotation with emphasis on Gene Ontology annotation. Also, the chapter reviews currently available mainstream GO browsers and methods to access GO annotations from some of the more specialized GO browsers, as well as the effect of functional gene annotation on biological data analysis. Chapter 9, *Methods for improving genome annotation*, focuses on recent progress in automated and manual annotations and their application to produce the human consensus coding sequence gene set, and also describes various types of non-coding loci found within the human genome. Chapter 10, *Sequences from prokaryotic, eukaryotic, and viral genomes available clustered according to phylotype on a Self-Organizing Map*, demonstrates a novel bioinformatics tool for large-scale comprehensive studies of phylotype-specific sequence characteristics for a wide range of genomes. The chapter discusses this interesting method of genome analysis that could provide a new systematic strategy for revealing microbial diversity, relative abundance of different phylotype members of uncultured microorganisms, and unveil the genome signatures.

In the fourth and last section, the book moves to biomolecular relationships and meta-relationships. Chapter 11, *Molecular network analysis and applications*, provides an overview of current methods for analyzing large-scale biomolecular networks and major applications on biological problems using these network approaches. Also, this chapter addresses the current and next-generation network visualization and analysis tools and future challenges in analyzing the biomolecular networks. Chapter 12, *Biological pathway analysis: an overview of Reactome and other integrative pathway knowledge bases*, provides further insight into the use of pathway analysis tools to identify relevant biological pathways within large and complex data sets derived from various high-throughput technology platforms. The focus of the review is on the Reactome database and several closely related pathway knowledge bases. Chapter 13, *Methods and challenges of identifying biomolecular relationships and networks associated with complex diseases/phenotypes, and their application to drug treatments*, explores various interesting methods to infer regulatory biomolecular interactions as well as meta-relationships and molecular relationships in complex disorders and drug treatments. The chapter addresses the challenges involved in the mapping of disease symptoms, identifying novel drug targets, and tailoring patient treatments.

The book, *Knowledge-Based Bioinformatics: From Analysis to Interpretation*, is the outcome of an international effort, including contributors from 19 institutions located in 7 countries. It brings into light the pioneering research and cutting-edge technologies developed and used by leading experts, and their combined efforts to deal with large volumes of data and derive functional knowledge to enhance biomedical research. The extensive coverage of topics from fundamental methods to application make it a vital reference for researchers and industry professionals, and an essential text for upper level undergraduate/first year graduate students studying the subject.

For the publication of this book, the contribution of many people from this cross-disciplinary field of bioinformatics has been significant. The editors would like to thank the contributing authors including: Eric Karl Neumann (Ch. 1), Hannah Tipney (Ch. 2), Lawrence Hunter (Ch. 2), Mikel Egaña Aranguren (Ch. 3), Robert Stevens (Ch. 3), Erick Antezana (Ch. 3), Jesualdo Tomás Fernández-Breis (Ch. 3), Martin Kuiper (Ch. 3), Vladimir Mironov (Ch. 3), Sarah Hunter (Ch. 4), Rolf Apweiler (Ch. 4), Maria Jesus Martin (Ch. 4), Mark Reimers (Ch. 5), Ning Sun (Ch. 6), Hongyu Zhao (Ch. 6), Dietrich Rebholz-Schuhmann (Ch. 7), Jungjae Kim (Ch. 7), Varsha K. Khodiyar (Ch. 8), Emily C. Dimmer (Ch. 8), Rachael P. Huntley (Ch. 8), Ruth C. Lovering (Ch. 8), Jonathan Mudge (Ch. 9), Jennifer Harrow (Ch. 9), Takashi Abe (Ch. 10), Shigehiko Kanaya (Ch. 10), Toshimichi Ikemura (Ch. 10), Minlu Zhang (Ch. 11), Jingyuan Deng (Ch. 11), Chunsheng V. Fang (Ch. 11), Xiao Zhang (Ch. 11), Long Jason Lu (Ch. 11), Robin A. Haw (Ch. 12), Marc E. Gillespie (Ch. 12), Michael A. Caudy (Ch. 12) and Mie Rizig (Ch. 13). The editors would also like to thank the book proposal and book draft anonymous reviewers. The editors would like to thank all the people who helped in reviewing the manuscript. The editors would like to acknowledge and thank Alpa Bajpai for her important role in editing this book.

Gil Alterovitz, Ph.D.
Marco Ramoni, Ph.D.

List of Contributors

Takashi Abe
Nagahama Institute of Bio-science
and Technology, Japan
takaabe@nagahama-i-bio.ac.jp

Erick Antezana
Norwegian University of Science
and Technology, Norway
erick.antezana@gmail.com

Rolf Apweiler
European Bioinformatics Institute,
Cambridge, UK
apweiler@ebi.ac.uk

Mikel Egaña Aranguren
University of Murcia, Spain
mikel.egana.aranguren@gmail.com

Michael A. Caudy
Gnomics Web Services
New York, USA
mcaudy@gmail.com

Jingyuan Deng
Division of Biomedical Informatics
Cincinnati Children's Hospital
Medical Center, USA
dengjn@mail.uc.edu

Emily C. Dimmer
European Bioinformatics Institute
Cambridge, UK
edimmer@ebi.ac.uk

Chunsheng V. Fang
Division of Biomedical Informatics
Cincinnati Children's Hospital
Medical Center, USA
fangcg@mail.uc.edu

Jesualdo Tomás Fernández-Breis
University of Murcia, Spain
jfernand@um.es

Marc E. Gillespie
College of Pharmacy and Allied
Health Professions
St. John's University, New York,
USA
gillespm@gmail.com

Jennifer Harrow
Wellcome Trust Sanger Institute
Cambridge, UK
jla1@sanger.ac.uk

Robin A. Haw
Department of Informatics and
Bio-computing, Ontario Institute for
Cancer Research, Canada
robinhaw@gmail.com

Lawrence Hunter
University of Colorado Denver
School of Medicine, USA
Larry.Hunter@ucdenver.edu

Sarah Hunter
European Bioinformatics Institute
Cambridge, UK
hunter@ebi.ac.uk

Rachael P. Huntley
European Bioinformatics Institute
Cambridge, UK
huntley@ebi.ac.uk

Toshimichi Ikemura
Nagahama Institute of Bio-science
and Technology, Japan
t_ikemura@nagahama-i-bio.ac.jp

Shigehiko Kanaya
Department of Bioinformatics and
Genomes, Nara Institute of Science
and Technology, Japan
skanaya@gtc.naist.jp

Varsha K. Khodiyar
Centre for Cardiovascular Genetics,
University College London, UK
v.khodiyar@ucl.ac.uk

Jung-jae Kim
School of Computer Engineering
Nanyang Technological University
Singapore
jungjae.kim@ntu.edu.sg

Martin Kuiper
Norwegian University of Science
and Technology, Norway
martin.kuiper@bio.ntnu.no

Ruth C. Lovering
Centre for Cardiovascular Genetics
University College London, UK
r.lovering@ucl.ac.uk

Long Jason Lu
Division of Biomedical Informatics,
Cincinnati Children's Hospital
Medical Center, USA
long.lu@cchmc.org

Maria Jesus Martin
European Bioinformatics Institute
Cambridge, UK
martin@ebi.ac.uk

Vladimir Mironov
Norwegian University of Science
and Technology, Norway
mironov@bio.ntnu.no

Jonathan Mudge
Wellcome Trust Sanger Institute
Cambridge, UK
jm12@sanger.ac.uk

Eric Karl Neumann
Clinical Semantics Group
Lexington, MA, USA
ekneumann@gmail.com

Dietrich Rebholz-Schuhmann
European Bioinformatics Institute
Cambridge, UK
rebholz@ebi.ac.uk

Mark Reimers
Department of Biostatistics, Virginia
Commonwealth University, USA
mreimers@vcu.edu

Mie Rizig
Department of Mental Health
Sciences, Windeyer Institute
London, UK
rejumar@ucl.ac.uk

Robert Stevens
University of Manchester, UK
robert.stevens@manchester.ac.uk

Ning Sun
Department of Epidemiology and
Public Health, Yale University
School of Medicine, USA
ning.sun@yale.edu

Hannah Tipney
University of Colorado Denver
School of Medicine, USA
Hannah.Tipney@ucdenver.edu

Minlu Zhang
Division of Biomedical Informatics
Cincinnati Children's Hospital
Medical Center, USA
zhangml@mail.uc.edu

Xiao Zhang
Division of Biomedical Informatics
Cincinnati Children's Hospital
Medical Center, USA
zhang2xh@mail.uc.edu

Hongyu Zhao
Department of Epidemiology and
Public Health, Yale University
School of Medicine, USA
hongyu.zhao@yale.edu

PART I

FUNDAMENTALS

Section 1

Knowledge-Driven Approaches

1

Knowledge-based bioinformatics

Eric Karl Neumann

1.1 Introduction

Each day, biomedical researchers discover new insights about our biological knowledge, augmenting by leaps our collective understanding of how our bodies work and why they fail us at times. Today, in one minute we accumulate as much information as we would have from an entire year just three decades ago. Much of it is made available through publishing and databases. However, any group's effective comprehension of this full complement of knowledge is not possible today; the stream of real-time publications and database uploads cannot be parsed and indexed as accessible and application-ready knowledge yet. This has become a major goal for the research community, so that we can utilize the gains made through the all the funded research initiatives. This is what we mean by biomedical knowledge-driven applications (KDAs).

Knowledge is a powerful concept and is central to our scientific pursuits. However, knowledge is a term that too often has been loosely used to help sell an idea or a technology. One group argues that knowledge is a human asset, and that all attempts to digitally capture it are fruitless; another side argues that any specialized database containing curated information is a knowledge system. The label 'knowledge' comes to connote information contained by an agent or system that (we wish) appears to have significant value (enough to be purchased). Although the freedom to use labels and ideas should not be impeded, an agreed

Knowledge-Based Bioinformatics: From Analysis to Interpretation   Edited by Gil Alterovitz and Marco Ramoni
© 2010 John Wiley & Sons, Ltd

use of concepts like knowledge would help align community efforts, rather than obfuscate them. Without this consensus, we will not be able to define and apply principles of knowledge to relevant research and development issues that would serve the public. The definition for knowledge needs to be clear, uncomplicated, and practical:

(1) Some aspects of Knowledge can be digitized, since much of our lives depends on the use of computers and the Internet.

(2) Knowledge is different from data or stored information; it must include context and sufficient embedded semantics so that its relevancy to a problem can be determined.

(3) Information becomes Knowledge when it is applicable to more general problems.

Knowledge is about understanding acquired and annotated (sometimes validated) information in conjunction with the context in which it was originally observed and where it had significance. The basic elements in the content need to be appropriately abstracted (classification) into corresponding concepts (usually existing) so that they can be efficiently reapplied in more general situations. A future medical challenge may deal with different items (humans vs. animals), but nonetheless share some of the situational characteristics and generalized ideas of a previously captured biomedical insight. Finding this piece of knowledge at the right time so that it can be applied to an analogous but distinct situation is what separates knowledge from information. Since this is something humans have been doing by themselves for a long time, we have typically been associating knowledge exclusively with human endeavors and interactions (e.g., 'sticky, local, and contextual,' Prusak and Davenport, 2000).

KDA is essential for both industrial and academic biomedical research; the need to create and apply knowledge effectively is driven by economic incentives and the nature of how the world works together. In industry, the access to public and enterprise knowledge needs to be both available and in a form that allows for seamless combinations of the two sets. Concepts must enable the bridging between different sources, such that the connected union set provides a business advantage over competitors. Academic research is not that different in having internal and external knowledge, but once a novel combination has been found, validated and expounded, the knowledge is then submitted to peer review and published in an open community. Here, rather than supporting business drivers, scientific advancement occurs when researchers strive to be recognized for their contribution of novel and relevant scientific insights. The free and efficient (and sometimes open) flow of knowledge is key in both cases (Neumann and Prusak, 2007).

In preparation for the subsequent discussions, it is worth clarifying what will be meant by data, information, and knowledge. The experimentalists' definition of data will be used for the most part unless otherwise noted, and that is information measured or generated by experiments. Information will refer to all forms of digitized resources (aka data by other definitions) that can be stored and recalled from a program; it may or may not be structured. Finally, based on the above discussion, knowledge refers to information that can be applied to specific problems, usually separate from the sources and experiments from which they were derived. Knowledge can exist in both humans and digital systems, the former being more flexible to interpretation; the latter relies on the application of formal logic and well-defined semantics.

This chapter begins by providing a review of historical and contemporary knowledge discovery in bioinformatics, ranging from formal reasoning, to knowledge representation, to the issues surrounding common knowledge, and to the capture of new knowledge. Using this initial background as a framework, it then focuses on individual current knowledge discovery applications, organized by the various components and approaches: ontologies, text information extraction, gene expression analysis, pathways, and genotype–phenotype mappings. The chapter finishes by discussing the increasing relevance of the Web and the emerging use of Linked Data (Semantic Web) 'data aggregative' and 'data articulative' approaches. The potential impact of these new technologies on the ongoing pursuit of knowledge discovery in bioinformatics is described, and offered as practical direction for the research community.

1.2   Formal reasoning for bioinformatics

Computationally based knowledge applications originate from AI projects back in the late 1950s that were designed to perform reasoning and inferencing based on forms of first-order logic (FOL). Specifically, inferencing is the processing of available information to draw a conclusion that is either logically plausible (inconclusive support) or logically necessary (fully sufficient and necessary). This typically involves a large set of chained reasoning tasks that attempt to exhaustively infer precise conclusions by looking at all available information and applying specified rules.

Logical reasoning is divided into three main forms: deduction, induction, and abduction. These all involve working with preconditions (antecedents), conclusions (consequents), and the rules that associate these two parts. Each one tries to solve for one of these as unknowns given the other two knowns. Deduction is about solving for the consequent given the antecedent and the rule; induction is about finding the rule that determines the consequent based on the known precondition; and abduction is about determining the precondition based on the conclusions and the rules followed. Abduction is more prone to problems since multiple preconditions can give rise to the same conclusions, and is not as frequently employed; we will therefore focus only on deduction and induction here.

Deduction is what most people are familiar with, and is the basis for syllogisms: 'All men are mortal; Socrates is a man: Therefore Socrates is mortal!' Deductive reasoning requires no further observations; it simply requires applying rules to information on preconditions. The difficulty is that in order to perform some useful reasoning, one must have a lot of deep knowledge in the form of rules so that one can produce solid conclusions. Mathematics lends itself well here, but attempts to do this in biology are limited to simple problems: 'P53 plays a role in cancer regulation; Gene X affects P53: Therefore Gene X may play a role in a cancer.' The rule may be sound and generalized, but the main shortcoming here is that most people could have performed this kind of inference without invoking a computational reasoner. Evidence is still scant that such reasoning can be usefully applied to areas such as genetics and molecular biology.

Induction is more computationally challenging, but may have more real-world applications. It benefits from having lots of evidence and observations on which to create rules or entailments, which, of course, there is plenty of in research. Induction works on looking for patterns that are consistent, but can be relaxed using statistical significance to allow for imperfect data. For instance, if one regularly observes that most kinases downstream of NF-kB are up-regulated in certain lymphomas, one can propose a rule that specifies this up-regulation relation in these cancers. Induction produces rule statements that have antecedents and consequents. For induction to work effectively one must have (1) sufficient data, including negative facts (when things didn't happen); (2) sufficient associated data (metadata), describing the context and conditions (experimental design) under which the data were created; and (3) a listing of currently known associations which one can use to specifically focus on novel relations and avoid duplication. Induction by itself cannot determine cause and effect, but with sufficient experimental control, one can determine which rules are indeed causal. Indeed, induction can be used to generate hypotheses from previous data in order to design testable experiments.

Induction relies heavily on the available facts present in sources of knowledge. These change with time, and consequently inductive reasoning may yield different results depending on what information has recently been assimilated. In other words, as new facts come to light, new conclusions will arise out of induction, thereby extending knowledge. Indeed, a key reason that standardized databases such as Gene Expression Omnibus (GEO, www.ncbi.nlm.nih.gov/geo/) exist is so we can discover new knowledge by looking across many sets of experimental data, longitudinally and laterally.

Often, reasoning requires one to make 'open world assumptions' (OWAs) of the information (e.g., *Ling-Ling is a panda*), which means that if a relevant statement is missing (*Ling-Ling is human* is absent), it **must** be assumed plausible unless (1) proven false (*Ling-Ling's parents are not human*), (2) shown to be inconsistent (*pandas and humans are disjoint*), or (3) the negation of the statement is provided (*Ling-Ling is not human*). OWAs affect deduction by expanding the potential solution space, since some preconditions are unknown and therefore unbounded (not yet able to be fixed). Hence, a receptor with no discovered ligand should be treated as a potential receptor for many different signaling processes (ligands are often associated with biological processes). Once a ligand is determined, the signaling consequences of the receptor are narrowed according to the ligand.

With induction, inference under OWAs will usually be incomplete, since a rule cannot be exactly determined if relevant variables are unknown. Hence some partial patterns may be observed, but they will appear to have exceptions to the rule. For example, a drug target for colon cancer may not respond to inhibitors reliably due to regulation escape through an unbeknownst alternative pathway branch. Once such a cross-talk path is uncovered, it becomes obvious to try and inhibit two targets together, one in each pathway, to prevent any regulatory escape (aka *combinatoric therapy*).

Another relevant illustration is the inclusion of Gene Ontology (GO) terms within gene records. Their presence suggests that evidence exists to recommend assigning a role or location to the gene. However, the absence of the attribute 'regulation of cell communication' could signify a few things: (1) the gene has yet to be assessed for involvement in 'regulation of cell communication'; (2) the gene has been briefly reviewed, and no obvious evidence was found; and (3) the gene has been thoroughly assessed by a sufficient inclusionary criteria. Since there is no way to determine, today, what the absence of a term implies, this would suggest that knowledge mining based on presence or absence of GO terms will often be misleading.

OWAs often cannot be automatically applied to relational database management systems (RDBMSs), since the absence of an entry or fact in a record may indeed mean it was measured but not found. A relational database's logical consistency could be improved if it explicitly indicated which facts were *always* measured (i.e., lack of fact implies measured and not observed), and which ones were *sometimes* measured (i.e., if measured, always stated, therefore lack of fact implies not measured). The measurement attribute would need to include this semantic constraint in an accessible metamodel, such as an ontology.

Together, deduction and induction are the basis for most knowledge discovery systems, and can be invoked in a number of ways, including non-formal logic approaches, for example SQL (structured query language) in relational databases, or Bayesian statistical methods. Applying inference effectively to large corpora of knowledge requires careful planning and optimization, since the size of information can easily outpace the computation resources required due to combinatorial explosion. It should be noted that biology is notoriously difficult to generalize completely into rules; for example, the statement 'P is a protein iff P is triplet-encoded by a Gene' is almost always true, but not in the case of gramicidin D, a linear pentadecapeptide that is synthesized *de novo* by a multi-enzyme complex (Kessler *et al.*, 2004). The failure of AI, 25 years ago, was in part due to not realizing this kind of real-world logic problem. We hope to have learned our lessons from this episode, and to apply logical reasoning to large sets of bioinformatic information more prudently.

1.3   Knowledge representations

Knowledge Representations (KRs) are essential for the application of reasoning methodologies, providing a precise, formal structure (ontology) to describe instances or individuals, their relations to each other, and their classification into classes or kinds. In addition to these ontological elements, general axioms such as subsumption (class–subclass hierarchies) and property restrictions (e.g., *P has Child C iff P is a Father* ∨ *P is a Mother*) can be defined using common elements of logic. The emergence of the OWL Web ontology language from the W3C (World Wide Web Consortium) means that such logic expressions can be defined and applied to information resources (IRs) across the Web, enabling the establishment of KRs that span many sites over the Internet and many kinds of information resources. This is an attractive vision and could generate enormous benefits, but in order for all KRs to work together, there still needs to be coherence and consistency between the ontologies defined (in OWL) and used. Efforts such as the OBO (Open Biomedical Ontologies) Foundry are attempting to do this, but also illustrate how difficult this process is.

In the remainder of this chapter, we will take advantage of a W3C standard format known as N3 (www.w3.org/TeamSubmission/n3/) for describing knowledge representations and factual relations; the triple predicate form 'A B_{rel} C' is to be interpreted as 'Entity A has relation B_{rel} with entity C.' Any term of the form '?B' signifies a named variable that can be anything that makes the predicate true; for example '?g a Gene' means ?g could be any gene, and the double clause '?p a Protein. ?p is_expressed_in Liver' means any protein is expressed in liver. Furthermore, ';' signifies a conjunction between phrases with the same subject but multiple predicates ('?p a Protein ; is_expressed_in Liver' as in the above). Lastly, '[]' brackets are used to specify any entity whose name is unknown (or doesn't matter) but which has relations contained within the brackets: '?p is_expressed_in [a Neural_Tissue; stage Embryonic].' One should recognize that such sets of triples result in the formation of a system of entity nodes related to other entity nodes, better known as a graph.

1.4   Collecting explicit knowledge

A major prerequisite of knowledge-driven approaches is the need to collect and structure digital resources as KRs (a subset of IRs), to be stored in knowledge bases (KBs) and used in knowledge applications. Resources can include digital data, text-mined relations, common axioms (subsumption, transitivity), common knowledge, domain knowledge, specialized rules, and the Web in general. Such resources will often come from Internet-accessible sources, and it is assumed that they can be referenced similarly from different systems. Web accessibility requires the use of common and uniform resource identifiers (URIs) for each entity as well as the source system; the additional restriction of uniqueness is not as easy to implement, and can be deferred as long as it is possible to determine whether two or more identifiers refer to the same thing (e.g., owl:sameAs).

In biomedical research, recognizing where knowledge comes from is just as important as knowing it. Phenomena in biology cannot be rigorously proven as in mathematics, but rather are supported by layers of hypotheses and combinations of models. Since these are advanced by researchers with different working assumptions and based on evidence that often is local, keeping track of the context surrounding each hypothesis is essential for proper reasoning and knowledge management. Scientists have been working this way for centuries, and much of this has been done through the use of references in publications whenever (hypothetical) claims are compared, corroborated, or refuted. One recent activity that is bridging between the traditional publication model and the emerging KR approach is the SWAN project (Ciccarese *et al.*, 2008), which has a strong focus on supporting evidence-based reasoning for the molecular and genetic causes of Alzheimer's disease.

Knowledge provenance is necessary when managing hypotheses as they either acquire additional supporting evidence (accumulating but never conclusive), or are disproved by a single critical fact that comes to light (single point of failure). Modal logic (see below), which allows one to define hypotheses (beliefs) based on partial and open world assumptions (Fagin *et al.*, 1995), can dramatically alter a given knowledge base when a new assumption or fact is introduced to the reasoner (or researcher). As we begin to accumulate more hypotheses while at the same time having to review new information, our knowledge base will be subject to major and frequent inference-driven updates. This dependency argues strongly for employing a common and robust provenance framework for both scientific facts and (hypotheses) models. Without this capability, one will never know for sure on what specific arguments or facts a model is based, hence impeding effective Knowledge Discovery (KD). It goes without saying that this capability will need to work on and across the Web.

The biomedical research community has, to a large extent, a vast set of common knowledge that is openly shared. New abstracts and new data are put on public sites daily whenever they are approved or accepted, and many are indexed by search engines and associated with controlled vocabulary (e.g., MeSH). However, this collection is not automatically or easily assimilated into individual applications using knowledge representations, so that researchers cannot compare or infer new findings against their existing knowledge. This barrier to knowledge discovery could be removed by ensuring that new published reports and data are organized following principles of *common knowledge*.

1.5   Representing common knowledge

Common knowledge refers to knowledge that is generally known (and accessible) by everyone in a given community, and which can be formally described. Common knowledge usually differs from *tacit knowledge* (Prusak and Davenport, 2000) and *common sense*, both of which are virtually impossible to explicitly codify and which require assumptions that are non-deducible[1]. For these reasons we will focus specifically on explicit common knowledge as it applies to bioinformatic applications.

An example of explicit common knowledge is 'all living things require an energy source to live.' More relevant to bioinformaticists is the central dogma of biology which states: 'genes are transcribed into mRNA which translate into proteins; implying protein information cannot flow back to DNA,' or formally:

∀ Protein ∃ Gene (Gene transcribes_into mRNA ∧ mRNA
        translates_into
Protein) ⇒ ¬ (Protein reverse_translate Gene).

This is a very relevant chunk of common knowledge that not only maps proteins to genes, but even constrains the gene and protein sequences (up to codon ambiguity). In fact, it is so common, that it has been (for many years) hard-wired into most bioinformatic applications. The knowledge is therefore not only common, but pervasive and embedded, to the point where we have no further need to recode this in formal logic. However, this is not the case for more recent insights such as SNP (single nucleotide polymorphism) associations with diseases, where the polymorphism does not alter the codons directly, but the protein is either truncated or spliced differently. Since the set of SNPs is constantly evolving, it is essential to make these available using formal common knowledge. The following (simplified) example captures this at a high level:

∀ Genetic_Disease ∃ Gene ∃ Protein ∃ SNP (SNP within
        Gene ∧ Gene
expresses Protein ∧ SNP modifies Protein ∧ SNP associated
Genetic_Disease) ⇒ SNP root_cause_of Genetic_Disease.

Most of these relations (protein structure and expression changes) are being curated into databases along with their disease and gene (and sequence) associations. It would be a powerful supplement if such knowledge rules could be available as well to researchers and their applications. An immediate benefit would be to allow for application to extend their functionality without need for software updates by vendors; simply download the new rules based on common understanding to reason with local knowledge.

Due to the vastness of common knowledge around all biomedical domains (including all instances of genes, diseases, and genotypes), it is very difficult to explicitly formalize all of it and place it in a single KB. However, if one considers public data sources as references of knowledge, then the amount of digitally

---

[1] Tacit knowledge is often related to human habit and know-how, and is not typically encodable and therefore not easily shared; common sense does involve logic but requires assumptions that cannot be formally defined.

encoded knowledge can be quickly and greatly augmented. This does require some mechanism for wrapping these sources with formal logic, for example associating entities with classes. Fortunately, the OWL-RDF (resource description framework) model is a standard that supports this kind of information system wrapping, whereby entities become identified with URIs and can be typed by classes defined in separate OWL documents. Any logical constraints presumed on database content (e.g., no GO process attribute means no evidence found to date for gene) can be explicitly defined using OWL (and other axiomatic descriptions); these would also be publicly accessible from the main source site.

Common knowledge is useful for most forms of reasoning, since it facilitates making connections between specific instances of (local) problems and generalized rules or facts. Novel relations could be deduced on a regular basis from the latest new findings, and deeper patterns induced from increasing numbers of data sets. Many believe that true inference is not possible without the proper encoding of complete common knowledge. Though it will take time to reach this level of common knowledge, it appears that there is interest in heading towards such open knowledge environments (see www.esi-bethesda.com/ncrrworkshops/kebr/index.aspx). If enough benefits are realized in biomedicine along the way, more organized support will emerge to accelerate the process.

The process for establishing common knowledge can be handled by a form of logic known as modal logic (Fagin *et al.*, 1995), which allows different agents (or scientists) to be able to reason with each other though they may have different subsets of knowledge at a given time (i.e., each knows only part of the story). The goal here is to somehow make this disjoint knowledge become common to all. Here, common knowledge is (1) knowledge (φ) all members know about ($E_G\varphi$), and importantly (2) something known by all members to be known to the other members. The last item applies to itself as well, forming an infinite chain of 'he knows that she knows that he knows that. . .' signifying *complete awareness of held knowledge*

$$\left(C\varphi = \lim_{n \to \infty} E^{n\cdots} E^2 E_G^1 \varphi\right).$$

Another way to understand this, is that if Amy knows X about something, and Bob knows only Y, and X and Y are both required to solve a research problem (possibly unknown to Amy and Bob), then Amy and Bob need to combine their respective sets as common knowledge to solve a given problem. In the real world this manifests itself as experts (or expert systems) who are called upon when there is a gap in knowledge, such as when an oncologist calls on a bioinformatician to help analyze biomarker results. Automating this knowledge expert process could greatly improve the efficiency for any researcher when trying to deduce if their new experimental findings have uncovered new insights based on current knowledge.

In lieu of a formal method for accessing common knowledge, researchers typically resort to searching through local databases or using Google (discussed later) in hopes of filling their knowledge gaps. However, when searching a

RDBMS with a query, one must know how to pose the query explicitly. This often results in not uncovering any new significant knowledge, since one requires sufficient prior knowledge to enter the right query in the first place, in which case one is searching only 'under the street lamp.' More likely, only particular instances of facts are uncovered, such as dates, numeric attributes and instance qualia. This is a distinguishing feature that separates databases from knowledge bases, and illustrates that databases can support at best very focused and constrained knowledge discovery. Specifically, queries using SQL produce limited knowledge, since they typically do not uncover generalized relations between things. Ontological relations rely on the ability to infer classes, groups of relations, transitivity (multi-joins), and rule satisfiability; these are the instruments by which general and usable knowledge can be uncovered. Standard relational databases (by themselves) are too restrictive for this kind of reasoning and do not properly encode (class and relation) information for practical knowledge discovery.

In many cases, in the bioinformatics community this has come to be viewed as knowledge within databases. For example, the curated protein database Swiss-Prot/UniProt is accepted as a high quality source of reviewed and validated knowledge for proteins, including mutational and splice variants, relations to disorders, and the complexes which they constitute. In fact, it is often the case that curated sets of information are informally raised to the level of knowledge by the community. This definition is more about practice and interpretation than any formal logic definition. Nonetheless, it is relevant and valid for the community of researchers, who often do apply logic constraints on the application of this information: if a novel protein polymorphism not in Swiss-Prot is discovered and validated, it is accepted as real and eventually becomes included into Swiss-Prot.

Nonetheless, databases are full of valuable information and could be re-formatted or wrapped by an ontological layer that would support knowledge inference and discovery, defined here as *implicit knowledge resources* (*IR* → *KR*). If this were to happen, structured data stores could be federated into a system of biomedical common knowledge: knowledge agents could be created that apply modal logic reasoning to crawl across different *knowledge resources* on the Web in search of new insights. Suffice it to say, practical modal logic is still an emerging and incomplete area of research. Currently, humans are best at identifying which new facts should be incorporated into new knowledge regarding a specific subject or phenomenon; hence researchers would be best served by being provided with intelligent *knowledge assistants* that can help identify, review, compare, and assimilate new findings from these biomedical IRs. There is a lot of knowledge in biology that could be formally common, consequently there is a clear need to transform public biomedical information sources to work in concert with knowledge applications and practices. Furthermore, this includes the Web, which has already become a core component of all scientific research communities.

1.6   Capturing novel knowledge

Not everything can be considered common knowledge; there are large collections of local-domain knowledge consisting of works and models (published and pre-published) created by individual research groups. This is usually knowledge that has not been completely vetted or validated yet (hypotheses and beliefs), but nonetheless can be accessed by others who wish to refute or corroborate the proposed hypotheses as part of the scientific method. This knowledge is connected to and relies on the fundamentals of biology, which are themselves common knowledge (since they form the basis of scientific common ground). So this implies that we are looking for a model which allows connecting local knowledge easily with common knowledge.

Research information is knowledge that is in flux; it is comprised of assumptions and proposed models (mechanisms of action). In modal logic (Fagin *et al.*, 1995) this is comparable to the KD45 axioms: an agent (individual or system) can believe in something not yet proven true, but if shown to be false, the agent cannot believe in it anymore; that is, logic contradictions are not allowed. KD45 succinctly sums up how the scientific process works with competing hypotheses, and how all parallel hypotheses can co-exist until evidence emerges that proves some to be incorrect.

Therefore, the finding (by a research group), that a mutation in the *BRCA2* gene is always associated with type 2 breast cancer, strongly argues against any other gene being the primary cause for type 2 susceptibility. Findings that have strong causal relations, such as nucleotide level changes and phenotypes of people always carrying these, are prime examples of how new-findings knowledge can work together with common knowledge. As more data is generated, this process will need to be streamlined and automated; and to prevent too many false positives from being retained, the balanced use of logic and statistics will be critical.

The onslaught of large volumes of information being generated by experiments and subsequent analyses requires proper data set tracking, including the capture of experimental conditions for each study. The key to managing all these associated facts is enforcing data provenance. Without context and provenance, most experimental data will be rendered unusable for other researchers, a problem already identified by research agencies (Nature Editorial, 2009). Provenance will ensure a reliable chain of evidence associated by conditions and working hypotheses that can be used to infer high-value knowledge associations from new findings.

1.7   Knowledge discovery applications

Once common and local knowledge are available to systems in a machine-interpretable form, the construction and use of knowledge-discovery applications that can work over these sources becomes practical and empowering. KDA has its roots in what has been labeled Knowledge Discovery and Data Mining (KDD, Fayyad *et al.*, 1996), which consists of several computational approaches that came together in the mid 1990s under a common goal. The main objective of KDD is to turn collected data into knowledge, where knowledge is something of high value that can be applied directly to specific problems. Specifically, it had become apparent that analysts were 'drowning in information, but starving for knowledge' (Naisbitt, 1982). It was hoped that KDD would evolve into a formal process that could be almost entirely computationally driven. It was to assist knowledge workers during exploratory data analysis (EDA) when confronted by large sets of data. The extraction of interesting insights via KDD follows more or less inductive reasoning models.

KDD utilizes approaches such as first-order logic and data mining (DM) to extract patterns from data, but distinguishes itself from DM in that the patterns must be validated, made intelligently interpretable, and applicable to a problem. More formally, KDD defines a process that attempts to find expression patterns (E) in sets of Facts (F) that have a degree of certainty (C) and novelty (N) associated with them, while also being useful (U) and simple (S) enough to be interpreted. Statistics plays a strong role in KDD, since finding patterns with significance requires the proper application and interpretation of statistical theories. Some basic KDD tools include Decision Trees, Classification and Regression, Probabilistic Graphs, and Relational Learning. Most of these can be divided into supervised learning and unsupervised learning. Some utilize propositional logic more than others.

Key issues that KDD was trying to address include:

- Data classification

- Interpretation of outcomes (uncovering relations, extraction of laws)

- Relevance and significance of data patterns

- Identification and removal of confounding effects (Simpson's paradox, http://plato.stanford.edu/entries/paradox-simpson/).

Patterns may be known (or hypothesized) in advance, but KDD is supposed to aid in the extraction of such patterns based on the statistical structure of the data and any available domain knowledge. Clearly, information comes in a few flavors: quantitative and qualitative (symbolic). KDD was intended to take advantage of both wherever possible. Symbolic relations embedded in both empirical data (e.g., *what conditions were different samples subjected to?*) and domain knowledge (e.g., *patient outcomes are affected by their genotypes*) begin to demonstrate the true symbolic nature of information. That is, data is about tying together relations and attributes, whether it is arranged as tables of values or sets of assertions. The question arises, how can we use this to more efficiently find patterns in data? The key here is understanding that relational data can be generalized as data graphs: collections of nodes connected by edges, analogous to how formal relational knowledge structures are to function (see above).

Indeed, all the KDD tools listed have some form of graph representation: decision trees, classification trees, regression weighted nodes, probabilistic graphs, and relational models (Getoor *et al.*, 2007). The linked nodes can represent samples, observations, background factors, modeling components (e.g., θ_i), outcomes, dependencies, and hidden variables. It would follow that a common way to represent data and relational properties using graphs could help generalize KDD approaches and allow them to be used in concert with each other. This is substantial, since we now have a generalized way for any application to access and handle knowledge and facts using a common format system, based on graph representation (and serialized by the W3C standard formats, RDF-XML or RDF-N3).

More recently, work by Koller and others has shown that the structure of data models (relations between different tables or sets of things) can be exploited to help identify significant relations within the data (Getoor *et al.*, 2007). That is, the data must already be in a graph-knowledge form in order to be effectively mined statistically. To give an example, if a table containing tested-subject responses for a treatment is linked to the treatment-dosing table and the genetic alleles table, then looking for causal response relations is a matter of following these links and calculating the appropriate aggregate statistics. Specifically, if one compares all the responses in conjunction with the drug and dosing used as well as the subject's genotype, then by applying Bayesian inference, strong interactions between both factors can be identified. Hence data graph structures can be viewed as first-order 'hypotheses' for potential interactions.

By the mid 1990's, the notion of publishing useful information on the Web began to take off, allowing it to be linked and accessed by other sites: the Web as a system of common knowledge took root and applications began to work with this. This was followed by efforts to define ontologies in a way that would work from anywhere on the Web, and with anything that was localized anywhere on the Web. A proposal was eventually submitted to the DARPA (Defense Advanced Research Projects Agency) program to support typing things on the Web. It was funded in 2000 and became known as the DAML (DARPA Agent Mark-up Language, www.daml.org/) project. This became the forerunner of the Semantic Web, and eventually transformed into the OWL ontology language that is based on Description Logic (DL).

Dozens of applications for KDD have been proposed in many different domains, but its effectiveness in any one area over the other is unclear. To this end, an open challenge has been initiated, called the KDDCup (www.kdnuggets .com/datasets/kddcup.html), to see how well KDD can be applied to different problem spaces. It has gained a large following in bioinformatics, addressing such diverse areas as:

- Prediction of gene/protein function and localization

- Prediction of molecular bioactivity for drug design

- Information extraction from biomedical articles

- Yeast gene regulation prediction

- Identification of pulmonary embolisms from three-dimensional computed tomography data

- Computer Aided Detection (CAD) of early stage breast cancer from X-ray images.

KDD was conceived during a time when the shortcomings of AI had surfaced, and the Web's potential was just emerging. We are now in an age where documents can be linked by other documents anywhere in the world; where communities can share knowledge and experiences; where data can be linked to meaning. The most recent rendition of this progress is the Semantic Web.

1.8   Semantic harmonization: the power and limitation of ontologies

One of the most important requirements for data integration or aggregation is for all data producers and consumers to utilize common semantics (Rubin *et al.*, 2006). In the past, it had been assumed that data integration was about common formats (syntax), but that assumed that if one *knows* the structure, one can *infer* the data meaning (semantics). This is now known to be grossly oversimplified, and semantics must also be clearly defined. RDF addressed the syntax issue by forcing all data relations to be binary based, therefore modeling all components as triples (subject, relation/property, object).

The emergence of the W3C OWL ontology standard has enabled the formal definition of many biological and medical concepts and relations. OWL is based on description logic, a FOL formalism that was developed in the 1980s to support class (concept) subsumption and relations between instances of classes. Using OWL, a knowledge engineer can create class hierarchies of concepts that map to real-world observations; for instance, 'Genes are encoded in DNA and themselves encode proteins.' OWL's other key feature is that it can be referenced from anywhere on the Web (e.g., used by a database) and incorporated into other non-local logical structures (ontology extension). It was designed to so that any defined ontological components are identifiable nodes on the Web; that is, all users can refer to the same defined Class. The most current version of OWL is OWL2, based on SROIQ logic supporting more expressive logic (Horrocks *et al.*, 2006). The OWL format is modeled after the Resource Description Framework (RDF) that will be described later.

The OWL standard allows knowledge systems to utilize ontologies defined by various groups, such as Gene Ontology, UniProt, BioPAX, and Disease Ontology. Data sets that one wishes to align with the ontologies now can apply a well-specified mechanism: simply reference the ontology URI from within a data documents and system. By doing so, all the data in the system is formally associated with concepts, as well as the relations concepts have with each other. Any third party also looking at the data can instantly find (over the Web) which ontologies were used to define the set.

Many of the current activities around developing ontologies in OWL are about defining common sets of concepts and relations for molecular biology (genes, proteins, and mechanisms) and biomedicine (diseases and symptoms). However, there is still no general agreement of how (completely) to define basic concepts (e.g., gene, protein) or what upper-level biological ontologies should look like or do. It is not inconceivable that this process will take many years still.

1.9   Text mining and extraction

One common source of knowledge that many scientists wish to access is from the unstructured text of scientific publications. Due to the increasing volume of published articles, it is widely recognized (Hunter and Cohen, 2006) that researchers are unable to keep up with the flow of new research findings. Much expectation is placed on using computers to help researchers deal with this imbalance by mining the content for the relevant findings of each paper. However, there are many different things that can be mined out of research papers, and to do so completely and accurately is not possible today. Therefore, we will focus here only on the extraction of specific subsets of embedded information, including gene and biomolecule causal effects, molecular interactions and compartments, phenotype–gene associations, and disease treatments.

One way to mine content is simply to index key words and phrases based on text patterns and usage frequency. This is all search engines do, including Google. This does quite well in finding significant occurrences of words; however it fails to find exactly what is being said about something, that is, its semantics. For instance, indexing the phrase '... cytokine modulation may be the basis for the therapeutic effects of both anti-estrogens in experimental SLE.' One can readily identify *cytokine modulation* (CM) and its association with *therapeutic effects* (TE) or *experimental SLE* (xSLE), but the assertion that 'CM is a TE for xSLE' cannot be inferred from co-occurrence. Hence, limited knowledge about things being mentioned can be obtained using indexing, such as two concepts occurring in the same sentence, but the relation between them (if there is one) remains ambiguous.

Word-phrase indexing is very practical for search, but for scientific knowledge inquiries it is insufficient; what is specifically needed is the extraction of relations R(A, B). Although more challenging, there has been a significant effort invested to mine relations about specific kinds of entities from natural language. This is referred to as Information Extraction (IE), and it relies much more heavily on understanding some aspects of phrase semantics. Clearly this hinges on predefining classes of entities and sets of relations that are semantically mapped to each other (ontology). The objective of this is to quickly glean key relations about things like biomolecules, biostructures, and bioprocesses from research articles, thereby permitting the rapid creation of accessible knowledge bases (KBs) about such entities.

As an example, if one wanted to find out (from published research) if a particular gene is associated with any known disease mechanisms, one would query the KB for any relation of the form ?gene ?rel [a Disease] (as well as [a Disease] ?rel ?gene). This form would allow any relation to a gene to be identified from the KB, where ?rel could mean 'is associated with,' 'influences,' 'suppresses,' or 'is over expressed in.' These relations should be defined in an ontology and the appropriate domain and range entity classes explicitly included. For IE to be most effective, it is useful to focus only on specific kinds of relations one is interested in, rather than trying to support a universal set. This helps reduce dealing with all the complexities of finding and interpreting specific relations from word-phrase patterns in natural languages, a problem that is far from being solved generically. Hence, it is desirable to have modules or cartridges for different IE target tasks, and which utilize different ontologies and controlled vocabularies.

Several open source or publicly accessible IE systems exist, including GATE, Geneways, OpenCalais, TextRunner, and OpenDMAP. OpenDMAP is specifically designed to extract predicates defined in the OBO system of ontologies (Relationship Ontology, RO), specifically those involved in protein transport, protein–protein interactions, and the cell-specific gene expression (Hunter *et al.*, 2008). They had applied it to over 4000 journals, where they extracted 72 460 transport, 265 795 interaction, and 176 153 expression statements, after accounting for errors (type 1 and type 2). Many of the errors are attributable to misidentification of genes and protein names. This issue will not be resolved by better semantic tools, since it is more basic and related to entity identification.

One possibility being considered by the community is that future publications may explicitly include formal identifiers for entities in the text, as well as controlled vocabularies, linked ontologies, and a specific predicate statement regarding the conclusion of the paper. Automated approaches that create such embedded assignments are being investigated throughout the research community, but so far show varying degrees of completeness and correctness, that is, both type 1 and type 2 errors. Much of this may be best avoided if the authors would include such embedding during the writing of their papers. Attractive as this sounds, it will require the development of easy to use and non-invasive tools for authors that do not impact on their writing practices. It will be quite interesting to follow the developments in this technology area over the next few years.

1.10   Gene expression

Gene Expression Analytics (GEA) is one of the most widely applied methodologies in bioinformatics, mixing data mining with knowledge discovery. Its advantage is that it combines experimentally controlled conditions with large-scale genomic measurements; as a technology platform has become commoditized so it can be applied cost effectively to large samples. Its weakness is that at best it is an average of many cells and cell types, which may be varied states, resulting in confounder effects; in addition, transcript levels usually do not correspond to protein levels. It has become one of a set of tools to investigate and identify biomarkers that can be applied to the research and treatment of diseases. There is great expectation here to successfully apply knowledge-driven approaches around these applications, justifying the enormous investment in funded research and development to create knowledge to support the plethora of next-generation research.

GEA works with experimentally derived data, but shows best results when used in conjunction with gene annotations and sample-associated information; in essence, the expression patterns for many genomic regions (including multiple transcript regions per gene in order to handle splice variants) under various sample conditions (multiple affected individuals, genotypes, therapeutic perturbations, dosing, time-course, recovery). The data construct produced is an $N \times K$ matrix, M, of expression levels, where N is the number of probes and K the number of samples. Much of the analytic enhancements have depended on performing appropriate statistics using replicate samples. This allowed the separation of variance arising from individual sample uncertainty and errors from the experimental factors that were being applied. The numeric data by itself can only support nearest neighbor comparisons, resulting in the construction of cluster trees: one for the relatedness of the expressing genes probed, the other for the relatedness of the tissue samples expressing the transcripts. Although many researchers try to find meaning in these cluster patterns (trees), they are for the most part artificial, due primarily to the experimental design, and have very little basis in normal biology.

More has been gained by associating additional knowledge to this matrix M, such that both genes and samples have linked attributes that can be utilized in deeper analyses. Examples include utilizing the GO ontology for each gene to see if a correlation exists between the nearness of genes within the M-derived gene cluster tree and the association of the genes with similar GO processes (Stoeckert et al., 2002). Other gene relations can be utilized as well common pathways (Slonim, 2002), common disease associations, and common tissue compartmentation. The same approach can be applied to the samples themselves, including variation in nutrients, genotypes (Cheung and Spielman, 2002), administered drugs, dosing, and clinical outcomes (Berger and Iyengar, 2009).

What is worthwhile remarking here is that all these different experimental design applications have a direct common correspondence with the mining of the numerically derived structures of the data in M. That is, all the attributes associated with genes or samples can be viewed as formal relations linked to each instance or a gene or of a sample. For example, all genes G_i with attribute A_k can be evaluated for their possible correlation with higher expression values ($M_{ij} > z$) over all samples S_j by testing for $P(M_{ij} > z | G_i.A_k)$, or even a subset of samples S_j with similar characteristics C_l, $P(M_{ij} > z | G_i.A_k, S_j.C_l)$. As described earlier, this generalization supports many forms of knowledge mining, and therefore opens any applications of microarrays or biomarkers as fertile ground for KDD.

1.11   Pathways and mechanistic knowledge

Pathways are the abstraction of molecular mechanisms that are the basis for molecular functions and processes. They appear as graphs with directional flow; however the edges can have multiple meanings, such as catalyzed stoichiometric reactions (substrate, product as input, output respectively), protein signaling cascades and transcription factor and binding site activation/repression of genes. These structures can be further broken down into all the interactions each of the molecules participates in, yielding interaction graphs that are not obviously mapped to the process-oriented pathways. In each of these cases, the structures can be represented as graph objects, that is, sets of nodes connected by edges (Figure 1.1).

Fundamental to note is that pathways are not data; that is, they are not derived from single experiments. Rather, they are the result of hypothesis building and

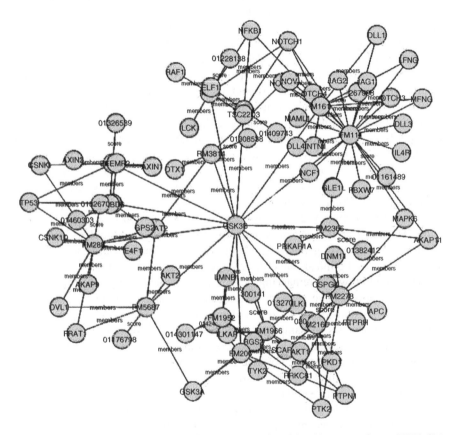

Figure 1.1   Assimilated knowledge around the GSK3b *gene from HEFalMp (Huttenhower et al., 2009), and its relations to other genes involved in similar processes.*

re-visioning over many experiments (and many years), and reflect our current knowledge of how biomolecules work together. In this regard, pathways are one manifestation of biological knowledge, and can be coded as a series of statements relating things to other things, that is, as graph objects with semantics (Losko and Heumann, 2009). The linked entities (nodes) can be mapped to existing protein and gene entities in other databases, thereby allowing pathways to be accessed and queried via entities from these other external sources (see Section 1.14.3).

Pathway structures lend themselves well to knowledge representations (*ADH1 catalyzes Ethanol_Acetylaldehyde_Reaction*), and can therefore aggregate additional facts and relations onto them. Annotations that describe possible regulations can be overlaid on top of these structures, possibly in another adjunct file (or graph). By using references to pathway models, it is possible to build layers of knowledge on top of more fundamental (canonical) pathways. These smart layers can include additional interactions, disease associations, and known polymorphic effects. Layers also provide a mechanism to describe and share additional knowledge that can be easily connected to existing pathway models via any software supporting the knowledge format. Reducing reliance on closed-vendor formats by increasing application independence is an important goal for making knowledge approaches successful.

The BioPax.owl ontology (currently as version 3) supports many kinds of relations within pathways (pathway steps, biochemical reactions, catalysis, modulation, part of complex, stoichiometric groups), and provides an exchange format supported by many pathway databases (Joshi-Tope *et al.*, 2005). Basing pathway information on an ontology means it is semantically structured, not just syntactically. Not only can it be used to find substructures of interest, but one can also infer new relations from it based on existing relations. As an example, if a kinase K is known to down-regulate another signaling protein S, one can infer that all the downstream proteins of $S(D)$ will also be affected by the K kinase; this can be expressed as a rule and applied whenever one wishes to find all downstream affected components:

K down-regulates S AND S up-regulates D $\Rightarrow$ K down-regulates D.

In addition, since molecular complexes are also modeled in BioPAX, a protein P which is part of a complex X, which is also part of another complex Y, is therefore also (P) part of complex Y, based on transitivity. Sets of these basic rules can be collected and applied strategically through a rule engine to help scientists find things such as potential disease mechanisms or candidate targets (Berger and Iyengar, 2009). Alternatively, mechanistic associations in conjunction with disease phenotypes can be used to explore novel drug applications (Qu *et al.*, 2009).This form of knowledge discovery has only begun to be explored, and since its application potential is great, it will be important to see how it evolves in the coming years. Its success depends largely on how much material is accessible as digital knowledge, which recently has begun to look promising (see Section 1.14.3).

Often, pathways are used in parallel with GEA to better understand the roles and dynamics of various components under different conditions, including progression of cancer. By clustering genes by expression changes one can investigate if some pathways show correspondence of the encoded proteins with their location within a pathway structure, for instance downstream of a key control point. The summarization of many sources of evidence can be compiled into a comprehensive knowledge graph and used by researchers to find potential functionally associated gene relations (Huttenhower *et al.*, 2009; see Figure 1.1). Pathway knowledge can also be combined with genotypic information to elucidate the effects gene variations have on mechanisms (Holford *et al.* 2009).

The tools of KDD can be applied here to help mine any possible correspondence between molecular interactions and classes of genes and their regulatory elements. Specifically, a pathway can be viewed as a proposed model structure S that can be the basis for analyzing expression dynamics using Bayesian network analysis $P(E|S, \theta)$. If one allows the model to evolve to better match expression data, the new relations can be interpreted as additional regulatory components on top of the pathway that were not known before.

1.12   Genotypes and phenotypes

As a final case study of knowledge-driven analysis, consider the complex problem of trying to identify relations between genotypes and phenotypes. This mapping has become increasingly relevant as researchers attempt to associate large numbers of human gene polymorphisms with observed traits, typically referred to as genome-wide associations (GWA). This analysis relies on the following existing resources in order to work: (1) detailed maps of genes in vicinities of polymorphisms, and (2) observed traits from familial studies that link sets of phenotypes and disorder symptoms to measured polymorphisms (Daly, 2010).

GWA by themselves cannot determine which specific gene is associated with a disorder, just that the influence seems to be nearby a polymorphism. Since there are multiple genes in these neighborhoods ($\sim$100 Mb), the actual affecting gene could be any one of these or even none of them. To further identify good gene candidates for the trait, one needs to analyze multiple sources of information and knowledge, including interactions and process membership (enrichment sets) with other proteins known to be associated with the related trait. Evidence may also come from animal models showing a related phenotype linked to the homologous gene. Then there might also be corroboration by way of common tissue expression for some of the genes, or even correlated expression changes due to a disease. In reality, there are many lines of reasoning with potential evidence, so the need for broad and flexible knowledge utilization is paramount.

Research into trait associations is ongoing and new evidence is continuously being generated. In order to validate an association, one must support an accumulative knowledge model that can work with many kinds of relations and assertions, implying no hard defined data schema. This is beyond the capability of standard databases, and illustrates that traditional IT storage approaches will not be sufficient to support most forms of knowledge discovery. Once again this argues strongly for a linked relational approach such as using RDF.

Where one hopes this is heading is that one day we can bring together all forms of information and knowledge to any set of experimental data and hypotheses, and efficiently identify all possible logical explanations of a biological phenomenon, or the causes of a disease, or the possible treatments of a cancer, or the prediction of responses to a therapy based on an individual's genotype and lifestyle.

1.13   The Web's role in knowledge mining

We have seen how search has evolved over the last few years, and become a staple of general computer use – clearly the simpler interface is a critical factor in the case of Google. With the focus on knowledge, we should not lose sight of how the interface of a knowledge system will determine its utility for different groups of users. As a common example, the appearance of faceted browsing enables users to better assess overall content and quickly direct focus onto relevant subsections of it. This highlights a shift away from the 'magic one-query' paradigm to one of quickly zooming onto the critical subset of assertion.

Biomedicine is about combining many forms of information effectively towards the formation of predictive and causal models. The models themselves define relations that are based on our abstracted knowledge, which applies to many instances of systems: common molecular mechanisms in cells of similar tissue types; altered regulation of cell growth in normal and neoplastic cells; pathological similarities between human disease and animal models; variations in cellular processes arising from genotypic differences between individuals. Furthermore, many of our recent discoveries come from cross-pollination between areas: mathematical models of evolving networks and neural development; Green Fluorescent Protein (GFP) from algae and microscopy of cellular changes.

The scope of the required KD solutions must be broad enough to handle all the sources of information that biomedicine and life science researchers need in their increasingly interdisciplinary activities. That implies the utilization of ontologies that can bridge the concepts and information sets related to these sources. However, such ontologies should be defined not to impose our current snapshot of how we think biology works, but rather serve as a set of components, by which we can effectively describe new phenomena and derived hypotheses; in other words, ontologies for constructions of new proposed models and views.

Since knowledge should not be thought of as having imposed boundaries, the KD approaches offered must support the combination of data and ideas.

1.14   New frontiers

1.14.1   Requirements for linked knowledge discovery

As previously described, most information potentially available for knowledge applications is embedded within relational databases (RDBMSs). This is the most common queriable storage format, upon which enormous infrastructures, both public and private, have been built. RDBMSs have the following desirable benefits:

Information can be structured as needed, and defined in a schema.

The content is selected and cleaned before uploading to meet functional quality.

SQL is a well-defined and validated query algebra that can be optimized.

Access control to databases is managed.

They also have the following limitations that arise out of these same advantages:

Their applications are limited by the original design goals.

The content cannot be easily expanded to include new kinds of information.

The database schema typically has no means to apply formal logic.

The query capability is limited to within a single database.

It is not straightforward to combine multiple databases over a network.

These limitations have hindered advancing knowledge discovery in life science research environments by making the cost to extend systems prohibitively expensive[2]. More so, in a world where scientific data arises from many locations and with very different structures, scientists have been impeded from taking full advantage of available knowledge from the community via the Internet. Bioinformatics has a strong hacker component that is constantly writing widgets and shims to bridge existing but incompatible resources (Hull *et al.*, 2004).

Several computer scientists have argued that ontologies are the solution to this problem, but as pointed to above, broad validated ontologies are not easy to create and do not trivially insert themselves into existing data systems. Ontologies represent a mechanism (along with a community process) for specifying content semantics, not a technology for bringing content together. The Linked Data model attempts to address most of these issues, providing a mapping to contemporary databases that supports federated queries and aggregation over the network. At the same time, linked data enable the incorporation of ontologies, even multiple ones. Two components necessary for data linking can be described using the concepts of data aggregation and data articulation, which now will be addressed.

1.14.2   Information aggregation

Aggregation is the process to efficiently find and correctly (semantically) link together information from multiple sources on the Web, based on relations

---

[2] Indeed this has spurred a culture of 'rebuild from scratch rather than build-upon.'

specified either in the content, or by a validated bridge source (look-up). This works best if the sources are already semantically defined and linking utilizes a referential linking model such as RDF triples. Aggregation also implies that not all information needs to be imported (i.e., massive data exchange) to build aggregate information, since the reference to a node is enough to state that all its other information is virtually connected as well. As an example, given the following two information sets:

```
PCKMz isa Kinase_Gene ; is_expressed_in Neural_Tissue  .

LTP is_necessary_for short-term_memory ; occurs_in
     Hippocampus_Tissue ;
is_caused_by local_Neural_Stimulation .
```

These two data sets (graphs) are disjoint (unconnected) except for the implied relation of `Hippocampus_Tissue` is a part of `Neural_Tissue`. If one now creates (or obtains from another researcher) the following statements:

```
PCKMz inhibited_by XC3751 . XC3751 blocks LTP . PCKMz
     has_role_in LTP .
```

Then once combined with the rest, these have the effect of connecting the previous statements (structurally and semantically) via the common `LTP` reference. Indeed, just knowing these facts is enough to connect the previous sets without having to explicitly transport any of the actual statements: their connections are available for discovery simply by accessing their URIs (e.g., all related information of LTP) through the Web.

I can formulate a query to ask: 'Which genes affect any kind of memory?' I could even do this using general concepts without using the specifics:

```
?aGene isa Gene; involved_in ?memProc . ?memProc isa
     Memory_Process .
```

Where the relation `involved_in` can be transitively inferred from the back-to-back `has_role_in` and `is_necessary_for` relations. The answer returned is:

```
PCKMz isa Gene .

PCKMz has_role_in LTP .

LTP is_necessary_for short-term_memory .

:= PCKMz involved_in short-term_memory .
```

These applications of aggregations do not happen trivially over the Web; reasoners require them to be pulled together in memory, either in one big batch, or incrementally as information is needed. Since there is no restriction (besides memory) on what one can aggregate, a researcher may wish to have all necessary information locally for computational efficiency. Aggregation simply makes the task of finding, transporting and building linked data much more efficient without needing to develop complex software architectures *de novo*; this has been well illustrated by the Linked Open Data initiative.

1.14.3   The Linked Open Data initiative

As a specific example of data linking and aggregation on a grand scale, the Linked Open Data (LOD, http://linkeddata.org/) initiative, which began in Banff at the WWW2007 conference, is continuously aggregating many public data sources. At the time of this writing, LOD has grown to over 7.7 billion triples over 130 different data sources (http://esw.w3.org/topic/TaskForces/CommunityProjects/ LinkingOpenData/DataSets/Statistics).

LOD addresses two points: (1) data from different sources are explicitly connected (rather than pages), and (2) the full set can be queried together based on their combined schema. LOD combines the notion of cloud computing and public access to connected and queriable data (not just Web pages). The URIs contained within serve as data identifiers that exactly pinpoint that record while at the same time are the points of linking between different data sets, for example ClinicalTrial and DrugBank. More recently, the Linked Open Drug Data (http:// esw.w3.org/topic/HCLSIG/LODD) component of LOD just won the 2009 triplification award for largest and most useful data sets converted to RDF (http:// triplify.org/Challenge/2009).

LOD is a live, Web-based proof-of-concept of what is possible if public data can be connected. Most of the original data relations are preserved, and new ones have been added between different sources when the relations where obvious. All these sets can now be queried using SPARQL (Figure 1.2, clinical trials data at http://linkedCT.org rendered with the Cytoscape S*QL plug-in). LOD is to serve as a starting point as to what is possible within the public Web. As discussed earlier, large-scale knowledge discovery requires robust and structured accessibility to large sets of information, complete with semantic relations and associated class definitions. LOD is helping point out how to move beyond our current reliance on local databases, and prepare us to begin using structured information on the Web to help solve complex, real-world challenges.

1.14.4   Information articulation

Pulling together information from different sources only gets you so far; information can be collected, but it may be in a form that does not provide significant value to specific inquiries, such as new insights that could arise from combined relations between instances. This is a limitation with the current forms of linked data, where data is available as a collected set, but does not necessarily offer any deeper logical insights. This limitation is primarily semantic, and indicates that there is more to knowledge mining than information models; additional logical relations are usually required to pose deeper inquiries.

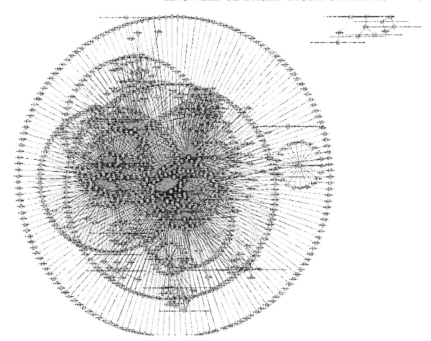

Figure 1.2   Results of a SPARQL query on LinkedCT, a LOD resource from Clin-icalTrials.gov, showing all the cancer clinical trials surrounding their respective specific cancer groups. Concentricity indicates studies investigating multiple cancers together.

For example, if a data set of customers can be linked to a data set of products from multiple vendors, one can readily query and find which customers tend to buy specific things from certain vendors, and which vendors offer products sold most frequently. However the insight into which vendor products are sold together (product–product relation) requires additional analytics and an expression to represent such novel associations (e.g., bought_with). The ability to infer new meaningful relations from existing semantics and incorporate these as additional assertions into the existing knowledge set is defined here as *information articulation*. The new relations increase the degrees of freedom of data nodes and thereby the reusability of data sets; in essence, the new relations *articulate the data*.

Information articulation can be best compared to the use of a spreadsheet whereby a user inserts a new column whose cell values are the direct result (computational output) of other existing columns from the same or different worksheets. However, rather than cells being created with new values, existing nodes (entities) are being linked by new (inferred) relations, or new nodes are being created to hold N-ary (fused) relations. This can be in the form of transitive closure (if A is part of B and B is part of C, then A is part of C) or other common rules (if Gene G is associated with Type 2 Diabetes, and Gene G is part of Pathway P, then Pathway P is involved in Type 2 Diabetes). The articulation takes the form of a rule: sets of specific relations create new deeper relations. The knowledge of when to apply the rule comes from a deeper domain understanding. Such knowledge may be defined within ontologies, but often a complex rule set cannot be coded using standard ontological frameworks.

Articulation is usually monotonic, only adding new relations to existing ones, without any deletions. The goal of information articulation is to create new usable insights by making novel connections between existing knowledge and new empirical data. Since novel relations must always be built from/on top of existing substrates, there needs to be flexibility for adding them onto existing structures, and this is what enables articulation. Another way to think about it is that when one performs inferences on knowledge, the resulting 'new findings' need to be connected with the input information so that they are retrievable and remain in context. These relations must make semantic sense for their respective domains.

Information articulation requires a sufficient degree of domain semantics to be known in order to make such inferences from initial entity sources (son-father-brother → uncle). This also implies that one cannot know everything about the information model in advance; the extension of knowledge needs to be done in such a way that it does not create inconsistencies (UNCLE ⊄ WOMAN). Information articulation also helps differentiate between *extensional sets* (data facts) and *intensional sets* (implied by semantics and reasoning). Intensional sets are often more endowed with usable meaning for future inferencing. Knowledge discovery will require that we can build upon facts by using tools and structures that can further articulate information with usable meaning.

1.14.5   Next-generation knowledge discovery

We have presented an overview of the relation between information and knowledge and how one can computationally apply knowledge to information to create new knowledge. Much of the theory has emerged from decades of research, but in the last several years, the combined expansion of the Web in conjunction with advances in semantic standards and technologies have dramatically opened the possibility of performing knowledge discovery on a global scale. Bioinformatics is one domain that can begin to take advantage of these resources to advance our understanding of biology and help us conquer diseases.

The key elements of knowledge discovery – logic, abstraction, relational graphs, and statistical induction – must be combined in consistent and flexible ways in order to address the richness of biomedical information. Once all of the biomedical information has been structurally and semantically normalized, many kinds of applications become possible. Comprehension of biological mechanisms will be critical for new therapeutics, and knowledge of one disease will often be key to the understanding of a very different disease. Biology itself is highly interlinked, so we will require information tools and logic that can match these structures.

In the near future, we will see new knowledge coming from multiple communities over many projects. Standards for information are important, but simply agreeing on the same formats and ontologies will not be sufficient. Applications that search, aggregate, and analyze diverse information will have to apply deeper logic fundamentals that cannot be addressed by ontologies alone. Information articulation will be key to performing deeper inductive-driven discovery across large volumes of information.

Pushing these new paradigms in the right direction is essential. New standards need to used and assessed, and if validated, the community must be encouraged to incorporate them into all our various resources, public as well as local. We still have a long way to go for increased awareness and general acceptance. For the most part, this will be driven by successful demonstrations of these information technologies in the context of scientific research and as part of the scientific process. In other cases they will become incorporated in commercial technologies driven by new opportunities in biotechnology research and personalized medicine. However, in all cases they will be chosen to realize benefits and improvements to knowledge discovery. A scientific community based on the sharing and commerce of knowledge discovery practices is closer at hand now than at any previous time in our history.

1.15 References

Berger, S.I. and Iyengar, R. (2009) Network analyses in systems pharmacology. *Bioinformatics*, **25**(19), 2466–72.

Cheung, V.G. and Spielman, R.S. (2002) The genetics of variation in gene expression. *Nat. Genet.*, **32**, 522–5.

Ciccarese, P., Wu, E., Wong, G., *et al.* (2008) The SWAN biomedical discourse ontology. *J Biomed Inform.*, **41**(5), 739–51.

Daly, A.K. (2010) Genome-wide association studies in pharmacogenomics. *Nat Rev Genet.*, **11**(4), 241–6.

Fagin, R., Halpern, J.Y., Moses, Y., and Vardi, M.Y. (1995) *Reasoning About Knowledge*, MIT Press.

Fayyad, U.M., Piatetsky-Shapiro, G., Smyth, P., and Uthurusamy, R. (eds) (1996) *Advances in Knowledge Discovery and Data Mining*, AAAI Press.

Getoor, L., Friedman, N., Koller, D., *et al.* (2007) Probabilistic relational models, in *Introduction to Statistical Relational Learning* (eds. L. Getoor and B. Taskar), MIT Press, pp. 129–74.

Holford, M.E., Rajeevan, H., Zhao, H., *et al.* (2009) Semantic web-based integration of cancer pathways and allele frequency data. *Cancer Inform.*, **8**, 19–30.

Horrocks, I., Kutz, O., and Sattler, U. (2006) The even more irresistible SROIQ, in *Proceedings of the 10th International Conference on Principles of Knowledge Representation and Reasoning (KR2006)*, AAAI Press, pp. 57–67.

Hull, D., Stevens, R., Lord, P., and Goble, C. (2004) Integrating bioinformatics resources using shims. Twelfth International Conference on Intelligent Systems for Molecular Biology (ISMB2004), Glasgow, UK.

Hunter, L. and Cohen, K.B. (2006) Biomedical language processing: what's beyond PubMed? *Mol Cell*, **21**, 589–94.

Hunter, L., Lu, Z., Firby, J., *et al.* (2008) OpenDMAP: an open source, ontology-driven concept analysis engine, with applications to capturing knowledge regarding protein transport, protein interactions and cell-type-specific gene expression. *BMC Bioinformatics*, **9**, 78.

Huttenhower, C., Haley, E.M., Hibbs, M.A., *et al.* (2009) *Exploring the Human Genome with Functional Maps*, Cold Spring Harbor Laboratory Press.

Joshi-Tope, G., Gillespie, M., Vastrik, I., *et al.* (2005) Reactome: a knowledgebase of biological pathways. *Nucleic Acids Res.*, **33**(Database issue), D428–32.

Kessler, N. Schuhmann, H., and Morneweg, S. (2004) The linear pentadecapeptide gramicidin is assembled by four multimodular nonribosomal peptide synthetases that comprise 16 modules with 56 catalytic domains. *J. Biol. Chem.*, **279**(9), 7413–19.

Losko, S. and Heumann, K. (2009) Semantic data integration and knowledge management to represent biological network associations. *Methods Mol. Biol.*, **563**, 241–58.

Naisbitt, J. (1982) *Megatrends*, Avon Books.

Nature Editorial (2009) Data's shameful neglect. *Nature*, **461**, 145.

Neumann, E.K. and Prusak, L. (2007) Knowledge networks in the age of the Semantic Web. *Brief. Bioinformatics*, **8**(3), 141–9.

Prusak, L. and Davenport, T. (2000) *Working Knowledge: How Organizations Manage What they Know*, Harvard University Business School Press.

Qu, X.A., Gudivada, R.C., Jegga, A.G., *et al.* (2009) Inferring novel disease indications for known drugs by semantically linking drug action and disease mechanism relationships. *BMC Bioinformatics*, **10**(Suppl 5), S4.

Rubin, D.L., Lewis, S.E., Mungall, C.J., *et al.* (2006) National Center for Biomedical Ontology: advancing biomedicine through structured organization of scientific knowledge. *OMICS* **10**(2), 185–98.

Slonim, D.K. (2002) From patterns to pathways: gene expression data analysis comes of age, *Nat. Genet.*, **32**, 502–508.

Stoeckert, C.J., Causton, H.C., and Ball, C.A. (2002) Microarray databases: standards and ontologies. *Nat. Genet.*, **32**, 469–73.

2

Knowledge-driven approaches to genome-scale analysis

Hannah Tipney and Lawrence Hunter

2.1 Fundamentals

2.1.1 The genomic era and systems biology

Revolutionary changes in technologies that interrogate biological preparations at genomic scale (high-throughput genomic and proteomic approaches, such as expression microarrays, GWAS (Genome Wide Association Studies), ribosome profiling/footprinting and ChIP-chips) have given molecular biologists the ability to comprehensively study biological function and process, and their associations to disease. However, this capacity to quickly and cheaply assay biological molecules on a large scale has also posed a profound challenge to the scientific ability to analyze and explain the resulting data.

The breadth of these high-throughput, genome-wide studies has contributed to a shift in how scientists view function. As most phenomena of interest to biomedical research involve the concerted activity of hundreds of genes and their products, the results of such high-throughput technologies are usually large lists of genes deemed 'interesting' and potentially implicated in the biological conditions under study. The definition of function has moved from that of a single entity and its behavior, to the much wider concept of multiple interactions between multiple entities and how their interactions may be altered in different

Knowledge-Based Bioinformatics: From Analysis to Interpretation Edited by Gil Alterovitz and Marco Ramoni
© 2010 John Wiley & Sons, Ltd

environments at different times. In molecular biology, the biological molecules of interest are genes and their products, so the focus has moved from single genes to groups of interacting genes, proteins, non-coding RNAs and other molecules.

This approach is intuitive. Life itself is beautifully integrative and inter-related; genes and their protein products function together in complex and dynamic networks; the cells where they reside are intimately associated with each other, forming intricate organs, which in turn contribute to systems, all of which function together harmoniously to produce a viable and living organism. The analysis of data from complete biological systems, and the study of how these interactions give rise to function is more generally and broadly known as systems biology (Hunter, 2009).

Understanding the importance of these large 'interesting' gene lists, how they behave and how they contribute to function, process or disorder is a critically important task. However, the analysis of such data that is both high in volume and complexity requires significant support.

2.1.2   The exponential growth of biomedical knowledge

In order to fully understand how and why a group of genes or proteins function together, one would like to be able to exploit the wealth of published biomedical knowledge. Unfortunately this is not a simple task. To fully understand a group of genes' or proteins' behavior, not only must knowledge associated with each gene of interest be accessed, but also knowledge relating to their relationships to each other and to the experimental scenario under study. For example, consider the results of a mRNA microarray study that identified 200 genes whose expression was deemed significantly different between breast cancer cells and normal breast cells. All previous knowledge associated with each of the 200 individual genes, all knowledge corresponding to the $40\,000$ (200^2) possible interacting pairs of genes or proteins, as well as the wealth of biomedical knowledge associated with breast cancer, normal breast development, and cancer in general provide the context for understanding the experimental result.

Biomedical knowledge is primarily captured and accessed in two ways: via databases or from papers published in peer-review journals. In 2009, more than 1170 peer-reviewed databases capturing gene- and protein-centric data were freely available to the public (Galperin and Cochrane, 2009). Such databases comprise a myriad of information, ranging from raw sequence data (GenBank (Benson et al., 2008); www.ncbi.nlm.nih.gov/Genbank), to knowledge related to disease associations (OMIM (Amberger et al., 2009); www.ncbi.nlm.nih.gov/omim), pathways (KEGG (Kanehisa et al., 2006); www.genome.jp/kegg/) and molecular interactions (iRefWeb (Razick et al., 2008); http://wodaklab.org/iRefWeb/), all of which may provide important context when investigating functional behavior. The type of knowledge captured in these databases generally falls into three categories: annotations, experimentally derived data, and the supporting biomedical literature. Annotations provide basic information associated with an individual gene or protein, such as function, location, structural components and roles in disease or disorder. Experimentally derived data encompasses everything from expression profiles to GWAS data and population frequencies. And finally, databases also frequently link directly to specific biomedical papers that support and report the annotations and experimental data presented by a database.

Peer-reviewed biomedical literature is the richest, most complete and most reliable source of data. However, it can also be the most overwhelming. PubMed (Sayers *et al.*, 2009; www.ncbi.nlm.nih.gov/pubmed), the online database of biomedical literature housed at the National Center for Biotechnology Information (NCBI) at the National Library of Medicine (NLM), has been growing exponentially since the early 1980s (Figure 2.1). Currently, PubMed comprises of over 19 million citations. In 2008, a total of 811 559 new entries were added to the database at a rate of 2220 per day; far too many papers to read manually (calculated from PubMed 2008 indexed entries. Accessed October 2009).

Furthermore, this flood of knowledge has been accompanied by the breakdown of traditional disciplinary boundaries, which historically had at least made it possible to keep up with new results in one's own area of specialty. The intentionally broad nature of genome-scale assays means relevant prior knowledge can arise from nearly any biomedical discipline, and it is becoming increasingly common to discover important and unsuspected roles for genes previously characterized elsewhere. Relaxin (*RLN1*, NCBI Entrez Gene GeneID: 6013 (www.ncbi.nlm.nih.gov/gene)), for example, was discovered in 1926 by Frederick Hisaw (Hisaw, 1926) and characterized as a key birth-related cervical ripening hormone in the 1950s (Graham and Dracy, 1953). More recently, relaxin has been identified as having additional roles in processes as diverse as osteoarthritis and heart failure (Kupari *et al.*, 2005; Santora *et al.*, 2007).

In summary, not only is there an overwhelming amount of knowledge available, but more and more of it is relevant to each and every biomedical scientist. The challenge facing biomedical researchers is no longer how to produce high-quality, system-wide experimental data, but how to analyze it in an efficient and thoughtful manner. The consequence of not being able to take advantage of this wealth of biomedical knowledge is hugely costly in terms of the wasted time, effort and money chasing weak leads, inadvertently duplicating already published results and missing important discoveries. Being a cross-disciplinary biomedical scientist is no longer a choice, but a necessity in order to effectively interpret this data.

2.1.3 The challenges of finding and interacting with biomedical knowledge

In biomedicine, knowledge completely surrounds us, permeating every resource and data set available for interrogation. Valuable efforts to warehouse biomedical information relevant to the interpretation of genome-scale data in an easy to use, integrated form have produced indispensible international multi-genome databases such as NCBI housed by the National Library of Medicine (www.ncbi.nlm.nih.gov), Ensembl (www.ensembl.org, a collaborative

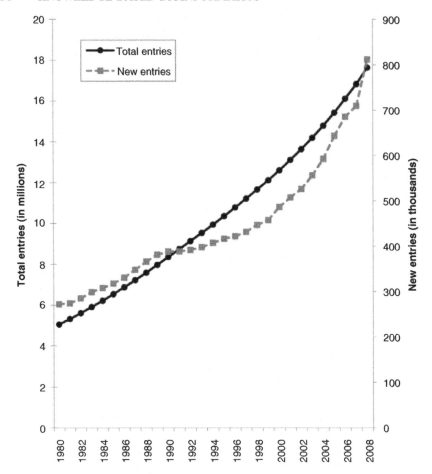

Figure 2.1   The growth of PubMed, 1980–2009. The dotted line indicates the number of new articles indexed in PubMed each year, while the solid line indicates the total number of articles indexed in PubMed at the end of each year. Data compiled from MEDLINE Citation Counts freely available at www.nlm.nih.gov/bsd/medline_cit_counts_yr_pub.html (accessed November 2009).

project between the European Bioinformatics Institute (EBI), and the Wellcome Trust Sanger Institute), and the University of California, Santa Cruz (UCSC) Genome Bioinformatics portal (http://genome.ucsc.edu), in addition to the various model organism databases such as MGI (www.informatics.jax.org) developed by The Jackson Laboratory, ZFIN at the University of Oregon (http://zfin.org) and WormBase International Consortium (www.wormbase.org).

Although essential to the modern biomedical scientist navigating genome-scale data, focusing only on the simple warehousing of biomedical data currently fails to adequately support knowledge-driven investigation for a number of reasons. Firstly, while undoubtedly wide ranging, the terabytes of data in these databases are inherently incomplete. Knowledge described in unstructured form (such as the natural language used in literature) or implied through inference is exceptionally difficult to capture, and so tends to be neglected. Secondly, the effective presentation of these large collections of data is arguably just as important as their assembly, and perhaps as difficult. These challenges are explored further in the sections below.

2.2   Challenges in knowledge-driven approaches

2.2.1   We need to read; development of automatic methods to extract data housed in the biomedical literature

While scientific methods, techniques and data sets have evolved, the publication of results, findings and hypotheses in peer-reviewed journals has remained the pre-eminent method of scientific discourse. PubMed continues to grow at the terrifying rate of 1.5 publications per minute (calculated from PubMed 2008 indexed entries, accessed October 2009), ensuring that the biomedical literature continues to be the richest source of available biomedical knowledge. However, Herbert Samuel's description of a library as 'thought in cold storage' (Hull *et al.*, 2008), conjuring images of a resource waiting, lying dormant until such time as it can truly be fully exploited, seems particularly apt when considering the knowledge captured by PubMed, and our current inability to thoroughly utilize it.

Recognizing the importance of the knowledge trapped within the biomedical literature, diligent groups of human experts manually curate key aspects of the biomedical literature and formally represent these assertions in databases, with pointers back to those documents that provided evidence for the assertion. For example, the mouse gene, caveolin 3 (*Cav3*, GeneID: 12391) is annotated with the term 'endocytosis' (Gene Ontology term identifier, GO:0006897 (http://amigo.geneontology.org)). This annotation was created due to the presence of a 'traceable author statement' in a publication, and so there is a hyperlink directly to the publication, via its PubMed ID (PMID: 10373486, (Das *et al.*, 1999)) from within the gene record. Manually identifying key features in text in a systematic and consistent manner is not an easy task, challenging even for trained annotators, and has spawned a myriad of metrics to assess the quality of manual annotations (Bada and Hunter, 2009; Artstein and Poesio, 2008).

Automatically parsing specific meaning and key terms from unstructured text is also incredibly difficult; in particular due to the ambiguous nature of terms and phrases frequently used in the natural written language of journal articles. For example, if you were searching the biomedical literature for gene products involved in ethanol synthesis, the synonyms (words of similar or identical meaning) of both 'synthesis' (e.g., biosynthesis, formation, anabolism) and of 'ethanol' (e.g., ethyl alcohol, hydroxyethane) would need to be considered to ensure a comprehensive search. Ambiguities, which although understood by humans, are difficult for computational systems to disambiguate, must be resolved (for example, alcohol may refer to ethanol or the more generic class of alcohols of which ethanol is one member), and a variety of lexical forms (e.g., anabolism of ethanol, ethyl alcohol is formed, synthesizes hydroxyethane) which can also include interspersed text (e.g., anabolism of a large amount of ethanol, ethyl and propyl alcohol are formed) all must be understood. When ambiguities are not resolved it becomes very difficult for computers to effectively identify and rationalize key pieces of information.

In an attempt to provide consistent descriptions of biomedical terms, relationships and entities, the biomedical community has turned to structured, controlled vocabularies, called *ontologies*. An ontology is a formal representation of a set of entities within a given domain, which specifies how such entities are related to each other within a hierarchy. There are many biomedical ontologies (www.obofoundry.org), but the most comprehensive is the Gene Ontology (GO), consisting of more than 28 000 terms and 5 relations (Ashburner *et al.*, 2000; www.geneontology.org). The GO comprises three separate ontologies that describe gene products in terms of their biological process, molecular function and the cellular components in which they reside. The terms and relationships within the GO are determined by community consensus and very carefully defined. Such precise and restricted definitions, that everyone in the community agrees on, enables uniform queries across different information sources, while the hierarchal structure allows queries to be undertaken at different levels. For example, one can use the GO to identify all the gene products in the human genome involved in transferase activity, or you can zoom in to all the protein serine kinases, or specify further only those products involved in transforming growth factor beta receptor activity (Figure 2.2). This structure also allows annotators to assign properties to genes or gene products at different levels, depending on the depth of knowledge about that entity.

Critically, the GO provides a controlled and structured vocabulary for describing genes and proteins that both humans and computers can understand, and therefore use when trying to describe and capture information from within the biomedical world. Recognizing this, there has been massive investment in biomedical ontologies; in the last six years alone, the US National Institutes of Health (NIH) has invested more than $52 million to support ontology development and use, including the Gene Ontology Consortium and the National Center for Biomedical Ontology (Leach *et al.*, 2009).

Yet, even with this large investment in ontological development and on-going human curation and annotation efforts, the biomedical community is still unable to keep pace with the tremendous rate at which knowledge is being published. Recent analysis showed that current manual curation rates will not be sufficient for the completion of all annotations, and even with the most optimistic of assumptions, decades will pass before annotations approach completion (Baumgartner *et al.*, 2007). Manual curation by humans is quite rightly highly regarded, producing highly accurate annotations essential for understanding the complexity

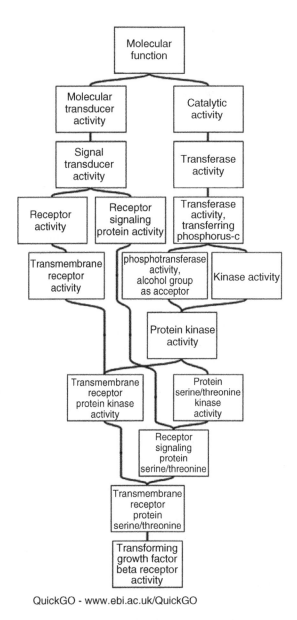

QuickGO - www.ebi.ac.uk/QuickGO

Figure 2.2   Visual representation of the Gene Ontology directed acyclic graph for the GO term 'transforming growth factor beta receptor activity' (GO:0005024). Terms are represented by boxes and linked to each other by the 'is a' relation. Graph extracted from the QuickGO website, November 2009 (www.ebi.ac.uk/QuickGO/).

of life. But it is expensive in terms of both time and money, and calls for researchers to voluntarily contribute to annotation efforts have been unsuccessful (Seringhaus and Gerstein, 2007). It is therefore critical to explore and invest in the development and use of automated methods able to effectively deal with, and extract information from, large amounts of text written in natural language that would otherwise be lost by human curation efforts.

The peer-reviewed literature is the ultimate repository of biomedical knowledge, and much of that knowledge has not been converted into structured, database form. Computational applications that process natural language (i.e., what people say or write for other people) have been developed to try to exploit the literature directly. One simple, yet surprisingly effective method is gene co-occurrence. The systematic overlap of publications that mention two genes can be used as evidence of some relationship between them and lead to the identification of involvement in common functions and processes (Schlitt *et al.*, 2003). To more comprehensively capture and extract the knowledge trapped within biomedical literature, more sophisticated methods are required, and two broad classes of approaches have been reviewed in depth in Cohen and Hunter (2004). Rule- or pattern-based approaches use background knowledge in the form of dictionaries, thesauri or ontologies, to create methods capable of distinguishing between different possible meanings in text. However, the development of such rules is largely undertaken by hand, which can be prohibitively labor intensive, and performance can be poor (Hunter and Cohen, 2006). Statistical techniques such as support vector machines have also been applied to biomedical text and generally are built by constructing models of the kinds of words that surround text of interest (Lee *et al.*, 2004). Statistical methods require large amounts of manually annotated text for training data, which is also challenging to create (Hunter and Cohen, 2006). Nevertheless, both sorts of natural language processing systems have had successes in identifying and extracting factual information from texts, in facilitating the retrieval of texts relevant to a particular query, and in creating summaries of large document collections. The broad field of application of natural language processing (NLP) techniques is sometimes called 'biomedical text mining.'

Since the general task of natural language processing is so challenging, researchers have focused in on very specific applications that are expected to be most useful. The same challenges in manually annotating texts for developing systems also make it difficult to fairly evaluate them. As slight differences in the definition of a 'useful task' and in the way the manual annotation is done can lead to significant differences in results, the biomedical text mining community has created a series of competitive evaluations to ensure that the results from different groups are comparable, and to track progress over time. For information retrieval (that is, finding documents with information relevant to a specific query), the TREC (Text REtrieval conference) Genomics Track, a yearly competition from 2003 to 2007, supported by the National Institute of Standards and Technology, was perhaps the most influential. The BioCreative competition (Critical Assessment of Information Extraction in Biology) has been held three times (2004, 2006 and 2009) and is focused on extraction of the information that could be entered into databases, with tasks oriented around identification of specific genes and evidence of protein–protein interactions. The broad computational linguistics community has also taken an interest in biomedical applications, with one of the 2009 North American Association for Computational Linguistics shared tasks focused on extraction of information about events (such as transport or phosphorylation) from the biomedical literature.

2.2.2 Implicit and implied knowledge; the forgotten data source

In addition to the wealth of biomedical knowledge explicitly stated in publications and databases, a significant amount is also hidden from view. Knowledge that is well known throughout a community is often viewed as so fundamental it no longer needs to be reiterated as it 'goes without saying'. For example, when a scientist reports the up-regulation of a protein in discussion of an mRNA microarray study, he or she assumes the reader knows that although the experiment measured mRNA levels, mRNA is being used as a proxy for protein levels (DNA makes RNA makes protein is the central dogma of biology (Alberts et al., 2008)). So the results are discussed in terms of proteins, without the author ever mentioning the mRNA–protein relationship, as they assume it to be so obvious, that it 'goes without saying.' Another example may involve the use of domain specific terms; researchers investigating the molecular mechanisms underlying facial clefting may be equally as interested in mouse models of micrognathia (mammalian phenotype ontology accession MP:0002639, (www.informatics.jax.org/searches/MP_form.shtml)), as those displaying abnormal mandibular morphology (MP:0000458) or cleft chin (MP:0000114). All three phenotypes describe abnormalities of the lower jaw. The mandible being the anatomical name for lower jaw, the chin being the common name for the region where the mandible fuses, while micrognathia is a medical term for describing an abnormally small jaw which tends to affect the mandible more frequently than the maxilla (upper jaw). It is reasonable to assume that craniofacial researchers are aware that mandible is another name for lower jaw, and that they are also familiar with medical terms used to describe common craniofacial abnormalities. Given these assumptions, when discussing or reporting these phenotypes, craniofacial researchers neglect to spell out the relationships between these phenotypes as they assume the knowledge that all three involve the mandible 'goes without saying.'

While this sort of shorthand is ubiquitous in human communication, it assumes that those who read publications or use knowledge have the same understanding or background as the author. Genome-wide technology has broken down disciplinary boundaries to such an extent that this is now a rather blinkered assumption, while this lack of boundaries also makes computational tasks much harder. The results of high-throughput experiments often include genes and proteins that the investigating biologist has never heard of, but that have been extensively studied in a different discipline in which they have no expertise. It is no longer reasonable to assume that all readers bring with them the same complement of background knowledge, so the challenge becomes to identify this high-level, so-called 'goes without saying' type of knowledge and ensure that it too is captured and represented in a way which can be used to help explore and interpret large 'interesting' gene lists.

A second and almost infinite untapped vein of knowledge is tied up in the process of thought, and is very challenging to capture. As molecular biologists analyze and think about a collection of information they make inferences about, for example, the function of a molecule, based on many different factors such as homology to other molecules, molecule interactions, expression patterns, and knockout phenotypes. Inference can come from shared physical characteristics; for example, if a novel protein is found to contain a EF-hand domain (InterPro identifier IPR002048, (www.ebi.ac.uk/interpro)), it is possible to infer that the protein has calcium binding ability, as a large number of proteins known to bind calcium do so via EF-hand domains. Or inference can be based on shared behavior; for example, if a collection of genes known to have a role in cell death display decreased expression in cancer tissues, it may be prudent to infer that genes of unknown function which also share the same expression pattern may also be involved in cell death. Another example of inference is based on shared outcome; for example, if two separate knockout mice generated by the ablation of activity of two distinct genes result in a common phenotype, such as heart hypoplasia (MP:0002740), it may be possible to infer that both genes are normally involved in the proliferation of cells during heart development. The degree to which any form of inference can be trusted can be highly variable. More weight, or importance, may be placed on an inference based on the knowledge that two genes are able to independently produce a rare phenotype, than on the knowledge that both proteins reside in the same cellular location, such as the cytoplasm.

The ability to capture knowledge which is either so pervasive in a community that it is deemed redundant and unnecessary to communicate, or which is inferred through shared attributes, behaviors or semantics, could potentially increase the depth and variety of knowledge available to researchers attempting to analyze complex data through computational tools supporting their analysis.

2.2.3 Humans are visual beings: so should their knowledge be

Once this enormous amount of knowledge has been gathered, the final but no less important consideration is how to present it to the investigating scientist. After all, a collection of hundreds of comprehensive gene summaries is no easier to digest than hundreds of journal publications, and may also not provide any clearer route to understanding a data set.

Humans are blessed with an innate ability to process large amounts of information visually. We are able to rapidly discern and identify patterns in data relatively easily when they are presented in visual form, but can struggle when the same information is presented to us in a more textual (or numerical) manner (Tao *et al.*, 2005). It is therefore unsurprising that visual representations of complicated biomedical data and results permeate almost every facet of biology. Molecular biologists, for example, are adept at looking at information in visual form; electrophoresis gels (visual representation of size), expression data as heatmaps (visualization of intensity), *in situ* hybridizations (identification of locations of functional interest), sketches of pathways and protein complexes (illustrative example of complex spatial and temporal process), and genome browsers (simplified and dynamic representation of dense and varied information organized spatially) are just some examples of how they routinely interact with information in a visual manner.

Currently, however, many representations of biological information are limited by either the type of data they depict (e.g., heatmaps and gels display numerical range) or by the lack of personalization and manipulation afforded to them during data display (e.g., KEGG is a collection of static pathway images, resistant to user manipulation). The many disparate types of knowledge which a biomedical researcher consults in order to undertake a comprehensive analysis of his or her genome-scale data posses a particularly challenging visualization problem. Not only is the breadth of the data wide (genome wide, encompassing all possible genes and all possible relationships between them), but it is also deep (large amounts of knowledge available for each gene or protein, and their relationships) and varied in both knowledge type (images, text, numerical data) and knowledge captured (interaction, structural, phenotype, and expression data to name a few).

Exploiting human visual ability should make it possible to present highly complex data, many orders of magnitude greater than would be possible with alphanumeric characters alone. The sheer volume and intricacy of data and information within the biomedical domain demands sophisticated visualization efforts if knowledge is to be made available in a systematic and intuitive manner. To date, the focus has remained on how to produce this data, with limited advances in leveraging creative visualizations to encourage and support insightful investigation of complicated biomedical data.

2.3   Current knowledge-based bioinformatics tools

As outlined above, finding ways to coherently gather, intelligently integrate and logically present biomedical knowledge of all data types, across many biomolecular entities, is a challenge, but one that is imperative to address. The output or results from high-throughput or genome-wide biomedical analyses are typically ranked lists of entities devoid of structure and lacking in context. A GWAS may produce a list of genes containing single nucleotide polymorphisms (SNPs) ranked by p-values dependent on their genetic association with a phenotype of interest. A microarray data set may yield a list of genes differentially expressed between two biomedical states ordered by statistical significance. In such instances it is not clear how (or even if) these genes and their protein products are related to each other, the biomedical scenario under study, or even what their 'normal' function may be.

Bioinformatics groups across the world are actively working to tackle this problem and a number of knowledge-based tools are freely available to assist in the analysis and interpretation of high-throughput results. The input for these tools is a set of genes or proteins identified as being of interest based on some high-throughput experiment, and typically the investigating bioscientist wants to understand how these genes and proteins are implicated in the biomedical scenario they are studying. Specifically, what are they doing, and how and with whom are they doing it? For example, if a collection of 400 genes and proteins has been identified as significantly differentially expressed during pancreatic development, questions to be addressed may include what are these genes doing during the formation of the pancreas, how do they interact with each other, what are the specific functions they undertake and how do they perform them?

2.3.1   Enrichment tools

Annotation enrichment is a strategy used by a number of bioinformatics tools to systematically map a large number of genes to their associated biological annotations and then statistically determine the most over-represented (enriched) biological annotation out of the hundreds of terms associated with the genes of interest (Huang da *et al.*, 2009). The result being the identification of a term (or terms) which describes some biological process or behavior common among a group of genes of interest, which also occurs more frequently in this group than expected by chance.

GO term enrichment analysis has become the default secondary analysis undertaken on any group of genes identified through high-throughput genomic methods, and there are a large number of tools that can be used to undertake the task (reviewed in Khatri and Draghici, 2005). Enrichment, as used during the analysis of GO terms associated with genes or proteins, is the idea that a specific term is present within a group of genes or proteins, at a frequency much higher (or lower) than would have been expected by chance. By comparing the distribution of terms within a gene set of interest to the background distribution of these terms (for example, across all genes represented on a microarray chip, or all proteins in a specific proteome), it is possible to identify terms which are under- or over-represented within a set of interest, and thus infer that these terms indicate an underlying biological function or process. For example, if 10% of the genes on the 'interesting' list are kinases, compared to 1% of the genes in the human genome (the population background) being kinases, by using common statistical methods (such as χ^2, Fisher's exact test, binomial probability or hypergeometric distribution), it is possible to determine that kinases are enriched in the gene list and therefore have important functions in the biological study undertaken (Huang da *et al.*, 2009). This type of enrichment analysis, the identification of functional classifications based on GO annotations attributed to the genes of interest, can help the development of explanations for the biological phenomena under study by moving the analysis of biological function from the level of single genes to the level of biological process (reviewed in Khatri and Draghici, 2005).

A drawback to this type of analysis is that only one knowledge or annotation source is utilized at a time, and in the case of the GO, this annotation source has some limitations. The GO is still a work in progress (Baumgartner *et al.*, 2007); its annotations remain incomplete and biased towards well-studied genes (Alterovitz *et al.*, 2007). In addition, the hierarchical structure of the GO results in many annotations that are very similar, but treated independently by enrichment methods, making it difficult to identify enrichment between different, but semantically similar terms (Khatri and Draghici, 2005). Many additional types of information or attributes can be used during enrichment analysis, such as membership to pathways, implication with disease or phenotype, and positional information such as chromosomal location. By considering many types of annotation data in concert, this type of secondary analysis can be more effective. Increased coverage increases analytical power when interpreting gene or protein lists.

The Database for Annotation, Visualization and Integrated Discovery (DAVID; Huang da *et al.*, 2009; http://david.abcc.ncifcrf.gov) developed by the Laboratory of Immunopathogenesis and Bioinformatics (LIB), NIH, does exactly that. DAVID classifies genes or proteins based on the co-occurrence of their annotation terms; however, it has progressed from using annotations from a single data source (for example GO terms), to flexibly integrating annotation terms from over 40 sources (including protein–protein interactions (PPIs), protein functional domains (e.g., InterPro), disease associations (e.g., OMIM), pathways (e.g., KEGG, BioCarta), sequence features, homology, tissue expression patterns and literature) for use during functional annotation enrichment. DAVID also undertakes functional annotation clustering. By measuring the relationship between annotation terms, DAVID is able to group similar, redundant and homogeneous annotation contents from the same or different resources into annotation groups. This reduces the burden of associating similar redundant terms, making biological interpretation more focused on a group level. DAVID also has intuitive visualizations that aid identification and investigation of the many-genes-to-many-terms associations reported, further supporting comprehensive understanding of how genes are associated with each other and their functional annotations. DAVID's ability to condense large gene lists into biologically meaningful modules greatly improves the 'ability to assimilate large amounts of information and thus switches functional annotation analysis from gene-centric to a biological module-centric analysis'(Khatri and Draghici, 2005).

The strategies outlined so far take a simple gene list as input, ignoring the qualitative information available in the data. The Gene Set Enrichment Analysis (GSEA) (Subramanian *et al.*, 2005; www.broadinstitute.org/gsea) method housed and developed by the Broad Institute, Cambridge, MA, USA, utilizes both prior biological knowledge and gene expression data. The expression data analyzed by GSEA is typically generated from samples belonging to two phenotypic classes (for example, tumor vs. normal or treated vs. untreated state), and the genes assayed are ranked based on their differential expression between these two classes. As the name implies, GSEA evaluates microarray data at the level of gene sets, which are defined on the basis of shared biological functions (e.g., a common GO category or biochemical pathway), physical position (e.g., chromosomal location), regulation (e.g., co-expression), or any other attribute for which prior knowledge is available. GSEA then determines if those genes in the gene set are randomly distributed throughout the larger ranked gene list (and therefore not significantly associated with either phenotypic class), or if they tend to be over-represented towards either the top or bottom of the larger ranked gene list, indicating an association between the gene set and the phenotypic classes under study (Figure 2.3). A phenotype-based permutation test is also used to determine the statistical significance of any enrichment, corrected to account for multiple hypothesis testing.

Focusing on sets of genes that share biologically important attributes can support the discovery of a biological function that may otherwise have been missed. For example, a small expression change in a single gene may seem inconsequential, but small increases in activity across all genes in a pathway can dramatically alter the flux of this pathway (Subramanian *et al.*, 2005). Combining expression data with *a priori* background knowledge allows the investigating scientist to identify unifying themes across data that give the results context.

All of these approaches, while helpful in discovering and understanding a biological theme common to a collection of genes or proteins, essentially produce a 'bag of genes' related by a common term. While useful, in itself this sheds little light on how the genes or proteins actually relate to each other or function together within a system. To fully understand a protein's function requires knowledge of all those entities with which it has an association (Jensen *et al.*, 2009).

2.3.2   Integration and expansion: from gene lists to networks

Life is a balanced process, in which genes, their products and various biomolecules work together in complex interacting groups and networks to create functioning systems. It is intuitive therefore to want to represent lists of genes or proteins as networks which more accurately represent the dynamic functional associations that exist between them *in vitro*. In lower, easily mutable organisms this is already a reality. In *Saccaromyces cerevisiae* (baker's yeast) PPI networks have been painstakingly constructed from systematic, genome-wide mutation experiments (Miller *et al.*, 2005; Uetz *et al.*, 2000). As of writing, the Database of Interacting Proteins (DIP; Salwinski *et al.*, 2004; http://dip.doe-mbi.ucla.edu/dip) contains records regarding 18 440 interactions amongst 4943 yeast proteins, derived from 23 034 experiments, likely a close-to-complete inventory. This is undoubtedly a valuable resource. Should a molecular biologist have a list of interesting proteins from a yeast experiment, they can instantly determine not only if these are implicated in a common process or pathway (using enrichment methods) but also exactly how each of the proteins interacts with each other, and additional proteins in the proteome. When considering experimentally validated PPIs from higher eukaryotes, however, the story is quite different.

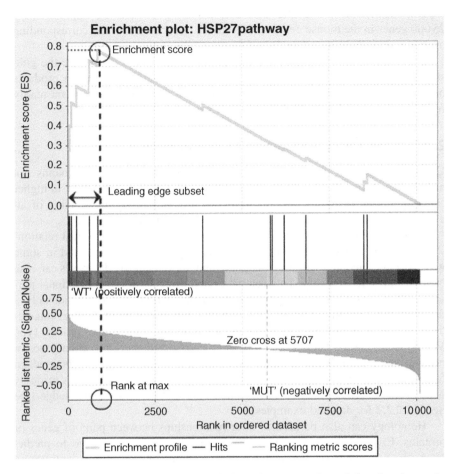

Figure 2.3   GSEA results: enrichment plot. The top section of the plot shows the running enrichment score (ES) as GSEA walks down the ranked gene list. The score at the peak of the plot (circled) is the ES for the gene set. A positive ES, as illustrated here, indicates enrichment at the top of the ranked gene list. The middle section of the plot shows where genes in the gene set appear with respect to the ranked list. The leading edge subset contains those genes which contribute most to the ES. The bottom third of the graph indicates the correlation between a gene in the ranked list and the observed phenotype. A positive value indicates a correlation with the first phenotype. Image produced by analyzing the p53 data set using the C2 (v2.5) functional human gene set as described in Subramanian et al., 2005, and illustrates enrichment in genes involved in the HSP27 pathway in NCI-60 cancer cell lines with normal vs. mutated p53. Figure produced by H. Kuehn at the Broad Institute of MIT and Harvard.

Currently, only 683 curated interactions between 502 mouse proteins from 917 experiments reside in the DIP database; assuming there to be approximately 25 000 genes in the mouse genome, this covers less than 2% of the corresponding proteome and likely less than 1% of the true PPIs in this species.

So while experimentally validated, high-coverage PPI networks are the gold standard, the 'answer' that can lay bare how genes and their protein products behave and interact to produce biological function, they are still a distant reality for the majority of biomedically important species, ourselves included.

2.3.3   Expanding the concept of an interaction

Considering just explicit physical interactions between genes or proteins not only frequently results in very sparse interaction networks (particularly in higher eukaryotes such as mouse and humans), but also utilizes a tiny fraction of all available knowledge associated with any gene or protein of interest.

While explicit or direct associations (such as PPIs) capture physical relationships between genes and proteins that can be experimentally measured in some way, attributes and annotations from gene- and protein-centric databases can also be used imply relationships between genes and proteins. From a functional perspective, an 'association' (as opposed to an interaction) can mean not only a direct physical binding between two proteins, but also a more indirect interaction, such as participation in the same metabolic pathway or cellular process (von Mering et al., 2005). These implicit or indirect associations occur when genes or proteins share a particular attribute, suggesting a possible common biomedical function. They are also examples of some of the more simple inferences made by biomedical scientists when reviewing their data and associated knowledge (see Section 2.2.2 for detailed examples).

Homology can also be used to infer relationships between pairs of genes or proteins. Experimentally validated PPIs in one species can be used to predict interactions in other species using orthology. Otholognes are genes in different species that are similar because they share a common ancestral gene; they were separated by a speciation event. A pair of interacting orthologs are known as interologs (Walhout et al., 2000), and the presence of a pair of interologs in a reference species is used to predict the presence of an interaction in the test species. The success of this method is highly dependent on the accurate prediction of orthology through the use of complex algorithms, rather than just sequence similarity (O'Brien et al., 2005). Interolog mapping has become an established method for predicating interactomes (Lehner and Fraser, 2004; Yu et al., 2004) and has been successfully used to predict prokaryote PPI networks (Arabidopsis) from interacting orthologs in eukaryotes (yeast, worm, fruit fly and human; Geisler-Lee et al., 2007).

By considering these implicit or indirect relationships between genes and their products, it is possible to create a much richer network, exploit a much greater proportion of the pre-existing biomedical knowledge, and also go some way towards capturing the often neglected source of implied and inferred knowledge.

STRING (Search Tool for the Retrieval of Interacting Genes/Proteins) is one of the next generation of integrated, multispecies databases which incorporate both direct and indirect protein–protein associations (Jensen *et al.*, 2009; http://string.embl.de). STRING aggregates, scores and weights data from four types of sources: genomic context (conserved genomic neighborhood, gene fusion events, co-occurrence of genes across genomes), high throughput experiments, co-expression/conservation, and prior knowledge (gene name co-mentions in PubMed abstracts). STRING behaves like a meta-database, mapping a wide variety of interaction data onto a common set of genomes and proteins, with its ultimate aim being to represent the union of all possible protein–protein associations (Jensen *et al.*, 2009). An interactive and dynamic interface presents and encourages the exploration of the evidence supporting each association, and also aids the comprehensive investigation of a group of interesting proteins and their associations with other biological entities (Figure 2.4).

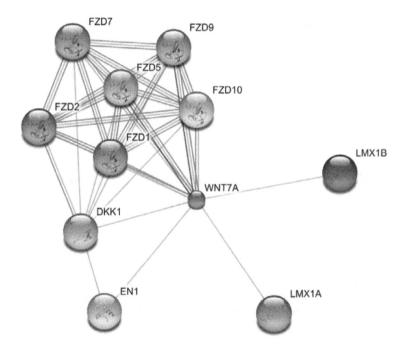

Figure 2.4   STRING integrated network – as produced by STRING 8.2 when searching for interactions involving the human WNT7A precursor. Interactions with 10 additional biomolecules are identified, with the number of arcs between any pair of entities indicating the different types of evidence supporting each interaction. Image produced by STRING 8.2, searching for Homo sapiens *WNT7A with default parameters, accessed November 2009 (http://string.embl.de; Jensen et al., 2009).*

2.3.4 A systematic failure to support advanced scientific reasoning

However, even when considering the most sophisticated of current bioinformatics tools, a disconnect remains between the support required by bioscientists to fully exploit their data sets, and the support these tools provide.

Scientists of any discipline, when conducting discovery-driven investigations, progressively and recursively apply three types of reasoning: classification, model-based, and narrative (Neressian, 2002). In the context of molecular biology, current bioinformatics tools excel in identifying surface-level insights associated with classification reasoning (for example classifying and identifying multidimensional relationships of interest). But they fail to support those deep causal insights essential for the development of novel and credible hypotheses, which can only be achieved by supporting high-order model-based and narrative reasoning activities (Mirel, 2009). Model-based reasoning requires the investigating scientist to interpret the data presented to them and engage in inference (for example in order to explain how and why a biological event may influence a particular disease mechanism), while narrative reasoning builds on, and refers to, both classification and mental modeling to create and develop stories which aid understanding of causal relationships (Mirel, 2009; Latour, 1999; Table 2.1). For scientists to progress from lists of genes and proteins, to novel hypotheses in which they are confident enough to invest their scarce research time and dollars, just delivering data is not good enough. To simulate hypothesis generation, bioinformatics tools must integrate and comprehensively support all three types of reasoning (Mirel, 2009).

2.4 3R systems: reading, reasoning and reporting the way towards biomedical discovery

Promising advances in tackling this complex challenge have recently been reported. Taking its name from the three broad classes of algorithms it utilizes (reading, reasoning and reporting), 3R systems focus specifically on assisting biologists to develop deep explanations for the biological phenomena they observe in their genome-scale data (Leach *et al.*, 2009). Introduced as a new approach to the exploration of high-throughput data capable of accelerating biomedical discovery, 3R systems integrate data from gene- and protein-centric databases and biomedical language ('reading'), with symbolic and quantitative network inference techniques ('reasoning') and visual approaches for presenting and dynamically exploring complex networks of information ('reporting'). In combination, this also addresses some of the difficulties associated with supporting the more complex forms of reasoning (model- and narrative-based) undertaken by scientists when carrying out discovery-based investigations. Significantly, the 3R approach focuses on the comparison and combination of two weighted networks. The 'knowledge network' is a representation of a large

Table 2.1   The stages of analysis and reasoning undertaken by scientists during discovery-driven investigation. Inspired by Table 1 of Mirel (2009).

Type of reasoning	Example question	Task description	Example task
Confirmation (validation)	Do I trust this tool? Do I trust its content?	Vetting the query results and the tool for accuracy, reliability, and timeliness	Search for familiar literature references
Classification and validation	Tell me what you know? Show me how my data fits into pre-existing categories, e.g., into a KEGG pathway	Classify relationships to find genes and protein interactions of interest	Find an association between a candidate gene from experimental findings and a protein associated with a disease
Model-based reasoning and validation	Why is this of interest to me? How does this relate to my work?	Place relationships of interest in context to mentally model explanatory biological events relevant to disease	Contextualize significant regulatory relationships in pathways
Narrative reasoning and validation	How does this fit with my work? Does it explain something I have observed? How does this data come together to tell a biological story? Is this a testable hypothesis?	Turning explanations about biological events into new, credible and plausible biological stories	No studied tasks exist, as observed scientists failed to achieve this level of reasoning using currently available bioinformatics tools

proportion of all available biomedical knowledge pertaining to genes, their protein products and their interactions. The 'data network' describes a particular high-throughput experimental data set (Leach *et al.*, 2009). In both graphs, biological entities that can be clearly and unambiguously defined are used as nodes, and arcs or lines between two nodes indicate some kind of association or relationship. The combination and display of both these graphs enables not only the exploration of existing knowledge in the context of a particular biological phenomenon, but also the development of novel and credible hypotheses.

2.4.1   3R knowledge networks populated by reading and reasoning

To be comprehensive, the knowledge network is populated with information 'read' from both biomedical databases and the biomedical literature. The gene- and protein-centric knowledge housed in biomedical databases comes in many different forms, but in nearly all instances is tied to a unique gene or protein identifier. The ability to unambiguously identify a gene or its protein products across a number of disparate resources allows the collation of diverse knowledge types, all relevant to a single gene or protein, in a single location. The 3R knowledge network uses NCBI Entrez Gene GeneID and Taxonomy Identifiers to disambiguate information extracted both at the gene and species level.

Explicit or direct associations (from experimentally validated PPIs), were retrieved from iRefWeb (Razick *et al.*, 2008), which itself integrates PPI information from the BIND, BioGRID, IntAct, MINT, MPPI, and OPHID databases. Because a single interaction between two proteins may be reported multiple times in different publications using different experimental methods, the 3R system uses PubMed identifiers and the Molecular Interaction Ontology (http://psidev.sourceforge.net/mi/rel25/data/psi-mi25.obo) to determine when PPIs are genuinely novel and when they are redundant. This is important because a bioscientist assessing PPI data between two proteins of interest may, for example, view multiple assertions of an interaction by a high-throughput method (such as yeast two hybrid) in two publications differently from an interaction which has been demonstrated using three different experimental methods, also reported by two independent publications. The provenance of these interactions and how they are reported is critical for a bioscientist to determine how much weight he or she wants to give a set of evidence. By capturing and displaying provenance data, such as the different data sources and biological methods used in determining a physical interaction, scientists are able to assess for themselves which information they wish to trust, and just how much they trust it. Additional explicit protein–DNA interactions were extracted from the TRANSFAC 10.2 (Wingender *et al.*, 1996) and PReMod (Ferretti *et al.*, 2007; http://genomequebec.mcgill.ca/PReMod) databases.

3R systems also 'read' the biomedical literature, and a simple but effective modification of co-occurrence, called the 'asymmetric co-occurrence fraction' (ACF) measure (Gabow *et al.*, 2008), was used to identify relationships between genes and/or proteins from Medline abstracts (Leach *et al.*, 2009). The pattern-based OpenDMAP system (Hunter *et al.*, 2008) is also being implemented to extract a wider variety of knowledge from abstracts (such as protein transportation events, PPIs, and expression locations). OpenDMAP is a general framework for recognizing instances of ontology terms and relationships among them in biomedical texts. It is dependent on sets of patterns specific to each ontology term and relationships among them in order to function. OpenDMAP has previously demonstrated its utility in extracting biologically important information during the 2006 BioCreative II evaluation by achieving the best accuracy at the protein interaction pairs subtask, which involved correctly identifying both of the interacting proteins and mapping them to database identifiers (Krallinger *et al.*, 2007). OpenDMAP is the first system developed to exploit community-consensus ontologies (e.g., the GO) as the central organizing principle of an information extraction system, and this has been credited for much of its observed success.

Additional arcs or edges can be added to the knowledge network based on simple inference. An arc was added if two genes were both annotated to the same metabolic pathway (Kanehisa *et al.*, 2006), the same Gene Ontology term (divided into molecular function, biological process and cellular component hierarchies), the same knockout phenotype, or shared at least one protein domain assignment. Currently, 13 different knowledge sources are used to infer associations between genes or proteins (including GAD, the GO, ChEBI, InterPro, KEGG, Reactome, and the MGI Phenotype database). For the knowledge resources that involve a nested hierarchy (such as the Gene Ontology), terms are merged when the information content score by the Jiang measure between them exceeds 19.0 (Lord *et al.*, 2003).

2.4.2   Implied association results in uncertainty

The resultant 3R knowledge network incorporates many different types of information from a variety of knowledge sources. However, not all associations asserted are equal. While the inclusion of implicit or inferred interactions drastically increases the number of interactions asserted and proportion of the proteome covered, it also contributes a large number of false positives due to computational predictions, experimental noise and the intentionally noisy nature of these inferred associations. For example, it is unlikely that all cytoplasmic proteins actually interact with each other as the GO cellular component knowledge source implies, but this co-localization information is useful when assessing a relationship if considered in conjunction with information obtained from other knowledge sources (Leach *et al.*, 2009). For this expanded coverage to be useful, there must be a way to assess the quality of the associations asserted.

While experimentally validated interactions between a pair of proteins can be seen as the gold standard of interaction data, meaningful assessment of the quality and reliability of associations asserted by other knowledge sources is challenging. When a large collection of interactions, asserted by a wide range of different knowledge sources, is available, a gold standard data set of validated interactions can be created and used to empirically assess the reliability of each interaction asserted by a given knowledge source. In well-studied organisms such as yeast, where the majority of interactions between genes and proteins have been validated, holding a subset of these interactions (or even all assertions from a particular knowledge source) as a gold standard set is a reasonable practice that enables the estimation of error rates for each knowledge source. In organisms such as human and mouse, however, there are already too few knowledge sources to justify withholding one as the gold standard. The inclusion of indirect associations between proteins or genes further compounds the issue by making it difficult to even determine which knowledge source would be an appropriate gold standard (Leach *et al.*, 2009).

The consensus reliability estimate (Leach *et al.*, 2007) does not require the designation of an explicit gold standard, and so was used to assign a reliability score to each knowledge source within the 3R system. Knowledge sources that explicitly name both genes and/or proteins (such as PPIs from iRefWeb, protein–DNA interactions from TRANSFAC and PReMod, and entities identified from the literature using the ACF) were used for the reliability calculation. The consensus estimate gives a higher reliability to a knowledge source if many other knowledge sources agree with its assertions on average (Leach *et al.*, 2007). All associations asserted by a given knowledge source are given the consensus reliability of that particular knowledge source.

It is entirely possible, and in fact common and desirable, to extract multiple assertions of a relationship between a pair of genes or proteins (and so multiple arcs) from different knowledge sources. In such instances, a summary arc with a higher reliability was generated using the 'noisy OR' function $P = 1 - \Pi_i (1 - r_i)$, where r_i is the reliability of knowledge source i (which can be scaled if necessary into the range 0 to 1 if probability is preferred; von Mering *et al.*, 2005; Li *et al.*, 2006; Sun *et al.*, 2007). Finally, in addition to calculating the overall reliability of each assertion from each knowledge source, 3R systems track the provenance of the sources of information used to generate the arcs in the knowledge network, for later display and exploration.

2.4.3   Reporting: using 3R knowledge networks to tell biological stories

Ultimately, the aim of 3R systems is to bring all this captured knowledge to bear on analyzing high-throughput experimental results, to explore large and complex data sets in light of all prior knowledge and to aid hypothesis generation. To tell a story, or develop a hypothesis, context is required, and so the identification of sub-networks containing just that knowledge pertinent to the investigating scientist is critical. The many protein complexes, interactions and functional pathways present within any biological system are all context-dependent, often transient occurrences dependent on particular temporal and spatial environments. No matter what the current state of the knowledge, at any given time only a subset will apply to the particular biological scenario under investigation and the rest may be considered noise. Using the results of the high-throughput experiment under study is therefore an intuitive way to achieve the context required for explanation development.

A simplistic approach to utilizing these experimental results is to visualize a subsection of the knowledge network that only includes nodes corresponding to those genes or proteins identified as 'interesting.' Such sub-networks, including the knowledge associated with the nodes and arcs, can then be visualized and explored using Cytoscape (Shannon *et al.*, 2003; Cline *et al.*, 2007; www.cytoscape.org/), an open source visualization tool. This approach has been successfully used to identify functional explanations of a gene list by exploiting inferences not present in any single knowledge source (Tipney *et al.*, 2009a).

A more effective method of harnessing this experimental data, however, is to create a second network from the experimental data and then combine both the knowledge and experimental data networks in ways that support intuitive and deep exploitation of the knowledge. In the application outlined below, genes were identified as interesting if they displayed differential expression during a large microarray study, then arcs were drawn between these genes of interest if they also displayed correlated expression (above a given threshold). The numerical weights on the arcs in this data network are the absolute correlation coefficients, while the nodes represent the genes of interest. Conceptually, data networks can be constructed from any high-throughput experimental data set of interest in which an arc can indicate an experimental association (for example, in-house protein or mRNA microarray studies, publically available data sets housed in GEO, GWAS, or metagenomic studies). In the instance of correlated expression, this indicates a possible functional relationship between the correlated genes; the assumption being that genes with similar biological behaviors (i.e., correlated expression) are likely involved in similar biological processes and functions.

Both the knowledge and experimental data networks have a common reference in the form of nodes representing genes or proteins, which supports the combination of both networks. Currently, 3R systems have successfully applied two different combination methods (Average and Hanisch Logit), and their utility is described in the following section.

2.5   The Hanalyzer: a proof of 3R concept

The Hanalyzer is a recently published implementation of a 3R system, which played a key role in understanding a craniofacial expression microarray time series, and in generating hypotheses about the function of four genes previously unsuspected to have a role in facial muscle development in the formation of the tongue (Leach *et al.*, 2009). The Hanalyzer, including Cytoscape plug-ins for visualization, is available as open source software via SourceForge at http://hanalyzer.sourceforge.net.

In this application, the mouse transcriptome was sampled at 12-hour intervals from E (embryonic day) 10.5 to 12.5, a period during development that begins with the formation of the facial prominences and ends when they fuse together to form the mature facial platform. Samples from three distinct facial regions (the frontonasal, maxillary, and mandibular prominences) were isolated at each time-point, with seven independent biological replicates prepared and analyzed for each sample. This data set and its initial analysis are described in detail in Feng *et al.*, 2009.

The knowledge network used in the Hanalyzer implementation was constructed (as described in Section 2.4.1) from nodes taken from a list of mouse genes that were significantly differentially expressed in this craniofacial development data set. The data network was also generated from this expression data set. Expression levels for all replicates were averaged, and for each gene the averages were normalized to the $\log_2$ ratio of the median expression level across all time points and tissues. Nodes representing each gene were then linked with arcs quantified by the Pearson correlation coefficient over time and tissue for all pairs of genes.

To fully investigate this craniofacial data set, the arcs of the knowledge and data networks were combined (since the nodes in both networks are mouse genes, it is trivial to align the nodes) using two different methods. A number of different methods for combining arcs in the data and knowledge networks are possible and are explored in detail in Leach *et al.*, 2009. However, since the goal when combining the networks was to highlight concurrence, the 'Average' and 'Hanisch Logit' methods were chosen due to their dependence on agreement between both knowledge and data sources in order to achieve a high combined score.

The Average method requires simply taking the mean of the reliability of the knowledge arc and the correlation coefficient in the corresponding data arc. This method gives equal weight to the contribution of information from both the background knowledge and the experimental data sources, providing scores that are an indication of how well observations from the experimental data set are supported by knowledge in the literature. The second method, a logistic combination function (Sohler *et al.*, 2004) which the authors named Hanisch Logit, averages the logistic function of each arc in the knowledge and corresponding data network. The distributions of scores in the Hanalyzer knowledge and data networks (the data scores were distributed closer to 1 than the knowledge scores), means that the Hanisch Logit metric tends to assert arcs with high scores in the data, but with less contribution from the knowledge network (Leach *et al.*, 2009).

To further prune each combined network, those arcs supported by less than three knowledge sources were excluded, and only the top 1000 highest scoring arcs from each combination metric were visualized in Cytoscape. In recognition that scientists rapidly became 'stymied when interface displays did not afford adequate interactivity' (Mirel, 2009), the Hanalyzer authors developed a number of Cytoscape plug-ins (freely downloadable from Cytoscape's plug-in menu) to aid exploration of the combined networks. The plug-ins were designed specifically to work with the unique collection of knowledge and data in these 3R networks, and support its exploration through the easy and interactive exposure and manipulation of information within the network (Tipney *et al.*, 2009b; Leach *et al.*, 2009).

The CommonAttributes plug-in (Leach *et al.*, 2009) enables the investigating scientist to quickly explore network provenance, exposing not just the knowledge sources supporting the assertion of an arc between two nodes (such as a common GO term, and co-occurrence in the literature), but also the specific details of the relationship. For example, that the term shared by two genes is GO:0006936 'muscle contraction' from the biological process GO, and that this GO term was attributed to each protein from separate publications, with direct links to these publications in PubMed; or in the instance of co-occurrence, the identification and hyperlinking to the exact abstracts in which a pair of proteins co-occur. The HiderSlider plug-in 'slides' over numerical values attributed to the arcs in the combined network, allowing the investigating scientist to determine which relationships they are exposed to, using arc weights as a filter (Tipney *et al.*, 2009b). The plug-in improves the experience of interacting with these complex networks. Being able to interact transparently with the network provenance dynamically allowed scientists to determine how willing they were to trust different assertions, as well as the network itself, while also providing access to detailed knowledge in context, which supports the creation and development of novel hypotheses (Mirel, 2009). Figure 2.5 outlines the full Hanalyzer system implementation.

The use of two different combination metrics allowed different reporting goals to be addressed during visualization. Arcs asserted by the Average method are strongly supported by both background knowledge and the experimental data, and so identify those relationships that are already understood, providing rapid orientation for the investigating bioscientist. Arcs asserted by the Hanisch Logit method (and not the Average method) indicate relationships which are strongly supported by the experimental data, but that have only modest support in the background knowledge. These arcs indicate associations that may be novel and unreported, and were used in this implementation to generate novel hypotheses.

By reviewing the combined 3R network, focusing on those arcs asserted by the Average metric, a sub-network of 20 genes of interest connected by 50 arcs was identified. Through investigation of the knowledge shared across these genes, it was observed that all 20 genes and their protein products were associated in some way to 'muscle', and more specifically that the network was involved in force generation and structural integrity of muscle. However, this network was intriguing not only because of its strong muscle theme but because the genes within this network displayed highly correlated expression with striking mandibular specificity (Figure 2.6). The expression of these 20 genes was consistently and exclusively up-regulated in the mandibular sample as development progresses from E10.5 to 12.5. Literature indicated that this expression profile was consistent with tongue muscle development; the tongue being the largest single muscle mass in the head and located within the mandible. At approximately E11, the migration of myogenic cells into the tongue primordia is complete, with myoblasts continuing to proliferate and differentiate until around E15, when they fuse and withdraw from the cell cycle (Yamane *et al.*, 2000). The same 20 genes were also up-regulated at the later E12–12.5 time point in the maxilla sample, consistent with the staggered onset of skeletal muscle cell

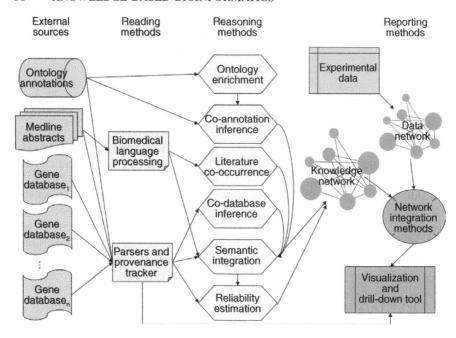

Figure 2.5   Hanalyzer system diagram, describing the modules of the Hanalyzer. Reading methods (rectangles) take external sources of knowledge (shapes on the far left) and extract information from them, either by parsing structured data or biomedical language processing to extract information from unstructured data. Reading modules are responsible for tracking the provenance of all knowledge. Reasoning methods (hexagons) enrich the knowledge that results from reading by, for example, noting two genes that are annotated to the same ontology term or database entry. All knowledge sources, read or reasoned, are assigned a reliability score, and all are combined using that score into a knowledge network that represents the integration of all sorts of relationships between a pair of genes and a combined reliability score. A data network is created from experimental results to be analyzed. The reporting modules (shapes in the bottom right) integrate the data and knowledge networks, producing visualizations that can be queried with the associated drill-down tool. (Adapted from Figure 1, Leach et al., 2009).

differentiation. The tongue matures approximately 1.5 days (in mice) earlier than all other skeletal muscles, and this is thought to correlate with the tongue's requirement for mammalian suckling immediately after birth (Amano *et al.*, 2002). The lack of significant muscle in the frontonasal prominence accounts for the low level of expression of these genes in that tissue.

Using the knowledge network to systematically report and explore the complete collection of relevant background knowledge made the interpretation of this complex set of evidence regarding the broad developmental function of a complex group of interacting genes a much more straightforward task. Once the

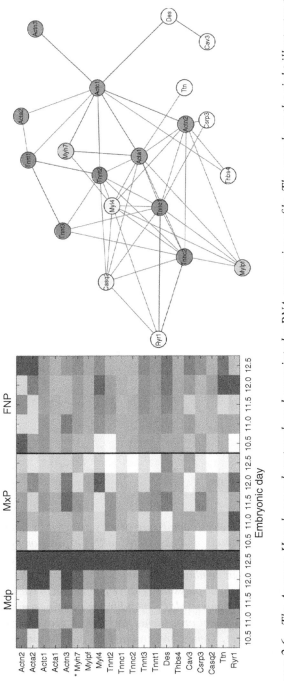

Figure 2.6 The Average Hanalyzer sub-network and associated mRNA expression profile. The graph on the right illustrates the sub-network generated by viewing only those edges asserted by the Average combinatorial metric. A total of 20 nodes and 50 edges are present. Nodes are labeled by gene symbol with shading indicating different protein families. White nodes indicate no common protein family. The heatmap on the left displays the relative expression of each gene in the Average sub-network across five time points and three tissues (MdP: mandibular prominence, MxP: maxillary prominence, FNP: frontonasal prominence), with dark gray indicating higher expression and white lower. Genes are grouped by protein family and clustered within these functional groups. Genes whose expression was classed as 'absent' in >99% of the samples are indicated by a * and are included here for completeness. (Adapted from Figures 7 and 8, Leach et al., 2009).

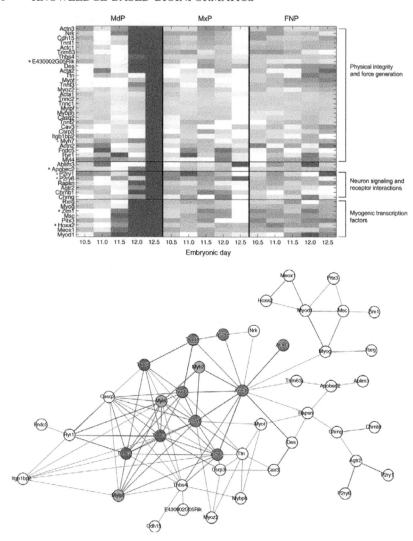

Figure 2.7   The Hanalyzer sub-network comprising of arcs asserted by both the Average and the Hanisch Logit methods, and associated mRNA expression profile. The graph illustrates the sub-network generated by viewing edges asserted by both the Average and Hanisch Logit combinatorial metrics. Nodes are colored as described in Figure 2.6. The heatmap displays the relative expression of each gene across five time points and three tissues (MdP: mandibular prominence, MxP: maxillary prominence, FNP: frontonasal prominence), with dark gray indicating higher expression and white lower. Genes are grouped by protein family and clustered within these functional groups. Genes whose expression was classed as 'absent' in >99% of the samples are indicated by a * and are included here for completeness. (Adapted from Figures 9 and 10, Leach et al., 2009).

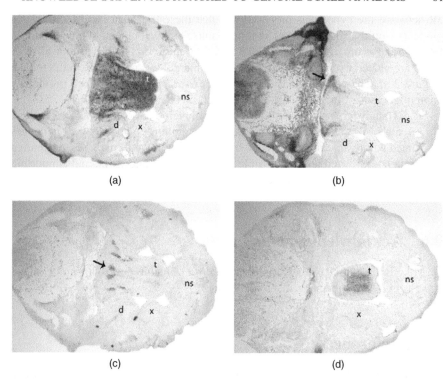

Figure 2.8   Gene expression in the developing mouse tongue. In situ *hybridiza-tion using anti-sense probes for* Zim1 (a), Hoxa2 (b), Apobec2 (c), *and* E430002G05Rik (d). *All panes are transverse sections of an E12.5 mouse head. Anterior is to the right on all panels. Dark staining represents hybridization signal from the probe. The arrows indicate areas of fainter staining. Control experiments using sense probes did not yield specific staining. d, mandible; ns, nasal septum; t, tongue; x, maxillary process. (Adapted from Figure 11, Leach et al., 2009).*

well-understood aspects of the sub-network had been explored and a biological explanation for the observations created, the arcs asserted only by the logistic metric (those with strong support in the experimental data and more modest support in prior knowledge) were added to the visualization, introducing an additional 25 genes to the network (total 45; Figure 2.7). The genes comprising this larger network displayed the same striking mandible-specific expression pattern seen in the Average combination network, suggesting that these additional genes were also implicated in tongue development. Exploring the collated knowledge associated with the additional genes indicated that many were indeed known to be involved in myogenic differentiation and synapse interactions. However, four genes (Riken clone *E430002G05Rik, Zim1, Apobec2* and *Hoxa2*) had no known relation to muscle development and were selected for experimental validation. Based on understanding developed during exploration of the combined

3R network, the hypothesis proposed was that these four genes and their protein products would also have a role in tongue muscle development. Whole-mount *in situ* hybridizations to E11.5 and E12.5 mouse embryos confirmed that all four genes are indeed expressed in the developing mandible, specifically in the tongue (Figure 2.8). These findings also implicated *Hoxa2* perturbation as a previously unreported cause of cleft palate (Leach *et al.*, 2009).

The Hanalyzer, as a proof of concept implementation, demonstrates how the 3R approach to knowledge integration is beginning to address the challenges associated with extracting knowledge from the literature and databases, presenting this complex knowledge in accessible and understandable ways, and supporting the higher order cognitive processes bioscientists engage in when they develop credible and novel hypotheses.

2.6   Acknowledgements

Dr Mike Bada and Dr Anis Karimpour-Fard for valuable discussions on content, and Justin Webb for image manipulation.

2.7   References

Alberts, B., Johnson, A., Lewis, J., *et al.* (2008) *Molecular Biology of the Cell*, 5th edn, Garland Science.

Alterovitz, G., Xiang, M., Mohan, M., and Ramoni, M.F. (2007) GO PaD: the Gene Ontology Partition Database. *Nucleic Acids Res.*, **35**, D322–7.

Amano, A., Yamane, A., Shimada, M., *et al.* (2002) Hepatocyte growth factor is essential for the migration of myogenic cells and promotes their proliferation during the early periods of tongue morphogenesis in mouse embryos. *Dev. Dyn.*, **223**, 169–79.

Amberger, J., Bocchini, C.A., Scott, A.F., and Hamosh, A. (2009) McKusick's Online Mendelian Inheritance in Man (OMIM). *Nucleic Acids Res.*, **37**, D793–6.

Artstein, R. and Poesio, M. (2008) Inter-coder agreement for computational linguistics. *Comput Linguist*, **34**, 555–96.

Ashburner, M., Ball, C.A., Blake, J.A., *et al.* (2000) Gene ontology: tool for the unification of biology. The Gene Ontology Consortium. *Nat. Genet.*, **25**, 25–9.

Bada, M. and Hunter, L. (2009) Using the Gene Ontology to annotate biomedical journal articles, in *Conference Proceedings: ICBO International Conference on Biomedical Ontologies July 24-26, 2009 Buffalo, New York, USA* (ed. B. Smith), University at Buffalo College of Arts and Sciences, Buffalo, NY, pp. 135–8.

Baumgartner, W.A., Jr., Cohen, K.B., Fox, L.M., *et al.* (2007) Manual curation is not sufficient for annotation of genomic databases. *Bioinformatics*, **23**(13), i41–8.

Benson, D.A., Karsch-Mizrachi, I., Lipman, D.J., *et al.* (2008) GenBank. *Nucleic Acids Res.*, **36**, D25–30.

Cline, M.S., Smoot, M., Cerami, E., *et al.* (2007) Integration of biological networks and gene expression data using Cytoscape. *Nat Protoc*, **2**, 2366–82.

Cohen, K.B. and Hunter, L. (2004) Natural language processing and systems biology, in *Artificial Intelligence Methods and Tools for Systems Biology* (eds W. Dubitzky and F. Azuaje), Springer, Norwell, MA, pp. 147–73.

Das, K., Lewis, R.Y., Scherer, P.E., and Lisanti, M.P. (1999) The membrane-spanning domains of caveolins-1 and -2 mediate the formation of caveolin hetero-oligomers. Implications for the assembly of caveolae membranes in vivo. *J. Biol. Chem.*, **274**, 18721–8.

Feng, W., Leach, S.M., Tipney, H., *et al.* (2009) Spatial and temporal analysis of gene expression during growth and fusion of the mouse facial prominences. *PLoS One*, **4**, e8066.

Ferretti, V., Poitras, C., Bergeron, D., *et al.* (2007) PReMod: a database of genome-wide mammalian cis-regulatory module predictions. *Nucleic Acids Res.*, **35**, D122–6.

Gabow, A.P., Leach, S.M., Baumgartner, W.A., *et al.* (2008) Improving protein function prediction methods with integrated literature data. *BMC Bioinformatics*, **9**, 198.

Galperin, M.Y. and Cochrane, G.R. (2009) Nucleic Acids Research annual Database Issue and the NAR online Molecular Biology Database Collection in 2009. *Nucleic Acids Res.*, **37**, D1–4.

Geisler-Lee, J., O'Toole, N., Ammar, R., *et al.* (2007) A predicted interactome for Arabidopsis. *Plant Physiol.*, **145**, 317–29.

Graham, E.F. and Dracy, A.E. (1953) The effect of relaxin – mechanical dilatation of the bovine cervix. *J. Dairy Sci.*, **36**, 772–7.

Hisaw, F.L. (1926) Experimental relaxation of the pubic ligament of the guinea pig. *Proc. Soc. Exp. Biol. Med.*, **23**, 661–3.

Huang da, W., Sherman, B.T., and Lempicki, R.A. (2009) Systematic and integrative analysis of large gene lists using DAVID bioinformatics resources. *Nat Protoc*, **4**, 44–57.

Hull, D., Pettifer, S.R., and Kell, D.B. (2008) Defrosting the digital library: bibliographic tools for the next generation web. *PLoS Comput. Biol.*, **4**, e1000204.

Hunter, L. (2009) Anatomy, physiology, and systems biology, in *The Processes of Life. An Introduction to Molecular Biology*. The MIT Press, Cambridge, MA, pp. 165–88.

Hunter, L. and Cohen, K.B. (2006) Biomedical language processing: what's beyond PubMed? *Mol. Cell*, **21**, 589–94.

Hunter, L., Lu, Z., Firby, J., *et al.* (2008) OpenDMAP: an open source, ontology-driven concept analysis engine, with applications to capturing knowledge regarding protein transport, protein interactions and cell-type-specific gene expression. *BMC Bioinformatics*, **9**, 78.

Jensen, L.J., Kuhn, M., Stark, M., *et al.* (2009) STRING 8 – a global view on proteins and their functional interactions in 630 organisms. *Nucleic Acids Res.*, **37**, D412–16.

Kanehisa, M., Goto, S., Hattori, M., *et al.* (2006) From genomics to chemical genomics: new developments in KEGG. *Nucleic Acids Res.*, **34**, D354–7.

Khatri, P. and Draghici, S. (2005) Ontological analysis of gene expression data: current tools, limitations, and open problems. *Bioinformatics*, **21**, 3587–95.

Krallinger, M., Leitner, F., and Valencia, A. (2007) Assessment of the second BioCreative PPI task: automatic extraction of protein-protein interactions, in *Proceedings of the Second BioCreative Challenge Evaluation Workshop (2007) Madrid, Spain*, Fundación CNIO Carlos III, pp. 41–54.

Kupari, M., Mikkola, T.S., Turto, H., and Lommi, J. (2005) Is the pregnancy hormone relaxin an important player in human heart failure? *Eur. J. Heart Fail.*, **7**, 195–8.

Latour, B. (1999) *Pandora's Hope: Essays on the Reality of Science*, Harvard University Press, Cambridge, MA.

Leach, S., Gabow, A., Hunter, L., and Goldberg, D.S. (2007) Assessing and combining reliability of protein interaction sources. *Pac Symp Biocomput.*, **2007**, 433–44.

Leach, S.M., Tipney, H., Feng, W., *et al.* (2009) Biomedical discovery acceleration, with applications to craniofacial development. *PLoS Comput Biol.*, **5**(3), e1000215.

Lee, K.J., Hwang, Y.S., Kim, S., and Rim, H.C. (2004) Biomedical named entity recognition using two-phase model based on SVMs. *J Biomed Inform*, **37**, 436–47.

Lehner, B. and Fraser, A.G. (2004) A first-draft human protein-interaction map. *Genome Biol.*, **5**, R63.

Li, J., Li, X., Su, H., *et al.* (2006) A framework of integrating gene relations from heterogeneous data sources: an experiment on Arabidopsis thaliana. *Bioinformatics*, **22**, 2037–43.

Lord, P.W., Stevens, R.D., Brass, A., and Goble, C.A. (2003) Investigating semantic similarity measures across the Gene Ontology: the relationship between sequence and annotation. *Bioinformatics*, **19**, 1275–83.

von Mering, C., Jensen, L.J., Snel, B., *et al.* (2005) STRING: known and predicted protein-protein associations, integrated and transferred across organisms. *Nucleic Acids Res.*, **33**, D433–7.

Miller, J.P., Lo, R.S., Ben-Hur, A., *et al.* (2005) Large-scale identification of yeast integral membrane protein interactions. *Proc. Natl. Acad. Sci. U.S.A.*, **102**, 12123–8.

Mirel, B. (2009) Supporting cognition in systems biology analysis: findings on users' processes and design implications. *J Biomed Discov Collab*, **4**, 2.

Neressian, N. (2002) The cognitive basis of model-based reasoning in science, in *The Cognitive Basis of Science* (eds P. Carruthers, S. Stich, and M. Siegal). Cambridge University Press, Cambridge, pp. 133–53.

O'Brien, K.P., Remm, M., and Sonnhammer, E.L. (2005) Inparanoid: a comprehensive database of eukaryotic orthologs. *Nucleic Acids Res.*, **33**, D476–80.

Razick, S., Magklaras, G., and Donaldson, I.M. (2008) iRefIndex: a consolidated protein interaction database with provenance. *BMC Bioinformatics*, **9**, 405.

Salwinski, L., Miller, C.S., Smith, A.J., *et al.* (2004) The Database of Interacting Proteins: 2004 update. *Nucleic Acids Res.*, **32**, D449–51.

Santora, K., Rasa, C., Visco, D., *et al.* (2007) Antiarthritic effects of relaxin, in combination with estrogen, in rat adjuvant-induced arthritis. *J. Pharmacol. Exp. Ther.*, **322**, 887–93.

Sayers, E.W., Barrett, T., Benson, D.A., *et al.* (2009) Database resources of the National Center for Biotechnology Information. *Nucleic Acids Res.*, **37**, D5–15.

Schlitt, T., Palin, K., Rung, J., *et al.* (2003) From gene networks to gene function. *Genome Res.*, **13**, 2568–76.

Seringhaus, M.R. and Gerstein, M.B. (2007) Publishing perishing? Towards tomorrow's information architecture. *BMC Bioinformatics*, **8**, 17.

Shannon, P., Markiel, A., Ozier, O., *et al.* (2003) Cytoscape: a software environment for integrated models of biomolecular interaction networks. *Genome Res.*, **13**, 2498–504.

Sohler, F., Hanisch, D., and Zimmer, R. (2004) New methods for joint analysis of biological networks and expression data. *Bioinformatics*, **20**, 1517–1521.

Subramanian, A., Tamayo, P., Mootha, V.K., *et al.* (2005) Gene set enrichment analysis: a knowledge-based approach for interpreting genome-wide expression profiles. *Proc. Natl. Acad. Sci. U.S.A.*, **102**, 15545–50.

Sun, J., Sun, Y., Ding, G., *et al.* (2007) InPrePPI: an integrated evaluation method based on genomic context for predicting protein-protein interactions in prokaryotic genomes. *BMC Bioinformatics*, **8**, 414.

Tao, Y., Friedman, C., and Lussier, Y.A. (2005) Visualizing information across multidimensional post-genomic structured and textual databases. *Bioinformatics*, **21**, 1659–67.

Tipney, H.J., Leach, S.M., Feng, W., *et al.* (2009a) Leveraging existing biological knowledge in the identification of candidate genes for facial dysmorphology. *BMC Bioinformatics*, **10**(Suppl 2), S12.

Tipney, H., Schuyler, R.P., and Hunter, L. (2009b) Consistent visualizations of changing knowledge. AMIA Summit on Translational Bioinformatics, San Francisco, CA, USA, 15–17 March 2009.

Uetz, P., Giot, L., Cagney, G., *et al.* (2000) A comprehensive analysis of protein-protein interactions in Saccharomyces cerevisiae. *Nature*, **403**, 623–7.

Walhout, A.J., Sordella, R., Lu, X., *et al.* (2000) Protein interaction mapping in C. elegans using proteins involved in vulval development. *Science*, **287**, 116–22.

Wingender, E., Dietze, P., Karas, H., and Knuppel, R. (1996) TRANSFAC: a database on transcription factors and their DNA binding sites. *Nucleic Acids Res.*, **24**, 238–41.

Yamane, A., Mayo, M., Shuler, C., *et al.* (2000) Expression of myogenic regulatory factors during the development of mouse tongue striated muscle. *Arch. Oral Biol.*, **45**, 71–8.

Yu, H., Luscombe, N.M., Lu, H.X., *et al.* (2004) Annotation transfer between genomes: protein-protein interologs and protein-DNA regulogs. *Genome Res.*, **14**, 1107–18.

3

Technologies and best practices for building bio-ontologies

Mikel Egaña Aranguren, Robert Stevens, Erick Antezana, Jesualdo Tomás Fernández-Breis, Martin Kuiper, and Vladimir Mironov

3.1 Introduction

Genomics technologies generate vast amounts of data of a wide variety of types and complexities, and at a growing pace. The analysis of such data and the mining of the resulting information is insufficient without a contextual interpretation, that is, biological knowledge deduced from the data. This knowledge states the data's biological meaning in terms of, for instance, molecular function, cellular location, or network interactions. Biological knowledge is diverse, vast, complex, and volatile. These factors, together with the nature of evolved systems, make the knowledge generated by the life sciences difficult to capture. As molecular biology has relatively recently included a systems approach, it has become increasingly important to have precise and rich representations of the catalogs that in turn form the basis of the networks and pathways that describe biological systems. Therefore, biological knowledge management is becoming essential for current research in life sciences (Antezana *et al.*, 2009a).

Biological knowledge has traditionally been represented in human interpretable formats like natural language in scientific literature, or somewhat more

Knowledge-Based Bioinformatics: From Analysis to Interpretation   Edited by Gil Alterovitz and Marco Ramoni
© 2010 John Wiley & Sons, Ltd

structured in database entries. The heterogeneous terminology used, together with the natural language form, has made it difficult to manage and use that knowledge, for both humans and, more importantly, computers. In order to use the computers' ability to handle complex and large amounts of information, it has become clear that biological knowledge should be codified in a machine interpretable form. Only in this way can biologists begin to exploit their hard-won data.

A widely used method for codifying knowledge in a machine interpretable form is to represent it in ontologies. Ontologies are computational formalizations of the concepts shared by a community of scientists. Thus, ontologies can be used to describe and define the entities of a domain, and their relations, axiomatically, with precise semantics. The expression of knowledge with precise semantics makes it possible for computers to perform, via automated reasoning, information management tasks that can save scarce human resources and retrieve more complete results from biological knowledge (e.g., new hypotheses).

Therefore, the use of bio-ontologies, that is, ontologies that represent biological knowledge, is essential in biological knowledge management and integration, and they have become mainstream within bioinformatics. Currently, there are established communities of bio-ontologists, like the Open Biomedical Ontologies (OBO) Foundry (Smith *et al.*, 2007; www.obofoundry.org/), which have produced important bio-ontologies such as the Gene Ontology (GO; Gene Ontology Consortium, 2000).

Many bio-ontologies exploit the very technology that will be used for building the Semantic Web (www.w3.org/standards/semanticweb/), which is the next 'smart' generation of the current Web, based on the automatic management of Web content. The W3C (www.w3.org/), the consortium responsible for the implantation of the Semantic Web and other open Web standards, has been fostering the Semantic Web Health Care and Life Sciences (HCLS) Interest Group (www.w3.org/blog/hcls) for working towards a Life Sciences Semantic Web (LSSW).

This chapter provides an introduction to the process of building bio-ontologies, analyzing the benefits and problems of modeling biological knowledge axiomatically, especially with regards to automated reasoning. Thus, the aspects that a biologist should consider in order to create a reusable, robust, rigorous, and axiomatically rich bio-ontology are briefly reviewed, providing pointers to successful engineering techniques and bio-ontologies. The aim of this chapter is not to provide a detailed methodology of the creation of bio-ontologies (the literature on the subject is vast); rather, the chapter highlights the elements that have to be taken into account, to help the reader to make informed decisions while building bio-ontologies.

3.2   Knowledge representation languages and tools for building bio-ontologies

An ontology represents knowledge through axioms. Axioms are used to describe the objects from the knowledge domain: their categories and the relationships between them. The axioms are written using a logical formalism, a Knowledge

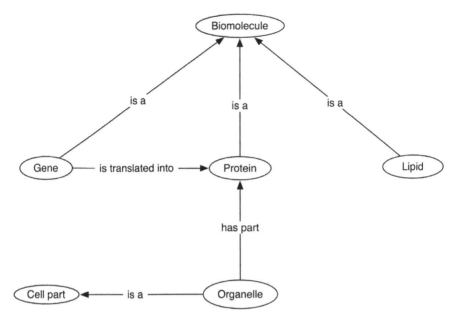

Figure 3.1   Simple bio-ontology, representing a 'toy' knowledge domain. The names of concepts – representing the categories or classes of objects in the domain – (e.g., Protein) and relations among the objects (e.g., has_part) are irrelevant for a computer; they are only 'understood' by humans. However, the structure of the ontology, expressed using axioms, is what the computer is able to manage, exploiting automated reasoning.

Representation (KR) language, which enables their computational interpretation (Figure 3.1).

The semantics of a KR language defines the computational interpretation of the statements (axioms) the ontologist makes in an ontology, thus, how the computer 'understands' such statements. The different KR languages offer different levels of expressivity (what can be said about a domain); therefore, ontologists are able to make statements at different complexity levels, depending on the expressivity of the language of choice. Expressivity is related to computational tractability: the more expressive a language, the less tractable; that is, the more computational resources are needed by a computer to operate on an ontology written in such a language.

Currently, the most used KR languages in life sciences are the Resource Description Framework (RDF)[1], the Web Ontology Language (OWL)[2], and the

---

[1] RDF is not strictly a language for creating ontologies. However, using a broad definition of ontology, and considering the widespread use of RDF in the LSSW and its close relation to OWL, it has been included in this chapter.

[2] RDF Schema (RDFS) offers functionality close to OWL. However, it has been left out of this review due to the fact that it is not widely used in the LSSW, and for the sake of brevity.

Table 3.1 Summary of features of RDF, OWL, and OBO. SW, Semantic Web oriented (Semantic Web stack of protocols, URIs, etc.); LS, Life Sciences; LSHC (IG/KB), W3C Life Sciences and Health Care (Interest Group/Knowledge Base).

	Repositories	SW	Reasoning	Strong points	Weak points	Editors	APIs	Communities of practice	Outstanding projects
RDF	OBO foundry, Bioportal[a]	Yes	No	Widely used outside LS Simple and intuitive SPARQL	Only triple like information	Protégé 3	Jena	LSHC IG All the SW communities	Bio2RDF, Biogateway, LSHC KB
OWL	OBO foundry, Bioportal	Yes	Yes	Widely used outside LS Expressive yet tractable	Difficult to understand and use	Protégé 3, 4 TopBraid composer[b]	OWL API	LSHC IG All the SW communities	OBI, BioPAX, CCO, PhosphaBase
OBO	OBO foundry, Bioportal, Ontology Lookup Service[c]	No	No	Widely used within LS Simple and intuitive	No formal semantics Not used outside LS	OBO-Edit COBrA-CT[d]	ONTO-Perl	OBO foundry	GO, CL

[a] www.bioportal.bioontology.org/.
[b] www.topbraidcomposer.com/.
[c] www.ebi.ac.uk/ontology-lookup/.
[d] www.aiai.ed.ac.uk/project/cobra-ct/.

OBO format. They mainly differ in terms of expressivity, tool availability, and communities of practice. Since RDF and OWL are official W3C recommendations to implement the Semantic Web, they are also used outside the life sciences domain, whereas the OBO Format is only used to represent life-sciences-related information. As RDF and OWL are part of the Semantic Web stack of technologies, OWL 'includes' RDF, and therefore an OWL ontology can be accessed with OWL-specific tools (OWL expressivity level) or RDF tools (RDF expressivity level). The following subsections describe the main features of each language, as summarized in Table 3.1.

3.2.1   RDF (resource description framework)

RDF (www.w3.org/TR/rdf-primer/) was designed to represent information about Web resources in the Semantic Web, thus to publish data in a basic machine processable form. The information in RDF is represented in statements formed by a subject, a predicate and an object, called triples. For example, a triple in RDF would read SWI4 participates_in G1/S_transition. SWI4 is the subject, participates_in the predicate, and G1/S transition the object (Figure 3.2). Triples can be combined to form a graph (Figure 3.3). In an RDF graph, the subject of a triple can be the object of another triple.

Figure 3.2 An RDF triple. A subject (SWI4) is related to an object (G1/S_transition) by a predicate (participates_in).

RDF uses URIs (Uniform Resource Identifiers; www.w3.org/standards/techs/uri) to identify entities (subjects, predicates, and objects). The use of URIs provides the possibility of referring to entities from different graphs that have been published in different resources on the Web. This enables a framework to combine graphs from different resources, or to combine graphs at query time.

RDF graphs can be queried using SPARQL (www.w3.org/TR/rdf-sparql-query/). SPARQL is a query language that can be used to retrieve smaller graphs from a target graph. In order to perform the retrieval, a user must define a query graph in which one or more entities are left as variables, and the query graph is matched against the target graph, returning the appropriate answer as a smaller sub-graph of the target graph.

RDF is based on a simple model that enables the representation of diverse information with very low computational costs, provided that such information can be captured as a set of subject–predicate–object triples. Therefore, the manipulation of RDF graphs through APIs (Application Programming Interfaces) like

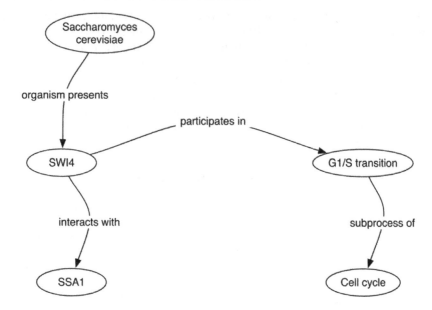

Figure 3.3   An RDF graph made by combining four triples. The triples share some common entities, such as SWI4, *which is the subject of two triples (*participates_in G1/S_transition *and* interacts_with SSA1*) and the object of another triple (*Saccharomyces_cerevisiae organism_presents*).*

Jena (http://jena.sourceforge.net/) is straightforward. This simplicity has made RDF the chosen language in several bioinformatics resources such as BioGateway (www.semantic-systems-biology.org/biogateway), Bio2RDF (http://bio2rdf.org/), and HCLS KB (www.w3.org/TR/hcls-kb/).

3.2.2   OWL (Web ontology language)

OWL (www.w3.org/TR/owl2-overview/) was designed as a language to publish machine processable and interoperable ontologies in the Web. OWL, compared to RDF, offers a semantic vocabulary to describe a knowledge domain. Such expressivity may have a higher computational cost. Nevertheless, OWL allows the representation of biological information with a finer granularity, opening up ample possibilities for interesting applications such as automated reasoning.

The OWL semantics is based on three elements: individuals, classes (sets of individuals), and properties (two individuals, or an individual and a data value, are linked in a pair along a property; Figure 3.4)[3]. Classes are built by specifying

---

[3] An OWL ontology that has classes, individuals and properties can be considered a Knowledge Base (KB). If there are no individuals, the artifact can be considered simply an ontology. An ontology describes a schema with which some entities of the domain (individuals) are described; a KB includes the schema (ontology) and the individuals.

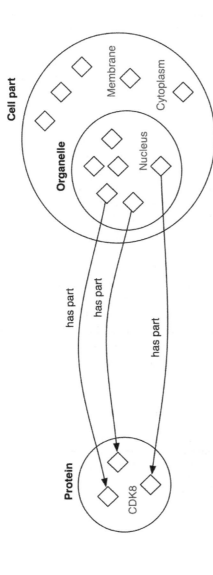

Figure 3.4 OWL classes (`Protein`, `Organelle`, `Cell_part`), individuals (`Nucleus`, `CDK8`, `Membrane`, `Cytoplasm`, and the rest of the diamonds), and an object property (`has_part`), used three times (in three pairs of individuals). The class `Organelle`, which has `Nucleus` as an individual (amongst others), is a subclass of `Cell_part`, which has other individuals (`Membrane`, `Cytoplasm`): every individual of `Organelle` is also an individual of `Cell_part`.

the conditions that the individuals should fulfill to belong to the class, in terms of which and how many relationships they should have, using class expressions. OWL offers universal (`only`) or existential (`some`) qualifiers and a plethora of typical logical constructs, such as negation (`not`), other Boolean operators (`or`, `and`), and more constructs, to create class expressions. Such constructs can be combined in complex (rich) class expressions. Class conditions can be either necessary (e.g., every nucleus is part of a cell, but being part of a cell is not enough to flag an organelle as nucleus) or necessary and sufficient (e.g., having a nucleolus as a part is a necessary and sufficient condition to flag an organelle as nucleus, as nuclei are the only organelles with nucleoli). The classes with at least one necessary and sufficient condition are called defined classes, whereas the classes with only necessary conditions are called primitive classes.

Classes can be subclasses of other classes, thus creating a taxonomy. The semantics of the subclass relation reads that, given a superclass S, every individual I of a given subclass of S is also an individual of S; for instance, all the organelles are cell parts, but not all the cell parts are organelles (membrane and cytoplasm are cell parts but are not organelles), therefore `Organelle` is a subclass of `Cell_part` (instead of an equivalent class).

There are three types of properties in OWL: properties that link pairs of individuals (object properties), properties that link individuals with data values (data type properties), and properties that can be used to add natural language information to axioms and entities, without affecting automated reasoning (annotation properties). Object properties can be arranged in hierarchies, and features of properties (such as transitivity) can be defined.

OWL can be expressed in various syntaxes. The most common computer readable syntax is RDF/XML (Figure 3.5). The Manchester OWL Syntax (MOS) offers a human-readable OWL syntax (Horridge *et al.*, 2006). For example, the expression from Figure 3.5 would read as follows in MOS: `Nucleus subClassOf has_part some Protein`.

```
<owl:Class rdf:about="#Nucleus">
<rdfs:subClassOf>
  <owl:Restriction>
   <owl:onProperty rdf:resources="#has_part"/>
   <owl:someValuesFrom rdf:resource="#Protein"/>
  </owl:Restriction>
 </rdfs:subClassOf>
</owl:Class>
```

Figure 3.5   OWL RDF/XML syntax of the MOS expression `Nucleus subClassOf has_part some Protein`.

OWL is based on Description Logics (DLs; Baader *et al.*, 2003), a well known logical formalism. OWL offers an optimal balance between expressivity and tractability, allowing the efficient application of automated reasoning on OWL ontologies. Automated reasoning consists of using a program to infer axioms from the axioms asserted in the ontology. The asserted axioms entail the inferred axioms. Thus, an automated reasoner makes axioms that were implicit explicit, showing further information to the bio-ontologist. For instance, let us consider the following two classes as being entities of our bio-ontology:

- `Nucleus`, with the axiom `Nucleus subClassOf part_of some Cell`. In order to be a nucleus it is necessary to be part of a cell, but being part of a cell is not enough on its own to be a nucleus (there are other organelles that are also part of a cell, but are not nuclei). Therefore, `Nucleus` is a primitive class.

- `Organelle`, with the axiom `Organelle equivalentTo part_of some Cell`. Anything that is part of a cell is an organelle. Therefore, `Organelle` is a defined class.

An automated reasoner will infer that `Nucleus` is a type of `Organelle`, thus the axiom `Nucleus subClassOf Organelle` will be made explicit or 'added' into the bio-ontology by the automated reasoner[4]. This is so because the axioms `Nucleus subClassOf part_of some Cell` and `Organelle equivalentTo part_of some Cell` entail the axiom `Nucleus subClassOf Organelle` (if all nuclei are part of a cell, and anything that is a part of a cell is an organelle, then nuclei are organelles).

The outcome of an automated reasoning process depends strongly on the axiomatic richness of the bio-ontology. It should also be noted that an automated reasoner acts in a 'ruthless' manner, showing the axioms that our modeling entails; in the above reasoning example, plasma membrane and cytoplasm should not be classified as organelles, indicating a likely modeling error on our side. It is necessary to regularly run an automated reasoner while building a bio-ontology, either to be reminded that our modeling is wrong or to highlight new information that was implicit ('hidden') in our modeling, entailed by the asserted axioms[5]. The more axioms we express in an ontology, the better; it is better to be axiomatically wrong (the automated reasoner tells us why we are wrong) than axiomatically correct and conceptually wrong (because we have not added those axioms). The automated reasoner shows the contradictions in our conceptual world.

In more concrete terms, automated reasoning can be used in the following ways:

(1) Perform complex queries against the knowledge stored in the ontology.

(2) Infer the class–subclass relationships from the class expressions; that is, build automatically the class hierarchy (taxonomy). For example, the

---

[4] This modeling (incorrectly) assumes that plasma membrane and cytoplasm should be classified as organelles; simplified for the sake of the example clarity.

[5] The automated reasoner will infer all the information entailed by the asserted axioms, including the information that a human would miss because of the extent or complexity of such information. That is why, among other reasons, automated reasoners can be so useful in knowledge-intensive disciplines like life sciences.

Normalization technique allows one to maintain a multiple inheritance in an ontology relying solely on the automated reasoner, provided that the appropriate class expressions are added to the ontology. In another example, an automated reasoner was used to check the completeness of the GO class hierarchy, in the Gene Ontology Next Generation (GONG) project (Egaña Aranguren *et al.*, 2008a).

(3) Given an individual and its relationships to other individuals, the automated reasoner can infer to which class(es) it belongs.

(4) Check the consistency of the asserted axioms, as the automated reasoner can flag contradictory axioms. Such a procedure is used, for example, to ensure that the information gathered from different resources commits to the same schema (Miñarro-Gimenez *et al.*, 2009).

Some OWL features stand out, apart from its expressivity, in terms of information integration:

- OWL (as well as RDF) relies on URIs to identify entities. Therefore, the Web machinery is also available for OWL.

- OWL is self-descriptive, that is, the schema and the data described using such schema are expressed in the same language: schema reconciliation is not needed, and the reconciliation problem is shifted to a more abstract (conceptual) level.

- Open World Assumption (OWA): OWL semantics interpret the absence of information as unknown rather than false. OWL assumes that, as the knowledge of the world we have is by definition incomplete, we cannot infer negation from the absence of information. Therefore, new information can be added to our bio-ontology and prior inferences remain valid, for example when importing entities from another OWL ontology (however, a new inconsistency may be triggered). This model fits with the biological knowledge domain, always being extended by different agents.

- Lack of Unique Name Assumption (UNA): in OWL, the fact that two entities have different names does not mean that they are different. Such entities need to be explicitly asserted to be different with the axioms `differentFrom` and `disjointWith`. On the other hand, different entities can also be asserted to be the same entity with the axioms `sameAs` and `equivalentTo`. For example, an OWL ontology can describe a gene with the name `CYC8`, and the same gene can be described in another OWL ontology with the name `SSN6`: they can be asserted to be the same entity (e.g., `CYC8 sameAs SSN6`), easing integration as no mapping must be created.

The expressivity and integrative features that OWL provides enable the representation of a considerable amount of biological concepts in a computationally accessible manner (Stevens *et al.*, 2007). Such features have promoted the use of OWL in several domains, and many tools supporting it have been also developed (www.w3.org/2007/OWL/wiki/Implementations), amongst which Protégé (http://protege.stanford.edu/) stands out as the most used OWL editor. Moreover, there are automated reasoners available for OWL, like Pellet (http://clarkparsia.com/pellet/) or FaCT++ (http://code.google.com/p/factplusplus/), and APIs like the OWL API (www.owlapi.sourceforge.net/). OWL has been successfully employed in projects such as OBI (www.purl.obolibrary.org/obo/obi), CCO (www.cellcycleontology.org/), BioPAX (www.biopax.org/), and PhosphaBase (www.bioinf.manchester.ac.uk/phosphabase/).

3.2.3   OBO format

The OBO format (www.geneontology.org/GO.format.shtml) has become the *de facto* KR language to model biological concepts for most of the OBO bio-ontologies, which are the most widely used bio-ontologies. Its development has been mainly fostered by the GO consortium (www.geneontology.org/). Figure 3.6 shows a sample entry of a term from the GO.

```
[Term]
id: GO:0005634
name: nucleus
def: "A membrane-bounded organelle of eukaryotic cells in which chromosomes
are housed and replicated. In most cells, the nucleus contains all of the cell's
chromosomes except the organellar chromosomes, and is the site of RNA synthesis
and processing. In some species, or in specialized cell types, RNA metabolism or
DNA replication may be absent." [GOG:go_curators]
synonym: "cell nucleus" EXACT []
xref: Wikipedia:Cell_nucleus
is_a: GO:0043231 ! intracellular membrane-bounded organelle
```

Figure 3.6   An OBO entry describing the term Nucleus *from the GO.*

In contrast to languages such as OWL, OBO has been tailored to the needs of the bio-ontologists (e.g., OBO offers an efficient mechanism for fine-grained annotations on ontology terms), resulting in the perception that it is more intuitive and more appropriate for biological knowledge modeling. Although OBO does not rely on any formal semantics, OBO algorithmic processing tools have been implemented, like the OBO-Edit reasoner (www.oboedit.org/docs/html/The_OBO_Edit_Reasoner.htm), the OBO Language (OBOL; Mungall, 2004), and the OBD-SQL reasoner (Mungall *et al.*, 2010). OBO ontologies can also be translated into OWL to exploit automated reasoning, but such translation is not completely free of problems (Golbreich *et al.*, 2007). In terms of expressivity, OBO can be used to represent relatively complex axioms, but composite expressions like Nucleus subClassOf (part_of some Cell) and (has_part only (Nucleus_membrane or Nucleolus and not Ribosome) cannot be expressed.

OBO is relatively human readable and easy to manipulate programmatically, with APIs like ONTO-PERL (Antezana *et al.*, 2008), or graphically, with ontology editors like OBO-Edit (http://oboedit.org/). OBO has been successfully employed in very influential projects such as the GO or the Cell Type Ontology (CL; Bard *et al.*, 2005). The GO is used for annotation by many current bioinformatics resources (www.ebi.ac.uk/GOA/). The CL is used in projects like XSPAN (www.xspan.org/).

3.3   Best practices for building bio-ontologies

Ontology building is still in a transition state from a 'craft' to a fully industrial engineering discipline (Bodenreider and Stevens, 2006). Therefore, there are neither established methodologies nor fully accepted principles. There are, however, practices that have already demonstrated their utility, and they are agreed to be important by the bio-ontologist community, explained as follows. Figure 3.7 summarizes such practices and the place they occupy in the bio-ontology development process.

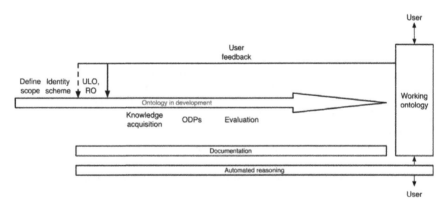

Figure 3.7   Diagram of the development cycle of a bio-ontology, with the best practices described in Section 3.3. The bio-ontology development starts by defining the scope, and it is repeated as necessary. User feedback is used to improve the bio-ontology, but generally without changing the scope and identity scheme, and barely changing the used ULO or set of relations. Documentation should be provided through the whole process. Automated reasoning should be used at development time (e.g., for consistency checking) and also users can exploit automated reasoning to query the ontology. Users can also interact with the ontology without using automated reasoning.

3.3.1   Define the scope of the bio-ontology

Bio-ontologies are able to perform a whole range of functions (Stevens and Lord, 2008). The function(s) of a bio-ontology will determine its scope and 'shape.' Therefore, explicitly and clearly defining the function (and hence the scope) of an ontology in early development stages, and sticking to such definition, is important to avoid spending too much effort in extending the ontology endlessly.

3.3.2   Identity of the represented entities

One of the most important elements of a LSSW is the identity of entities that form the biological knowledge domain, such as genes and proteins (Good and Wilkinson, 2006). Thus, many current bioinformatics resources describe the same entity with different identities (e.g., many resources give different names to the same gene). Different global identity schemes have been proposed to address the problem, but none has prevailed. The latest proposal is the Shared Names endeavour (http://sharedname.org/).

It is important to use an explicit identity scheme for the bio-ontology being built, and be consistent in its application. It might be that the identity scheme chosen does not 'succeed' and be used in the future by other resources, but nonetheless it will facilitate internal knowledge management, and if another identity scheme succeeds later on, it will be possible to map to it.

3.3.3   Commit to agreed ontological principles

There are ontological principles that are useful in order to make the bio-ontology interoperable with other bio-ontologies and resources. Such principles, however, impose a certain structure on our bio-ontology, and they determine strongly the subsequent modeling (Schulz *et al.*, 2008). Therefore, the bio-ontologist must maintain an equilibrium between using such principles and being too influenced by such principles in the modeling process. Thus, the bio-ontology development should follow a minimal commitment policy.

In the case of OBO bio-ontologies, there is a set of relationships, collected in the Relation Ontology (RO; Smith *et al.*, 2005), that can be used in our bio-ontology. The use of such relations favors the integration with other bio-ontologies that also use RO, as, for example, the participates_in relation in our bio-ontology will be the same participates_in relation present in such other bio-ontologies. Therefore, bio-ontologies using such relations can be efficiently integrated and queried. Also, the RO relations have a precise semantic definition, saving time for the bio-ontologist, as there is no need to define the relations of the bio-ontology (if satisfied with the RO definition).

The use of an Upper Level Ontology (ULO), deeply related with the use of a set of relationships like RO, is also a recommended ontological practice. A ULO is generally an ontology with a few concepts that sits on the upper levels of the bio-ontology we are building, providing basic distinctions of types of concepts, like process vs. thing, self standing vs. refining entity, and so on. A ULO not only helps in integration with other bio-ontologies that are based in the same ULO, but also helps in building a sound and modular bio-ontology by creating a cleaner structure with explicit distinctions. One of the most used ULOs in bio-ontologies is the Basic Formal Ontology (BFO; Grenon *et al.*, 2004).

3.3.4 Knowledge acquisition

There are different ways of populating our bio-ontology with knowledge, described as follows. These methods are not disjoint; they can be used in a complementary manner.

An ideal method for obtaining the knowledge is to elicit it directly from the domain experts or the prospective users of our bio-ontology. Knowledge can also be obtained from extant resources. For example, data can be integrated from different resources in our bio-ontology, or knowledge from other bio-ontologies can be reused. Reusing content of other bio-ontologies is important to ease development and create a useful bio-ontology, since such a bio-ontology will be more interoperable with other resources. The OBO foundry ontologies offer a wealth of content that can be reused and extended with new axioms and entities. For example, that is the strategy followed in the creation of CCO (Antezana *et al.*, 2009b).

3.3.5 Ontology Design Patterns (ODPs)

ODPs are solutions for common modeling problems that appear when building ontologies (Egaña Aranguren *et al.*, 2008b). Thus, an ODP solves a concrete problem efficiently, as the ODP has been tested by a community of ontologists, and agreed to be an efficient modeling solution. Each ODP is thoroughly documented, clearly stating the requirements that the use of the ODP fulfills; that is, the problem that it solves. An ODP is like a 'cooking recipe' of how to create axioms that perform a given function within an ontology. Therefore, a bio-ontologist need only explore ODPs and apply the appropriate one in the bio-ontology being built. For example, in the case of the Value Partition ODP (Figures 3.8 and 3.9), such an ODP solves the problem of how to represent a feature that has only certain values (e.g., the height of a person can only be tall, medium or short). Ideally, if a bio-ontologist is confronted with the problem of representing such structure in a bio-ontology, he or she will explore ODP catalogs (see below), read the documentation, and, as the Value Partition ODP fulfills his or her requirements, apply it in the bio-ontology. Following such a procedure the bio-ontologist saves a lot of time, as many axioms are applied automatically in the bio-ontology.

ODPs are presented as fragments of ontologies that solve a concrete modeling problem, as a concrete set of axioms, but with an abstract structure: when applied in the ontology, such axioms relate the actual entities of the ontology. Therefore, ODPs can also be regarded as modules of ontologies to be applied 'off the shelf': an ontology can rapidly be built by applying a collection of ODPs.

Using ODPs in the development of an ontology makes such development faster, more consistent, and explicit. The resulting bio-ontologies have a richer

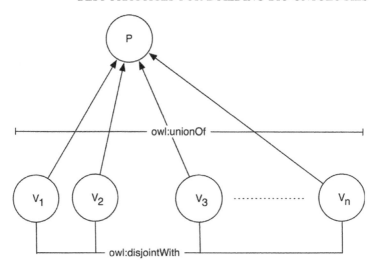

Figure 3.8   Abstract representation of the Value Partition ODP. This ODP solves a concrete problem; namely, how to represent exhaustive sets of values in OWL. P can be any feature (regulation, color, height, etc.), and V any value (positive or negative; red, blue or white; etc.). This abstract structure is presented with documentation that explains how the ODP can be used (e.g., motivation, structure, elements, implementation, and result).

axiomization, obtained with less effort, enhancing automated reasoning. They are also more reusable and interoperable with other bio-ontologies.

There are two main catalogs where ODPs can be obtained (http://odps.sf.net/, http://ontologydesignpatterns.org). Once an ODP has been chosen, there are different methods for applying it. The ODP can be directly imported into the ontology, manually recreated, or applied with ODP-oriented tools like the NeOn toolkit (http://neon-toolkit.org) or the Ontology PreProcessor Language (OPPL; http://oppl.sourceforge.net/).

3.3.6   Ontology evaluation

Ontology evaluation is a controversial issue, and there is a wealth of methodologies to choose from, depending on the needs of the project. Three main and complementary categories can be identified, according to the aims of the evaluation process: ranking, correctness, and quality.

Ranking approaches pursue the selection of the best ontology for a particular task, so they apply criteria that focus on that particular task. Ranking strategies may be driven by users, experts, and so on. Bio-ontologies can get different results using different ranking strategies, as different quality aspects are measured. For example, in Aktiverank (Alani *et al.*, 2006), ontologies are ranked against search terms, so that the best ontology is the one that best matches the query. For this

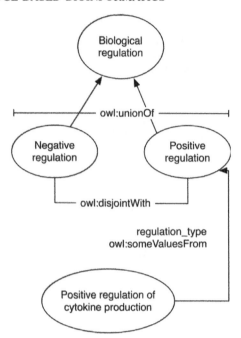

Figure 3.9   Application of the Value Partition ODP in GO. Thus, the abstract structure from Figure 3.8 is converted to a concrete structure with concrete entities, and linked to the rest of the bio-ontology by the `regulation_type` *relation (with the existential qualifier).*

purpose, quantitative metrics such as the coverage of an ontology for the given search term, the number of connections (relations, subclasses, superclasses, and siblings), or the closeness of the classes that matches the search terms in the ontology are used.

Correctness approaches determine the quality of a bio-ontology by applying formal theories. The most relevant approach is provided by Ontoclean (Guarino and Welty, 2004), which checks for the formal correctness of the taxonomy, based on rigidity, identity, unity, and dependence principles.

Quality approaches provide frameworks that are based on a series of qualitative and quantitative criteria that can be organized in quality dimensions. The goal of such approaches is to provide an overview of the strengths and weaknesses of the bio-ontologies in the particular quality dimensions rather than finding the best one for a particular task. Quality approaches are likely to include criteria that cannot always be optimized simultaneously, and this makes their application more complex. In (Fernández-Breis *et al.*, 2009), an ISO 9126-based framework was proposed, comprised of seven quality dimensions: structural, functionality, reliability, usability, efficiency, maintainability, and quality in use.

3.3.7 Documentation

Most KR languages allow the inclusion of information to axioms and entities in the form of annotations that are not processed by the automated reasoner[6]. For example, OWL allows one to create custom annotation properties or use the already defined `rdfs:comment`, `rdfs:label`, or the Dublin Core (http://dublincore.org/) annotation properties. The OBO format has its own set of annotations tailored to the OBO community needs. Annotations are usually used to capture information that cannot be represented in axioms, to capture information that should not be represented in axioms (e.g., the name of an entity in different languages) or to express facts about the modeling in natural language (e.g., the rationale for modeling decisions).

It is important to capture as much information as possible in annotations, as it will be used by other developers or users. Such annotations should also be as structured as possible: for example, the GO term names are syntactically very repetitive (Ogren *et al.*, 2004), which helps in computationally processing them (Egaña Aranguren *et al.*, 2008a).

3.4 Conclusion

The Life Sciences Semantic Web (LSSW) faces many challenges. KR languages with precise semantics like OWL, being powerful and robust solutions for a truly distributed and automatic knowledge management, are not free of problems. The increasing volume of available data and supporting bio-ontologies reveals limitations in terms of performance, especially regarding automated reasoning and the management of KBs. Performance issues are expected to be solved as the technology evolves. However, there are also problems in the 'social' side of bio-ontology creation, the main one being the lack of agreement in modeling principles: for example, there is not even a consensus on how to represent a concept as important and basic as the one of species (Schulz *et al.*, 2008). Such lack of agreement is a community problem, but there are practices, like the use of ODPs, that can contribute to its solution.

Even taking into account these problems, the LSSW offers an increasing number of examples that make good on its promise to help in the information management of biological knowledge, and to support advanced queries that demonstrate the power of semantic data integration.

The adoption of a precise semantics opens new paradigms of biological research, like the Semantic Systems Biology (SSB) approach (Antezana *et al.*, 2009c). SSB is a systems biology approach that combines Semantic Web

---

[6] The term 'annotation' has a somewhat different meaning in bioinformatics and KR. In bioinformatics, an annotation is information attached to biological data, such as the molecular function of a gene product. In KR, an annotation is extralogical information added to an axiom or an entity of an ontology, usually using natural language. We are using the KR meaning throughout the chapter.

technologies for analyzing data and formalized knowledge to engineer biological system models. Kitano's Systems Biology paradigm (Kitano, 2002) hinges on mathematical model-based system behavior predictions, or hypotheses, and validation in new experiments. In SSB, data and new knowledge are (automatically) checked for consistency against existing knowledge, and queries and automated reasoning on semantically integrated knowledge are used to extract new knowledge and hypotheses.

Post *et al.* applied such an approach to study the role of histone modification in gene expression regulation (Post *et al.*, 2007). In that use case as well as in other efforts such as the YeastHub (Cheung *et al.*, 2005), CViT (Deisboeck *et al.*, 2007), and the Cell Cycle Ontology (Antezana *et al.*, 2009b), the workflow of an SSB approach was followed. Some other initiatives are NeuroCommons (http://neurocommons.org), focused on neuroscience, and the SSB portal (www.semantic-systems-biology.org). All these initiatives demonstrate the added value that the SSB approach can offer to the understanding of biological systems.

This chapter has provided a brief overview of the extant technologies and tools to build bio-ontologies, as well as real bio-ontology examples and pointers to the future of the LSSW, like SSB. Also, it has highlighted the most important issues and practices that should be taken into account in order to create a useful bio-ontology with the least possible distress. Creating proper bio-ontologies is a very hard task; however, it is even harder to manage biological data, information, and knowledge efficiently without them.

3.5   Acknowledgements

Mikel Egaña Aranguren was funded by the Autonomous Community of the Region of Murcia, Spain (BIO-TEC 06/01-0005). Vladimir Mironov was funded by FUGE Mid-Norway.

3.6   References

Alani, H., Brewster, C., and Shadbolt, N. (2006) Ranking ontologies with AKTive-Rank. International Semantic Web Conference (ISWC 2006), Athens, GA, USA, 5–9 November 2006.

Antezana, E., Egaña, M., De Baets, B., *et al.* (2008) ONTO-PERL: an API supporting the development and analysis of bio-ontologies. *Bioinformatics*, **24**(6), 885–7.

Antezana, E., Kuiper, M., and Mironov, V. (2009a) Biological knowledge management: the emerging role of the Semantic Web technologies. *Brief. Bioinformatics*, **10**(4), 392–407.

Antezana, E., Egaña, M., De Baets, B., *et al.* (2009b) The Cell Cycle Ontology: an application ontology for the representation and integrated analysis of the cell cycle process. *Genome Biol.*, **10**, R58.

Antezana, E., Blondé, W., Egaña, M., *et al.* (2009c) BioGateway: a Semantic Systems Biology tool for the life sciences. *BMC Bioinformatics*, **10**(Suppl 10), S11.

Baader, F., Calvanese, D., McGuinness, D., *et al.* (eds) (2003) *The Description Logic Handbook*. Cambridge University Press, Cambridge.

Bard, J.B.L, Rhee, S.Y., and Ashburner, M. (2005). An ontology for cell types. *Genome Biol.*, **6**, R21.

Bodenreider, O. and Stevens, R. (2006) Bio-ontologies: current trends and future directions. *Brief. Bioinformatics*, **7**(3), 256–74.

Cheung, K.H., Yip K.Y., Smith, A., *et al.* (2005) YeastHub: a semantic web use case for integrating data in the life sciences domain. *Bioinformatics*, **21**(Suppl 1), i85–96.

Deisboeck, T.S., Zhang, L., and Martin, S. (2007) Advancing cancer systems biology: introducing the Center for the Development of a Virtual Tumor, CViT. *Cancer Inform*, **5**, 1–8.

Egaña Aranguren, M., Wroe, C., Goble, C., and Stevens, R. (2008a) In situ migration of handcrafted ontologies to Reason-able Forms. *Data Knowl. Eng.*, **66**(1), 147–62.

Egaña Aranguren, M., Antezana, E., Kuiper, M., and Stevens, R. (2008b) Ontology Design Patterns for bio-ontologies: a case study on the Cell Cycle Ontology. *BMC Bioinformatics*, **9**(Suppl 5), S1.

Fernández-Breis, J.T., Egaña Aranguren, M., and Stevens, R. (2009) A quality evaluation framework for bio-ontologies. International Conference on Biomedical Ontology (ICBO 2009), Buffalo, USA, 24–26 July 2009.

Gene Ontology Consortium (2000) Gene Ontology: tool for the unification of biology. *Nat. Genet.* **25**, 25–9.

Golbreich, G., Horridge, M., Horrocks, I., *et al.* (2007) OBO and OWL: Leveraging Semantic Web technologies for the life sciences. International Semantic Web Conference (ISWC 2007), Busan, Korea, 11–15 November 2007.

Good, B.M. and Wilkinson, M.D. (2006) The life sciences Semantic Web is full of creeps! *Brief. Bioinformatics*, **7**(3), 275–86.

Grenon, P., Smith, B., and Goldberg, L. (2004) Biodynamic Ontology: applying BFO in the biomedical domain, in *Ontologies in Medicine*, (ed. D.M. Pisanelli), IOS Press, pp. 20–38.

Guarino, N. and Welty, C.A. (2004) An overview of OntoClean, in *Handbook on Ontologies* (eds S. Staab and R. Studer), Springer, pp. 151–72.

Horridge, M., Drummond, N., Goodwin, J., *et al.* (2006) The Manchester OWL syntax. OWL: Experiences and Directions (OWLED 06), Athens, GA, USA, 10–11 November 2006.

Kitano, H. (2002) Systems biology: a brief overview. *Science*, **295**(5560), 1662–4.

Miñarro-Gimenez, J.A., Madrid, M., and Fernández-Breis, J.T. (2009) OGO: an ontological approach for integrating knowledge about orthology. *BMC Bioinformatics*, **10**(Suppl 10), S13.

Mungall, C.J. (2004) OBOL: integrating language and meaning in bio-ontologies. *Comp. Funct. Genomics*, **5**(6-7), 509–20.

Mungall, C.J., Gkoutos, G.V., Smith, C.L. *et al.* (2010) Integrating phenotype ontologies across multiple species. *Genome Biol.*, **11**, R2.

Ogren, P.V., Cohen, K.B., Acquaah-Mensah, G.K., *et al.* (2004) The compositional structure of Gene Ontology terms. Pacific Symposium on Biocomputing (PSB 04), Big Island, Hawaii, USA, 6–10 January 2004.

Post, L.J.G., Roos, M., Marshall, M.S., *et al.* (2007) A semantic web approach applied to integrative bioinformatics experimentation: a biological use case with genomics data. *Bioinformatics*, **23**(22), 3080–7.

Schulz, S., Stenzhorn, H., and Boeker, M. (2008) The ontology of biological taxa. *Bioinformatics*, **24**(13), i313–21.

Smith, B., Ceusters, W., Klagges, B., *et al.* (2005) Relations in Biomedical Ontologies. *Genome Biol.*, **6**, R46.

Smith, B., Ashburner, M., Rosse, C., *et al.* (2007) The OBO Foundry: coordinated evolution of ontologies to support biomedical data integration. *Nat. Biotechnol.*, **25**, 1251–5.

Stevens, R. and Lord, P. (2008) Application of ontologies in bioinformatics, in *Handbook on Ontologies in Information Systems*, 2nd edn (eds S. Staab and R. Studer), Springer, pp. 735–56.

Stevens, R., Egaña Aranguren, M., Wolstencroft, K., *et al.* (2007) Using OWL to model biological knowledge. *Int J Hum Comput Stud.*, **65**(7), 583–94.

4

Design, implementation and updating of knowledge bases

Sarah Hunter, Rolf Apweiler, and Maria Jesus Martin

4.1 Introduction

A clear message emanates from the major bioinformatics infrastructure institutes and initiatives: *the development and maintenance of biological knowledge bases is core to their missions*. At the European Bioinformatics Institute (EBI), the aim is '...to provide freely available data and bioinformatics services to all facets of the scientific community in ways that promote scientific progress.' Similarly, the National Center for Biotechnology Information (NCBI) 'creates public databases... develops software tools... and disseminates biomedical information... for the better understanding of human health and disease.' Biology and biological research is becoming inexorably linked with computation as a consequence of both the digitization of biological data and increasingly vast quantities of these data. The organization and provision of biological data in knowledge bases is therefore critical for related research to be able to continue progressing in a manageable and effective way.

The first biological knowledge bases housed data in relatively simple formats, and one of the first to be widely recognized was the Protein Data Bank (PDB), which stored protein structures from an international consortium of crystallographers and structural biologists (Bernstein *et al.*, 1977). The motivations behind

Knowledge-Based Bioinformatics: From Analysis to Interpretation   Edited by Gil Alterovitz and Marco Ramoni
© 2010 John Wiley & Sons, Ltd

forming the PDB are as relevant today as they were at its inception almost 40 years ago; that is, the concept of sharing data via a single resource to mutual scientific benefit. Other databases followed suit, and by the 1980s scientists were depositing novel genome and EST nucleotide sequences in EMBL-bank, DDBJ, and GenBank, and protein sequences in Swiss-Prot.

The technological advances in genome sequencing over the past few years have led to a massive decrease in cost and increase in throughput of genome sequencing projects. A consequence of this is that the number of nucleotide sequences (and, by extension, protein sequences) which have been deposited into public repositories has also grown at a rapid rate, with almost 10 million protein sequences currently existing in the UniProtKB repository, compared with under a million sequences 10 years ago and around 10 thousand sequences 20 years ago. Biology and its data are regularly stated to be expanding in size faster than Moore's Law, and this can generally be attributed to these improvements in technology and a widening of their availability. The prevalence of the trend is apparent in Figure 4.1, which displays data depositions in structure, nucleotide, and protein knowledge bases over the past decade.

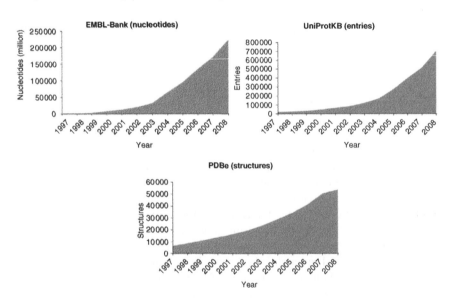

Figure 4.1   Trends in biological knowledge-base sizes over the past decade. Data from EMBL-Bank (for DNA data), UniProtKB (for protein data), and PDBe (for 3D structure data).

The expanding numbers of protein sequences and related data demand new approaches in data storage, data structuring, and data analysis. In this chapter we will discuss how biological knowledge bases are developed and maintained, mainly using the EBI databases InterPro (Hunter *et al.*, 2009) and UniProtKB (UniProt Consortium, 2009) as illustrative examples.

In 2002, the Universal Protein Resource (UniProt) was established as a central resource of protein sequences and functional information, providing essential services to a wide range of scientists in biological research. UniProt is the result of a collaborative effort between well established, but until then, distinct protein databases: Swiss-Prot and TrEMBL, operated by the Swiss Institute of Bioinformatics (SIB) and the European Bioinformatics Institute (EBI); and PIR-PSD, produced by the Protein Information Resource (PIR). The consolidation of these resources under UniProt facilitated the introduction and development of new database components, each of which was specifically designed to address a key need in protein bioinformatics. The UniProt Knowledgebase (UniProtKB) provides protein sequences with extensive annotation and cross-references and it is an essential resource for scientists working in biological research. The UniProt Archive (UniParc) is the main sequence storehouse which is used extensively as a source of protein sequences for other biological resources, including InterPro. The UniProt Reference Clusters (UniRef) condense sequence information and annotation to facilitate both sequence similarity searches and protein sequence analysis. The UniProt Metagenomics and Environmental Sequences database (UniMES) focuses on storing the increasing volume of data from environmental and metagenomics studies.

The manual curation of sequences in databases such as UniProtKB/Swiss-Prot is of undisputed importance; however, it is painstakingly slow work and, as such, cannot hope to keep up with the explosion in sequence data. In the 1990s, several groups recognized that automated methods of classifying and annotating sequences would be necessary in order to deal with the increasing volume of uncharacterized sequences. The resources produced by these groups aimed to model current knowledge about protein domains, families, and sites and worked under the assumption that proteins with similar sequences and/or structures could possibly also share similarity of biological function. Each of the resources use a slightly different, complementary technique for modeling, variously using profile hidden Markov models (HMMs), motif 'fingerprints,' scoring matrices, and regular expressions. In 1999 the InterPro database was created, and initially amalgamated four resources (PRINTS, PROSITE, ProDom and Pfam) together. Since then, seven other databases have also joined the InterPro consortium; SMART, PIR SuperFamily, Panther, HAMAP, TIGRFAMs, CATH-Gene3D, and SUPERFAMILY. Each member database of InterPro has its own niche in methodology and/or biological focus, making InterPro the most comprehensive protein classification resource available.

The amount of data is not the only aspect which has grown rapidly; increasingly, a greater number of biologists are accessing these resources and using them in their daily work. Biology is becoming more digitized and computer-centric, and this places other pressures on knowledge bases, where they must cope not only with an audience that is significantly larger but also which has a widening range of experiences and expectations.

The combination of factors presented above has presented us with numerous challenges, including the question of how to create a knowledge base that is optimally designed for maintenance as well as for delivering its contents to its users. In this chapter we will outline the methodologies we have used to overcome these challenges, explaining how they have influenced the design of these knowledge bases and the infrastructure that supports them.

4.2   Sources of data in bioinformatics knowledge bases

Data feeding into public biological repositories can generally be subdivided into three types: data submitted by external users and collaborators; data generated by internal database curators; and data generated by automatic means.

4.2.1   Data added by internal curators

Central to the biological knowledge-base concept is data curation. Biological data curation typically involves experts (such as PhD level scientists) interpreting data and literature and using this information to populate the database. It is a highly time-consuming and manual process, yet it is highly appreciated as it adds value and quality to a resource.

The range of work performed by a curator can be quite wide and can include the addition of information stating whether data is ready to be released to the public or not; the composition of text culled from literature sources to describe an aspect of a database entity; the association of ontological terms and classification systems to an entry and the structuring of relationships between data in the database (e.g., 'This protein entity inhibits this enzyme's activity.' or 'This structural fold relates to these functional families.').

The size of a database curation team can vary considerably from a single curator to multiple curators spread over multiple sites (UniProt has 60 curators working at 3 different sites). The most obvious factor influencing this is the type and volume of data these curators are handling.

4.2.2   Data submitted by external users and collaborators

Knowledge bases often allow submission of data by users who are not part of the institution where the resource was developed and maintained. In fact, the *raison d'être* of several knowledge bases is to be the central archive of a particular kind of data (for example, the INSDC's nucleotide archives EMBL, DDBJ, and GenBank or ArrayExpress, the repository for microarray data held at EBI). These databanks provide tools to allow submission of data on both a small and a large scale. For small-scale submissions, Web-based forms tend to be provided where a user can interactively enter data which is subsequently validated (automatically and/or manually) before it is allowed into the repository. In the case of large-scale submissions, a standardized data format is typically defined, usually in a language such as XML (eXtensible Markup Language), although proprietary file formats also exist (these proprietary formats may be a throw-back to the early days of bioinformatics repositories; many were flat file-based, initially).

A more recent addition to the repertoire of user data submission tools is the *Wiki* concept, where any user can edit pages containing data from a resource. A few databases, such as the RNA family database Rfam (Gardner *et al.*, 2009), use Wikis to allow users to contribute annotations to their resource. This approach has become more popular within Bioinformatics resources in recent years and it is likely that other knowledge bases will follow this trend and use Wikis as another method for obtaining data to supplement existing content.

4.2.3   Data added automatically

A significant proportion of data contained in Bioinformatics knowledge bases is the result of automatic associations or predictive calculations. In order to be able to cope with large increases in data volumes and complexity, data providers are resorting to automatic predictions to populate areas that would otherwise be 'thin' on information. The information that is be added in this way is not as in-depth as that added by a human curator; however, these processes' throughput is much greater.

Additionally, rather than relying on manual processes to load and transform data, automated pipelines are implemented (where possible) that require minimal human intervention. Again, this is as a consequence of the need to handle very large amounts of data in an efficient way.

Finally, in an environment when quality and accountability is critical, data auditing has a central role and is highly recommended. Auditing is of major importance in knowledge bases, as it tells the user and knowledge-base provider how the data has changed, when it changed and who or what made those changes.

As will be demonstrated in the rest of this chapter, understanding the data which are captured by knowledge bases is the key to understanding the design choices made in their implementation.

4.3   Design of knowledge bases

Designing bioinformatics knowledge bases is not an easy task; where many have succeeded, many more have failed. All too often, these failures occur because the resultant software does not adequately address the needs of the end user. One reason for this is that technology becomes a central driver of the design, rather than the focus being kept on what the software will eventually be used for. Another cause is a lack of communication between resource developers and end users. For these reasons, design is arguably the most critical phase in a knowledge base's development. In this section we will describe steps that can be taken during the design phase of a knowledge base that ought to increase the likelihood of a successful outcome.

4.3.1 Understanding your end users and understanding their data

An obvious major factor that has an influence on the design of a knowledge base is the type of data that is being stored in it and what questions will be asked of that data by users. Knowing the kinds of user who will be accessing your resource (and how they will be accessing it) is therefore critical if you are to produce something that is subsequently adopted and used by your target community. Knowledge base developers frequently become too inwardly focused during the design and development phase, forgetting to prototype and iterate designs with end user input. This lack of external input is damaging as it can lead to a project taking a tangential path and producing a final product which does not fulfill requirements.

The technical people designing and building knowledge bases are not always experts in the particular data domain that will be represented. Therefore, a first challenge is to ensure that a dialog exists between the knowledge base developers and domain experts throughout the design process. An understanding needs to be reached between both parties regarding what the important data to capture is, what terminology to use during design discussions, and what the eventual scenarios for usage will be. Many software development projects therefore start with a requirements gathering phase, where the data that are to be represented in the knowledge base are cataloged in consultation with developers, the domain expert, and the resource's intended user base. It may be that the domain expert or developer already feels she or he has a good idea of what will be contained within the knowledge base, but it is generally a good idea to canvass the community for opinions, so that the data catalog is as comprehensive as is necessary. However, a danger exists that trying to take all ideas and opinions into the final design leads to a loss of focus in the functionality, and development teams should remain mindful of this.

Domain experts and developers work together to identify which data components are central to the resource, with the intention that a model of the data can be created which adequately represents it. The advantage of creating a data model is that a description of the data then exists that can be referred to throughout the design process. Hopefully, this means that the chances of superfluous features being added or something important being overlooked or omitted are diminished. Data models should attempt to represent the data in a way that is meaningful to both the users of the resource and the developers working on it. Typically, a data model will consist of entities (i.e., the data objects that have been cataloged as being core to the resource) and the relationships between these entities; data representations are therefore often called Entity Relationship Models (ERMs). This type of representation is frequently encountered in object-oriented approaches to software design or where a relational database is under development. Developers tend to use standard formats to describe a data model; examples of this are DDL (Data Definition Language), DTD (Document Type Definition) used with XML documents, or UML (Unified Modeling Language) which can offer additional benefits, such as the ability to map to other languages, such as Java or XML.

4.3.2   Interactions and interfaces: their impact on design

Broadly speaking, there are two main ways that users access bioinformatics knowledge bases: *programmatically* and *interactively*. Programmatic access tends to be used where a user wishes to do some sort of large-scale analysis of data which is too complex or time-consuming to perform manually, or where they want to seamlessly display information from a resource in conjunction with data from their own, without having to worry about keeping a local copy of that data up to date. Interactive access is the more traditional method of using a Web interface or GUI (Graphical User Interface) to allow a user to search and browse the data in your knowledge base. Both require careful design as they each need to behave in a way that is expected by the user.

In the past, bioinformatics knowledge bases perhaps did not place as much emphasis on usability as they do currently. This is possibly because the data they were storing and visualizing initially were not as complicated as they are now. This increase in complexity, combined with higher expectations from a more sophisticated, Web-aware user base has made usability testing a priority in knowledge-base design.

4.4   Implementation of knowledge bases

4.4.1   Choosing a database architecture

In biological repositories in the 1970s, punch cards were still used for data storage and exchange; in the 1980s the first releases of the Swiss-Prot database used a text file-based catalog of protein sequences; in the 1990s, moves were made towards utilizing database management systems to organize increasing amounts of data, some of which had not been encountered before (e.g., from MicroArray technologies). Now, in the 2000s, we are faced with handling and analyzing enormous quantities of genomic and proteomic data resulting from new projects sequencing many thousands of genomes.

The vast majority of modern bioinformatics databases have complex data structures and consequently rely on relational database management systems (RDBMSs) to ensure data integrity. For early, simple versions of knowledge bases, a file system-based approach might have sufficed; however, this is no longer sustainable. The main RDBMSs used in Bioinformatics knowledge bases are typically ANSI SQL compliant, and, at EBI, are mainly Oracle (InterPro, UniProt and ChEBI, for example) or MySQL (Ensembl), although PostGres is occasionally also used.

There are three principle factors to take into account when deciding upon your database architecture.

(1) **Performance:** as databases get larger and more widely utilized, performance of the system becomes a key factor. Using a well-established DBMS and optimizing both the hardware and software set-up should

help limit negative effects on performance and allow fast updating and querying of the data contained within. The use of appropriate indexes also contributes a positive effect.

(2) **Back-up and recovery:** being able to recover from losses of data or disasters is crucial if the amount of up-time of a database is to be maximized and users' faith in the resource is to be maintained. Inefficient back-up strategies can not only delay the recovery of data but also impact negatively on performance.

(3) **Purpose:** the role of a particular database within a larger system will obviously affect how it is configured. If a database is constantly updated by multiple concurrent users and the data in it is dynamic, a normalized, relational database would be well suited in this role. However, if the data is more static and the primary function of the database is for querying rather than editing, a denormalized, query-optimized schema might be chosen or, alternatively, an indexed file-based system.

At the heart of UniProtKB and InterPro is a series of databases (both relational and flat-file) which are used for development and testing of software changes, production of data, and serving of that data to the public. Each database has a different purpose, outlined in Figure 4.2.

Being able to uniquely identify entities within a biological database is important not just for ensuring referential integrity within the database itself but also to allow users to unambiguously refer to entities within it. Within UniProtKB, there are multiple ways to identify protein sequence entities:

(1) An entry name *ID*, which is a unique but non-stable human-readable identifier, often containing biologically relevant information. It consists of up to eleven uppercase alphanumeric characters with a naming convention that can be symbolized as X_Y, where X is the mnemonic protein identification code and Y is the code for the species identification. For example, SRPK1_HUMAN.

(2) An *Accession* which uniquely identifies the protein entity. For example, Q96SB4.

(3) A *Checksum* which is calculated on the sequence and uniquely identifies it. For instance, 900E980FE1C16B9A.

(4) A numerical identifier which is only used internally within the database, as primary keys for enforcing referential integrity of the data.

An often overlooked issue within biological databases is how to make sure that users understand which identifiers are stable and therefore suitable for long-term identification of a particular entity. In UniProtKB, the accession uniquely identifies a specific protein, usually the longest variant of the sequence and typically one from a particular species. However, with time, the underlying sequence

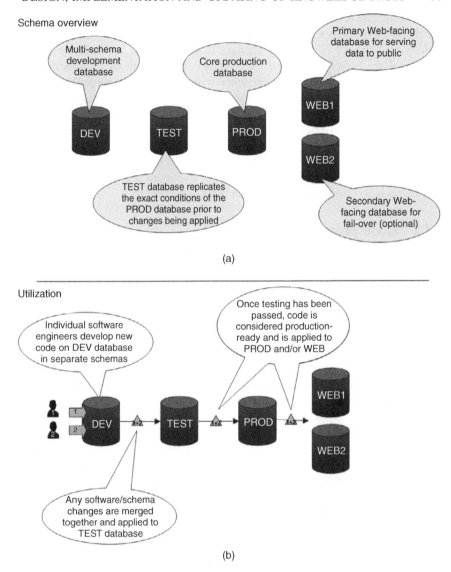

Figure 4.2   Outline of how database instances are arranged in InterPro and UniProtKB, and what purposes they serve.

itself may change (due to corrections of sequencing or gene prediction errors, for example), or other protein entries in UniProtKB may be merged in with that entry. In this case, the accession may become 'secondary' to the entry; that is, a user will still be able to use their accession to access the information around the protein but it may no longer be the primary way that the protein is identified. If the exact sequence of the protein is important for the purposes of the user, the accession on its own is probably not sufficient. Instead, the user should use the version number provided with the accession, as this only changes if the underlying sequence of the entry changes (e.g., Q96SB4.2 indicates the second version of the protein identified by Q96SB4). These methods of identification should therefore be used whenever users wish to refer to a particular protein, such as when they are creating a cross-reference from another resource or are publishing protein-related data in a scientific journal. The entry name ID should not be used in this manner as it is liable to change and is not tracked; it is only provided because of its readability (for example to ensure that related entries have consistent and similar names or when entries have manually annotated protein names). Later in the chapter, we discuss how important sequence and accession versioning is to internal processes at UniProt and InterPro.

4.4.2   Good programming practices

In any software development project, there should be two mantras: *Keep Things Simple* and *Use Good Programming Practices*. Generally speaking, the fewer technologies you use in a project, the easier it is going to be to maintain; this includes the programming languages that are going to be used. The prevalent programming language used within the UniProtKB and InterPro projects is Enterprise Java after a conscious decision was made to minimize the number of programming languages used overall, so that any software engineer on the team can interchangeably work on the code, insuring maintainability of the resource for the future. Other scripting languages, such as shell programs, SQL, Perl and python continue to be used by our teams but not extensively.

Where possible, developers should be encouraged to share development tasks, partly so that multiple software engineers can support, develop and maintain the code. It also reduces the risk of errors appearing because developers can review and check each other's code as it being written. Storing of code in a single, centralized code repository such as CVS (Concurrent Versioning System) or SVN (Subversion) allows multiple software engineers to access and edit software components of a single system in a manageable, robust manner. In both InterPro and UniProt, CVS is used to capture changes to the software accessing the database, together with SQL scripts that describe both the original database schema and the changes which have been applied to it. These schema evolution scripts are tagged when major software releases occur, to allow tracing of which schema worked with a particular version of the software and vice versa. In addition to a central repository, a continuous integration server (such as TeamCity or Cruise-Control) may be used to encourage frequent updates of code to the repository, with thorough overall testing and builds of the software.

In UniProt and InterPro, 'unit tests' are systematically used during the software development process to ensure that the code meets initial design requirements and would continue to perform as intended if changes were made at a later date. Each developer needs to write a test for every class as it is developed, making sure that this covers the entire functional requirement for that class. It is important that all code that is released into the source code repository includes tests that run at 100%. This ensures that all functionality always works. When there is a new code release, all tests are run and any failure can be quickly identified and fixed. The addition of tests in the code is a cost-effective way of ensuring software quality, in particular, when many developers are involved in the maintenance of the code.

When new features are being implemented in any data supply pipelines or interfaces to any knowledge base, we strongly recommend that initial development and testing of each feature should be performed by software engineers in a development environment. Once this preliminary stage is complete, all code changes should then be merged and re-tested as a whole on a test database, which is effectively a copy of the main, 'live' production database. The production database is most important because it is here that new data is loaded from external collaborators and where curators add and edit data; it is therefore important that the integrity of this database is always intact and rigorous testing of all code has taken place. In the event that data or code in this database becomes corrupted, backup and recovery strategies should allow rollback of the database to any point in time, reducing the likelihood of data loss.

4.4.3   Implementation of interfaces

The data contained in biological knowledge bases can be presented to end users in a variety of ways, and different users may prefer different methods of access. The most basic interface would be a dumping out of the contents of the knowledge base into a delimited, flat-file format and presenting it for download from an FTP site or similar. While this is relatively easy to do, the data provided in this way are not particularly easy to manipulate or query. As a result, knowledge-base teams may find that users begin requesting the production of multiple, different file formats which essentially contain the same data but organized in subtly different ways. This has an obvious maintenance overhead. However, if the flat file structure is formatted appropriately, such as in a recognized standard XML format, it should be possible to also provide indexing capabilities on the file(s) in order that appropriate data may be retrieved. A drawback to allowing users to download data locally is that there are potential difficulties keeping the data synchronized. As we discuss later in the chapter, a way to avoid this is by providing programmatic and graphical user interfaces to the knowledge bases, so that data is pulled directly from the database.

UniProt data can be downloaded via FTP in various formats including XML, RDF, fasta and flat file. In order to process the information contained in these formats, the users need to write a parser that translates the input into suitable data structures. Furthermore, data format changes within UniProt can lead to significant maintenance overhead. In order to facilitate access to our data, we maintain a number of programmatic interfaces which include protocols using the open standards REST and SOAP (Web services), Java applications (for example, UniProt-JAPI (Patient *et al.*, 2009)) and database federation approaches (such as BioMart).

The UniProt REST interface (www.uniprot.org) provides access to UniProt data, as individual records and queries, in a simple and consistent URL schema, and returns appropriate content type headers (e.g., application/xml for XML resources) and response codes. All URLs can be bookmarked and linked, allowing the user to obtain the most recent information available at the time.

UniProt data is also delivered through SOAP (www.w3.org/TR/soap/) services for both database information retrieval and sequence analysis. The Web services technology used in UniProt is built in open standards (SOAP as the messaging protocol for transporting information; WSDL (www.w3.org/TR/wsdl) as the standard method for describing the Web service; and UDDI (www.uddi.org/specification.html) as the platform-independent XML-based registry for services). Information about these services can be accessed from www.ebi.ac.uk/Tools/webservices/ (Labarga et al., 2007).

UniProt provides a Java application programming interface (UniProtJAPI) that allows remote access for Java applications processing UniProt and related data, such as scores and start and end positions of the signatures in InterPro. This API (Application Programming Interface) represents each UniProtKB record as a Java object and provides methods to access all of its information. For example, `getDescription().getProteinName()` returns the protein name associated with this entry. The UniProtJAPI also provides the ability to perform text and sequence similarity searches across this data, allowing users to access a single database entry with a given accession number, or whole entry sets matching a defined criteria.

4.5   Updating of knowledge bases

The release cycles of bioinformatics knowledge bases can vary in length from a matter of weeks to annual updates. Multiple factors can influence a release cycle, but the biggest factor by far is the availability of the data that populates the knowledge base. In this section, we outline how data in these databases is maintained by a variety of processes, each with its own advantages and drawbacks; a variety of approaches being required due to the changing nature and volume of the data that is captured and stored.

4.5.1   Manual curation and auto-annotation

Although centralized sequence repositories are an essential means of providing a user with the sequences themselves as quickly as possible, it is clear that associating additional information with a sequence greatly increases the scientific value of the resource. UniProt has earned an international reputation for high-quality manual and automated annotation of protein sequences. Manual annotation is a slow and labor-intensive process and is generally the rate-limiting step in the production of any curated biological database. Curation of protein sequences aims to enrich basic sequence data with additional information from a wide range of sources such as the scientific literature and specially selected high-quality sequence analysis tools. Proteins for which there exist published functional, biochemical, and/or structural data are the main targets for manual annotation. Curators add knowledge-derived information such as protein function, biologically relevant domains and sites, post-translational modifications, the subcellular location of the proteins and their complexes, developmental- and tissue-specific expression levels, splice-variant isoforms, and the publications used in the annotation process. These manually annotated protein sequences constitute the 'reviewed' section of the UniProt Knowledgebase (UniProtKB/Swiss-Prot).

Since the number of protein sequences continues to grow exponentially, sophisticated computational techniques are employed for their analysis, in particular for sequences that have not yet been manually characterized. Many of the automatic procedures used in community annotations are based on sequence similarity searches using tools such as BLAST (Altschul *et al.*, 1990). An alternative approach is to use protein signatures, such as those included in InterPro, which allow the identification of distant relationships to novel sequences, and hence the prediction of protein functions and structure. Often, multiple methods are combined together in annotation pipelines. All of these methods increase the amount of information associated with proteins, and provide a rapid, automated means of analysis, however, they all rely on the availability of accurately annotated sets of reference proteins from which to make predictions. Manual annotation is thus essential, not only to provide high-quality information to database users, but also to supply accurate information on which automated methods can be based. Thus, the combination of manual curation and automated methods is essential for providing high quality functional annotation to an increasing number of proteins. This remains the core of UniProt annotation activities.

The automatic annotation system developed in UniProt is based upon rules derived from the combination of (1) a protein family classification provided by the InterPro protein family and domains database, and (2) published experimental data, which is incorporated in the manually annotated section of UniProtKB, UniProtKB/Swiss-Prot. The use of protein family and domain classifications allows the characterization of proteins that are difficult to identify when using pair-wise alignment methods. It also provides an effective means to retrieve relevant biological information from vast amounts of data as well as reflecting underlying gene families. The analysis of these families is essential for subsequence comparative genomics and phylogenetic analysis. UniProtKB/Swiss-Prot represents a rich and consistent source of standardized functional annotation for the large unreviewed sections of UniProtKB: UniProtKB/TrEMBL and UniMES. A single annotation *item* (i.e., protein name, enzyme nomenclature, protein function or ontology relationships) can be triggered as a result of combining data from different annotation *sources* (i.e., automatic annotation rules, PDB structures, model organism databases, etc.), and/or multiple annotation items can result from a single annotation source. To distinguish all derived evidence of annotations, it has been necessary to establish a methodological system of evidence 'tags', where the description of the annotation procedure is described. This system of evidence tags is essential for the maintenance and updating of the database, and to assist the end user with the interpretation of annotations contained within the database.

Core to InterPro are the predictive signatures which classify protein sequences into protein families and domains. Each member database is continually generating new signatures based on data being published in the literature and novel sequences being deposited in the public domain. Update frequency can range from every three weeks to yearly. InterPro is often sent a pre-release of data, so that it can be prepared for inclusion into the resource and released to the public as quickly as possible.

Data provided by constituent InterPro member databases can vary considerably. Most obviously, the types of models which make up the predictive signatures and the associated data for interpreting results from these models (significance cut-offs, for example) are different from one database to the next. The breadth of annotation provided by member databases ranges from very limited (an identifier and model) to detailed (some databases provide in-depth annotation of the families and/or domains that they are modeling, with literature references and Gene Ontology terms included). InterPro also extracts relevant data from related databases. For example, in addition to the protein sequences themselves, information regarding the taxonomic spread of proteins in a particular InterPro entry must be extracted from UniProtKB and loaded into InterPro. Similarly, structural data is taken from wwPDB and utilized in InterPro entries, as is data regarding enzyme classifications (EC), where appropriate. All of these disparate data sources must then be molded into a consistent, high-quality set of annotation by the InterPro curation team and the database's pipelines.

In order to be included into InterPro, a signature has to be considered biologically significant by an InterPro curator (i.e., it is representing a 'real' protein family rather than a similar set of sequences). This decision is taken by a curator who then additionally checks whether or not a signature is similar to anything already in InterPro. If it appears that two signatures are representing the same entity (a functional domain, for example), they are placed into the same entry. If a signature does not resemble any existing InterPro entry, a new entry will be created for it. If the signatures of an entry match a subset of proteins compared to another, such as when a more specific sub-family has been described, in this case, the entries are related to each other in a hierarchy called 'parent-child.' A descriptive abstract and name for the entry are written by curators and, if possible, Gene Ontology (GO) terms are mapped to it.

In order to be able to process the deluge of data, the curation process in all knowledge bases therefore needs to be intuitive and (consequently) efficient. InterPro and UniProtKB both provide powerful GUI-based curation tools which allow the speedy update of data in their respective repositories. Where possible, tasks are automated. For example, in UniProtKB, there are various *macros* – recorded procedures associated with keystrokes – which the curators can use to quickly short-cut to a particular functionality in the tool. In InterPro, prediction of certain data items, such as an abbreviated name, is performed to allow the curator to accept or edit the suggested information. Visualization tools showing the relationships between data in InterPro are also used by curators when they are making decisions about how to enter data into the database or assess its quality.

4.5.2   Clever pipelines and data flows

There are two scenarios to consider when dealing with a multi-source database:

(1) The design of clever pipelines and schemas that can accommodate different sources of conceptually identical data types.

(2) The implementation of an update mechanism that allows multiple processes to modify the data independently.

There are two good examples of such multi-source databases in UniProt. *UniParc* is an archive of all protein sequences – including new and revised protein sequences – from many public sources, including UniProtKB and various external sources, such as DDBJ/EMBL/GenBank CDS translations, RefSeq protein sequences, Ensembl predicted proteins, PDB protein sequences, protein sequences in patents, and other protein data (see Figure 4.3). It offers a single point of entry that allows access to all versions of sequences previously and currently available from the protein databases.

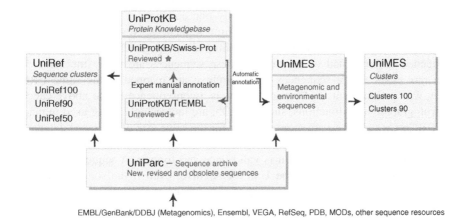

Figure 4.3   Schematic of the UniProt set of knowledge bases.

Populating a database with all protein sequences would potentially lead to a huge amount of data redundancy, since the same sequence could be found in multiple sources (UniProtKB, RefSeq, PDB, etc.). However, in UniParc, each unique sequence is assigned a unique identifier (UPI) and is stored only once. The basic information stored with the UPI and the sequence is a checksum (both

CRC64 cyclic redundancy check and MD5 numbers are stored internally and are used as a hash-key to speed up the identification of unique sequences); the accession number(s); version number(s); source database(s); and a time stamp. In this way, the UniProt Archive overcomes the unsatisfactory situation that arises when there is no single point of entry that allows access to all the available versions of protein sequences that occur in a variety of source databases, and leads to sequences being not easily traceable across databases. In essence, the UniProt Archive acts as a basic 'translation hub' indicating which sequence corresponds to which identifier in various databases.

Three basic aspects of the technical implementation of the UniProt Archive are of central importance:

- The hash coding of sequences to efficiently determine if a given sequence is already present in the archive.

- The assignment of increasing alphanumeric ID values to incoming sequences to guarantee that the archive's IDs preserve a chronological order.

- The immutability of the archive; making sure that sequences, once archived, cannot be deleted and inserted again into the archive under a new ID.

Every sequence that is imported into the archive receives a unique identifier and a hash value based on the CRC64 algorithm. Most proteins are guaranteed to receive unique checksums because only one collision of hash values can be expected in a database containing 2.3 trillion sequences. Nevertheless, for every sequence sharing the same CRC64, additional string matching of the hashed sequence is performed to ensure that these sequences are really the same. In the rare cases where sequences in the archive are mapped to the same CRC, they will get different UniParc identifiers as a consequence of the above check. Cryptographic hash functions, such as MD5, are increasingly being used in place of CRC64 as collisions are more unlikely for the same size of sequence database.

The immutability of the archive guarantees that a sequence, once archived, will have a stable identifier that is valid once and forever. A sequence will still exist in its original form in the UniProt Archive, even if the source sequence in, for example, WormPep has changed. The changed WormPep sequence will also enter the archive, but the change in sequence will lead to a new CRC64 hash value, and thus a new unique UniProt Archive Sequence Identifier will be assigned. The status of a sequence in the source database (i.e., whether the sequence still exists or has been deleted) is indicated for each source sequence record. These features of the archive allow us to use the UniProt Archive as a versioning server to retrieve 'historic' sequence data, which is particularly important in the context of patent claims. The incremental order of the ID numbers allows sequence-based computational analysis tools to perform a comparison of two identifiers using a relational operator and determine which is more recent (i.e., UPI0000123A > UPI00000456 will return 'true').

The immutable and sequential nature of UniParc is also exploited by the pipelines within InterPro. A significant proportion of the data in InterPro is calculated automatically by a tool called InterProScan, which searches member database signatures against protein sequences. InterProScan is implemented as a modular pipeline, which takes multiple sequences as input, uses the searching software associated with each member database to look for matches of signatures to these proteins, and filters the output according to various significance criteria, such as e-values. A schematic of how InterProScan works is shown in Figure 4.4.

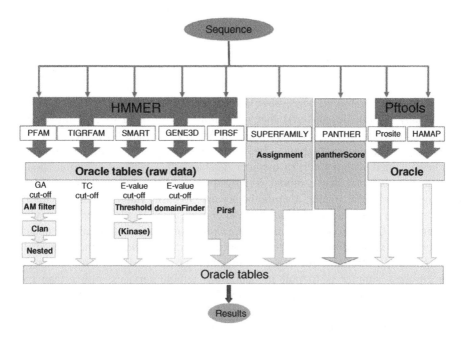

Figure 4.4   Schematic of InterProScan architecture. Results from the search algorithms are typically stored in raw data 'staging tables' in an Oracle database. The entire output from the search is then post-processed (i.e., cut-offs applied, etc.) altogether and persisted in a final set of main database tables. Not all searching algorithms are displayed in the diagram due to space limitations.

In order to keep the complement of all signature matches against all protein sequences up to date, InterProScan uses the sequence versioning capabilities of UniParc to ensure that only completely new sequences are calculated against the signature databases. The reason for this is that the majority of algorithms currently used within InterPro are relatively computationally expensive. To search an average-length protein sequence against an average-sized hidden Markov model (HMM) using the HMMER algorithm (v.2.3.2) takes around two seconds. This may not seem like a long time; however, when one considers that the size of UniProt is nearing 10 million proteins, and InterPro contains over 50 000 HMMs at the time of writing, the problem begins to look a little more daunting! Using the archive, it was possible to reduce the number of calculations by being careful to only recalculate data when the sequence and/or the model or algorithm had changed. Looking again at the diagram of InterProScan, one can see that the initial, compute-intensive, algorithm search step is completely divorced from the processing and filtering steps that further refine the raw results from these searches. By storing the raw results and only recalculating them where absolutely necessary, we estimate that we have saved many thousands of compute hours on our local cluster. Being able to generate this data as quickly as possible is critical if we are to have a sensible release cycle.

4.5.3   Lessening data maintenance overheads

To overcome the problems of inconsistent data formats and types it is a good idea for collaborators to agree upon a standard format for data exchange. There are many worldwide consortia that have done this in order to ease movement of data between their repositories (e.g., the MIAPE standard for proteomics data exchange (Martens *et al.*, 2007)). In InterPro it was agreed that all core information contained within InterPro member databases that were in common should be inventoried and that a model which represented this information would be designed. From this model, it would be possible to easily transform data from one format (e.g., objects in a database) to another (e.g., an XML file for exchange with others). By doing so, the amount of time and effort required to maintain update pipelines is reduced because a single format means a single pipeline is needed.

A way to minimize the effort of maintaining and updating data and make the process simpler is to pull in data from remote services, rather than trying to federate information locally. DAS (Distributed Annotation System) (Prlic *et al.*, 2007) is a client-server system in which a single client is able to integrate information from multiple, disparate servers. It allows producers of data to expose the contents of their repositories in a standardized and easy-to-discover fashion. The protocol itself is very simple and requires very little data download from the server to the client site.

The DAS route is used by UniProtKB to allow better exploitation of its data by other databases and to facilitate other groups' contribution of information to UniProt. The UniProt DAS server allows researchers to show their research results (for example, identified peptides or signal sequences) on the UniProt reference sequence server, in the context of UniProtKB annotation. This approach is more tailored towards Bioinformatics groups or labs with sufficient bioinformatics support.

The UniProt DAS Reference Server (www.ebi.ac.uk/uniprot-das/index.html) provides sequence and feature data from UniProtKB as well as sequence data from UniParc and IPI (International Protein Index; Kersey *et al.*, 2004). The server can be queried using any of the following:

- UniProtKB Accession numbers; for example, O35502.

- UniProtKB IDs; for example, A4_HUMAN.

- IPI IDs; for example, IPI00015171.

- UniParc IDs; for example, UPI0000125656.

Besides UniProtKB, many other knowledge bases expose their data in this way, including InterPro and its member databases, Ensembl, PDB and model organism databases.

4.6 Conclusions

This chapter has described how the changes to the biological data landscape have impacted on how knowledge bases have developed over the past decades. There is no doubt that the factors we describe – the volume and complexity of these data – will continue to shape the architecture design choices that are made for many years to come. A key challenge for bioinformatics in these coming years will be to harness new, appropriate technologies in order to ensure that the capture of data and its provision to the public does not lag behind advances in biology.

4.7 References

Altschul, S.F., Gish, W., Miller, W., *et al.* (1990) Basic local alignment search tool. *J Mol Biol*, **215**, 403–10.

Bernstein, F.C., Koetzle, T.F., Williams, G.J., *et al.* (1977) The Protein Data Bank: a computer-based archival file for macromolecular structures. *J Mol Biol.*, **112**, 535–42.

Gardner, P.P., Daub, J., Tate, J.G., *et al.* (2009) Rfam: updates to the RNA families database. *Nucleic Acids Res*, **37**, D136–40.

Hunter, S., Apweiler, R., Attwood, T.K., *et al.* (2009) InterPro: the integrative protein signature database. *Nucleic Acids Res*, **37**, D211–5.

Kersey, P.J., Duarte, J., Williams, A., *et al.* (2004) The International Protein Index: an integrated database for proteomics experiments. *Proteomics*, **4**(7), 1985–88.

Labarga, A., Valentin, F., Anderson, M., and Lopez, R. (2007) Web services at the European Bioinformatics Institute. *Nucleic Acids Res*, **35**, W6–11.

Martens, L., Orchard, S., Apweiler, R., and Hermjakob, H. (2007) Human proteome organization proteomics standards initiative: data standardization, a view on developments and policy. *Mol. Cell Proteomics*, **6**, 1666–7.

Patient, S., Wieser, D., Kleen, M., *et al.* (2009) UniProtJAPI: a remote API for accessing UniProt data. *Bioinformatics*, **24** (10), 1321–22.

Prlic, A., Down, T.A., Kulesha, E., *et al.* (2007) Integrating sequence and structural biology with DAS. *BMC Bioinformatics*, **8**, 333.

UniProt Consortium (2009) The Universal Protein Resource (UniProt). *Nucleic Acids Res*, **37**, D169–74.

Section 2

Data-Analysis Approaches

5

Classical statistical learning in bioinformatics

Mark Reimers

5.1 Introduction

Biology has been changed dramatically in the past decade by the emergence of powerful high-throughput assays, such as microarrays and highly parallel sequencing. These methods have brought issues of data analysis from the wings to the forefront of biology. In reciprocity, the research agendas of statistics have also been dramatically changed by the advent of high-throughput genomics. At recent Joint Statistical Meetings many of the applied sessions focused on high-throughput genomic data, and many of the theoretical discussions focused on 'Large P – Small N' problems, where there are many more variables than measured samples, which is now the norm in genomic data analysis.

As in any new field there are many claims. However, in many cases the important questions to ask about a new method are classical statistical questions; for example, how much evidence is there for apparent structure in the data? How can the accuracy of classification be tested using only a limited sample? These kinds of classical questions will inform this brief guide to classical statistical learning in bioinformatics.

5.2 Significance testing

The oldest form of statistical learning is deciding whether a particular measure differs between the populations from which two samples have been taken. Here

Knowledge-Based Bioinformatics: From Analysis to Interpretation   Edited by Gil Alterovitz and Marco Ramoni
© 2010 John Wiley & Sons, Ltd

I will assume that the reader has a basic grounding in statistics and discuss some issues peculiar to high-throughput data analysis.

5.2.1 Multiple testing and false discovery rate

Most genomic researchers compare many genes or loci across a set of samples; therefore multiple-comparisons issues come up immediately. Suppose that a researcher assays 20 000 gene expression levels between two conditions that (unknown to the experimenter) actually do not differ at all. If the researcher performs a t-test at a significance level of 1% then he or she might expect to find 200 genes that appear significant at a 1% threshold. However, suppose that the researcher anticipates this problem, and performs the test at a much stricter threshold, at which he or she might expect to find fewer than one false positive, for example a threshold of $p < 10^{-5}$; then the researcher is likely to find no significant changes at all, even when real differences exist, because the typically small sample sizes employed in genomic experiments make it difficult to achieve very large t-scores. A considerable statistical industry has sprung up addressing this dilemma. Let's start with a few definitions.

5.2.1.1 Definitions

The Null (default) Hypothesis about a gene is that no change occurs in that gene across the populations that are being compared. Suppose that genes are numbered 1 through M, and we denote the corresponding Null Hypotheses for each gene by H_1 through H_M. We suppose that $M_0 < M$ of these Null Hypotheses are actually true (i.e., no real differences in those genes). Table 5.1 will help keep track of the numbers.

Table 5.1 Symbols for how testing results come out in one experiment.

Hypotheses	Accepted	Rejected	Number
True	U	V	M_0
False	T	S	$M - M_0$
	W	R	M

The number of null hypotheses rejected wrongly ('false discoveries') is here denoted by V and the number of rejected null hypotheses is denoted by R. Note that R is determined by the data and the testing procedure, and is known after each experiment, but that S, T, U, and V are all unknown values, which may vary from experiment to experiment, while M_0 is a fixed number, depending on the experimental question, although the true value of M_0 is unknown to the investigator.

We distinguish two types of errors:

Type 1: false positives; when a true null hypothesis is rejected.

Type 2: false negatives; when a false null hypothesis is retained; that is, a true difference is not discovered.

Most testing procedures try to guarantee an upper bound on Type 1 errors, while keeping Type 2 errors to a minimum.

For any testing procedure, we may think about the false positive rate in several ways:

(1) The *family-wise error rate* (FWER) is the probability of (at least) one false positive somewhere among the R genes which appear significant: FWER $= P(V > 0)$; this can often be calculated from first principles.

(2) The *false discovery rate* (FDR) is the expected proportion of false positives among the R selected genes: FDR $= E(V/R)$, over many runs of the same experiment; this can only be estimated from the data.

Definition (2) above doesn't explicitly address how to handle the case when $R = 0$ (and therefore $V = 0$), that is, when the testing procedure produces no genes. There are two common approaches. The Benjamini and Hochberg (1995) approach counts cases where no genes are selected as having a false positive proportion of 0, in their sense. Storey (Storey and Tibshirani, 2003a; Storey and Tibshirani, 2003b) argued that this is not what most people mean by FDR, and that the expectation in definition (2) is meaningful only for $R > 0$, that is, there are some genes selected by the testing procedure; in their definition FDR $= E(V/R | R > 0)$.

I agree with John Storey that his version of the FDR is closer to our intuitive sense of false discovery rate. However, the Benjamini and Hochberg (BH) FDR is easier to work with, both to develop theory, and to compute, and in my experience the BH sense of FDR is more widely used (Reiner *et al.*, 2003). The relationship between estimates of these two versions of FDR is not always clear. Since there is always some chance of coming up empty handed, the FDR in Storey's sense must logically be somewhat larger than the FDR computed by the BH procedure. If a selection procedure identifies many highly significant genes, the FDRs estimated by the two approaches turn out to be similar. However if only a few genes are selected by a testing procedure, then the FDR in Storey's sense for that testing procedure will usually be much higher than the BH sense of FDR, and there may be no genes at all selected at the same level of Storey's FDR (Storey and Tibshirani, 2003b).

5.2.2   Correlated errors

The commonly used BH procedure only guarantees that the average proportion of false positives, over many repeated experiments, is bounded by the specified value. In practice the variability can be quite high. This is because in real data the errors are correlated, so that false positives tend to occur in groups. Figure 5.1 shows that the actual proportion of false positives selected by the BH procedure can be quite variable, in a simulation.

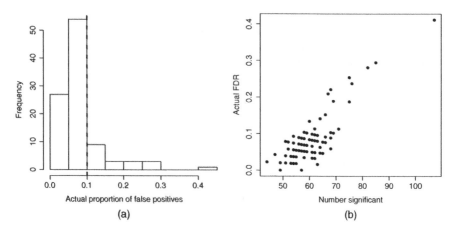

Figure 5.1   Results of a 100-simulations study of the Benjamini–Hochberg procedure. One hundred out of 1000 variables were different between 2 samples of 10 each. The errors were correlated: one-third of the variance in errors was accounted for by the first principal component. (a) Histogram of actual proportions of false positives selected by BH procedure aiming for an FDR of 0.1 (10% – shown by dashed line). (b) Actual proportion of false positives plotted against number of genes selected. The analyst might be inclined to think the reverse of what actually happens.

5.3   Exploratory analysis

The aim of exploratory analysis is to uncover unexpected patterns or relationships in data. In the jargon of machine learning, this is 'unsupervised learning.' Generally speaking, when a researcher has fairly substantial prior ideas about the relationships to be expected among the samples or genes, the statistical techniques of hypothesis testing are both more powerful and less prone to misinterpretation.

5.3.1   Clustering

Clustering has become the most widespread exploratory technique used in genomic data analysis. The clustering habit goes back to the yeast cell-cycle paper, Spellman *et al.* (1998), that brought genome-wide microarrays to the attention of researchers. That paper used a high-dimensional clustering of gene expression patterns to more than double the number of known cell-cycle regulated genes. This signal success had the unfortunate side effect that, for several years thereafter many researchers felt compelled to cluster every microarray data set. In recent years, however, researchers have recognized that clustering is not the appropriate tool for all tasks, in particular not for identifying genes differentially expressed between predefined conditions; clustering has been used with more discrimination recently, and is no longer an obligatory feature of all genomic papers.

Clustering was an appropriate methodology for the Spellman paper. Their explicit aim was to identify groups of genes with similar expression patterns across the cell cycle, and that is exactly what clustering does. Their research premise was that specific transcription factors would activate modules of genes, each module in turn turning 'on' as the cell cycle progressed. This type of modular structure can readily be identified by cluster analysis.

The aim of clustering is to identify similarities in items on which a large number of variables are measured (see Chapter 12 of Johnson and Wichern (2007)). Most people find it difficult to intuitively judge similarity based on multiple measures, and clustering methods provide an objective method of aggregating differences based on each variable measured into a single measure of dissimilarity between each pair of samples. Clustering may be performed on genomic samples in order to identify subtypes among the samples; for example clustering may be a first step towards identifying molecular subtypes of a disease such as cancer, which is known to be heterogeneous (Perou *et al.*, 2000), but whose substructure is unknown.

Clustering may also be performed on genes, as in Spellman *et al.* (1998), for discovering groups of genes with similar expression patterns across a wide range of biological conditions. Note the requirement for a wide range of conditions; clustering genes across a small number of conditions, or across many rather similar conditions, doesn't give very distinct clusters, in my experience. Cluster analysis needs a large number of contrasting conditions to isolate distinct modules.

There are several types of clustering algorithm. Some methods assume a pre-specified number of clusters, and try to fit each sample into one group. These 'k-means' methods, so called as they cluster all samples around k mean profiles, are fast and easy to compute; therefore k-means methods are often used for clustering large numbers of items such as genes or genomic measures. A common alternative method is to aggregate individuals into small clusters, and then to aggregate small clusters into larger clusters, and so on, to form a hierarchy of clusters. This so-called hierarchical method is often used for clustering samples.

Clustering should be viewed as a heuristic tool, which is useful for suggesting hypotheses, but clustering by itself does not quantify evidence to support these suggestions. Furthermore, clustering involves a number of arbitrary choices, some of which may drastically affect the resulting cluster diagram (Do and Choi, 2008; Garge *et al.*, 2005; Kerr *et al.*, 2008). Most clustering software offers several choices of distance metric; that is, how to summarize all differences between items in a single numerical measure. Furthermore, hierarchical clustering offers the user several choices of linkage; that is, what criteria to employ to join smaller clusters into larger clusters.

It is always possible to construct a clustering from any data set, and researchers may be misled by apparent deep branches. The hierarchical cluster diagrams in Figure 5.2 were constructed from completely random data with the same distribution characteristics as microarray data (i.e., variation was described by a t distribution with 5 d.f.). The diagrams were constructed from the same set of random numbers, using two metrics to define distance: Euclidean (L^2) distance, based on the sum of squared differences, and Manhattan (L^1) distance, based on the sum of absolute values of differences. There are two apparently distinct clusters in both versions; unfortunately they don't agree on which items belong together.

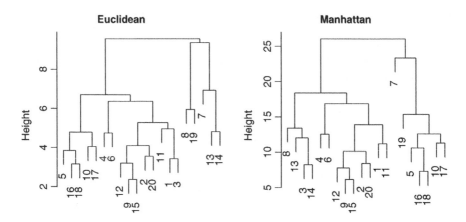

Figure 5.2   The same (random) data set clustered using two different metrics.

This artificial example illustrates an important point about clustering. There are many choices for clustering metrics, and in order to be confident that the clusters are real, it is wise to try several choices. I find Euclidean, Manhattan, and correlation[1] distances to be the most useful. If the resulting cluster diagrams are fairly similar, then the clustering is more likely to reflect real underlying relationships. In assembling a hierarchical cluster, I find complete linkage, average linkage and Ward's method to be the most useful.

Clustering can be reproducible: Figure 5.3 shows an example of real data where the same two metrics give very similar deep branches. Note that the leukemias cluster together and well apart from the other samples, as do the melanomas, in both versions of the clustering, but the relationships within these groups appear to change between versions.

---

[1] The correlation distance is defined as $1 - r(x, y)$, where $r(x, y)$ is the correlation between two items x and y. This distance is insensitive to changes of scale, and so is particularly useful if a researcher suspects normalization problems in data.

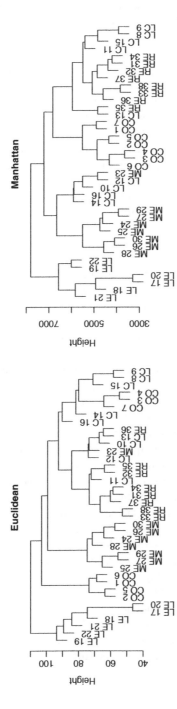

Figure 5.3   Two clusterings of the same data. The cell lines are numbered (arbitrarily) for clarity of the figure. Tissues of origin are indicated by the first two letters. LE, leukemia; ME, melanoma; CO, colon; RE, kidney; LC, lung cancer. This data set is available at http://discover.nci.nih.gov/datasets.jsp.

If a striking clustering appears, most researchers want more objective evidence to validate that clustering. Here a silhouette score is useful. A silhouette score measures how close are items within each cluster to each other, relative to distances between items in different clusters. Another approach is to 'bootstrap' the clustering. The idea is to ask whether a particular cluster depends on only one feature or on many. To check this, one may resample from the variables at random and rebuild the clustering based on only those selected variables, repeating both steps many times. If almost all the resulting clusters share a particular branch, then that branch is regarded as fairly well attested.

For all the effort spent on validating clusters, it is not uncommon to find that samples cluster by date of array preparation or by technician. In my experience these kinds of artifacts seem to show up more often when using the Manhattan metric. Principal components analysis (see below) also frequently arranges items by technical covariates rather than real biology.

5.3.2 Principal components

Principal components analysis (PCA) is another valuable exploratory tool, as is the closely related technique of multidimensional scaling (Johnson and Wichern, 2007; Mardia *et al.*, 1979). PCA constructs synthetic variables (components) as linear combinations (sums of multiples) of the measured variables, in order that the values of a small number of these synthetic variables can efficiently encapsulate most of the values of the measured variables across all the samples. The multiple of a gene measure that occurs in the sum for a particular component is called the *loading* of the component on that measure.

Sometimes PCA uncovers processes which coordinate several genes across the samples. In many cases, the first few components are linear combinations of gene measures heavily loaded on a distinct small subset of the genes; then there may be a ready biological interpretation to those components. Just as often the components are linear combinations with small loadings on a large number of measures, and these combinations have no easy biological interpretation. Furthermore PCA is not a robust technique; that means the results are easily distorted by a few outliers in data. One ready sign of outliers driving the PCA is a principal component which has large loadings for only one or two genes, or has large values in only one or two samples.

Even if there are no outliers, a researcher should think critically about the meaning of PCA results. A biologically meaningful component usually loads heavily on a relatively small subset of gene measures; furthermore, two different biologically meaningful components don't usually depend heavily on the same genes. A useful heuristic is to plot the loadings of different components against each other. A plot that looks like 'blob' usually suggests difficulty of interpretation. Figure 5.4 illustrates easy and difficult interpretations.

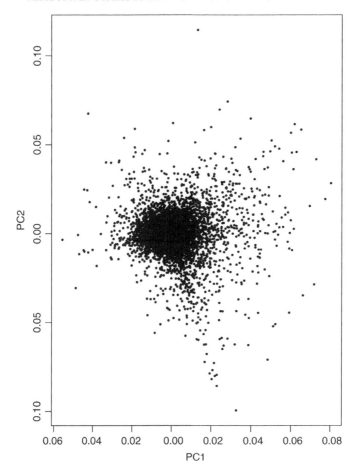

Figure 5.4   Plot of loadings of the first two principal components (PCs) from analysis of a subset of NCI 60 data. The second PC probably reflects some identifiable biology, since there are a few dozen genes which load heavily on PC2 but have almost no loading on PC1. The first PC may or may not reflect real biology – but compare Figure 5.5, which suggests that the first PC distinguishes leukemias from other samples. This data was obtained from http://discover.nci.nih.gov/datasets.jsp.

5.3.3   Multidimensional scaling (MDS)

There are several MDS techniques, and they aim to do much the same thing as PCA but more flexibly and for more types of data. MDS techniques differ from PCA in allowing a researcher to represent data for which Euclidean distance is not a meaningful measure of differences between samples (Johnson and Wichern, 2007). Furthermore, some variants of MDS allow one to iteratively improve a graphical representation beyond what PCA can achieve. The price for this added flexibility is that the synthetic variables constructed by MDS techniques are not related in any simple way to the original variables, and so it is often hard to interpret an MDS diagram. In practice MDS is often used as a kind of graphical clustering, which is useful for suggesting unsuspected relationships, but is not a precise representation.

Figure 5.5 shows an MDS plot for the same subset of the NCI 60 expression data, whose loadings are shown in Figure 5.4. We can see that mostly the leukemias and melanomas segregate nicely from the epithelial tumors, and within the epithelial tumors, the colon cancers are pretty distinctive, but that

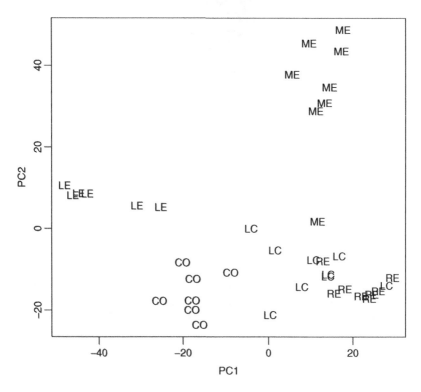

Figure 5.5   MDS plot of a subset of the NCI 60 data. The horizontal axis corresponds to the first PC; that component seems to distinguish blood-derived cancer cell lines from the others. The vertical axis corresponds to the second PC; that component seems to distinguish melanomas from the others, and represents coordinated expression of relatively few genes, as shown in Figure 5.4. Symbols: LE, leukemia; ME, melanoma; CO, colon; RE, kidney; LC, lung cancer.

the overall gene expression profiles of the lung tumors are not very distinct from those of the kidney tumors. In fact the other major epithelial tumor types in the NCI 60 – ovarian and breast (not shown) – also mix in on these broad dimensions. We see one puzzle: one of the melanomas seems indistinguishable from the epithelial lines at coordinates (10, 0). That cell line is LOX-IMVI; in fact LOX-IMVI doesn't express TYR, which is a distinctive marker of melanocytes, and so this cell line may very well be misclassified.

Recall that PCA doesn't handle skewed data very well. An alternative technique known as independent components analysis (ICA) is specifically designed to deal well with skewed data (Hyvärinen *et al.*, 2001). ICA differs from PCA in that linear combinations are sought that are uncorrelated but also are statistically independent, which is often not the case in PCA. Frequently the component values are skewed rather than Normal. This approach may work well when there are very distinct sets of genes up-regulated in distinct conditions. A nice example used the cell-cycle data (Lee and Batzoglou, 2003), where distinct groups of genes are expected to be very high for brief periods, so that a skewed distribution of the components would be expected.

There are other methods related to PCA, but which are more resistant to outliers, for instance Croux *et al.* (2006). In my opinion these other methods deserve more attention than they have received so far.

5.4   Classification and prediction

A common goal of clinical microarray studies is to predict clinical outcomes from genomic measures. In many recent clinical studies, researchers have employed modern machine learning (ML) algorithms, often methods developed or refined just a few years previously. However, there is a long tradition in statistics of predicting outcomes using multiple predictors. The advantage of many ML methods is that they are designed to work efficiently with large numbers of predictor variables. In contrast, most classical statistical procedures work efficiently with a modest number of variables and classical procedures struggle to sort out relationships among enormous numbers of variables. On the other hand, some caution about ML is needed, since many ML procedures were originally designed for data sets with very large numbers of samples. Most microarray studies have far fewer samples and far more variables than many ML methods were intended to handle. There is as yet little statistical theory for ML methods in this 'large P, small N' situation (many predictors but few samples). Although ML methods may get answers quickly, it is hard to assess their significance. It is quite easy in practice to 'over-fit' data using ML methods, resulting in excellent predictions on existing data but poor predictions on new data. On the other hand, most classical statistical algorithms have internal measures to help assess their predictive accuracy beyond the training sample (Hastie *et al.*, 2001). A general approach to reduce the danger of over-fitting is cross-validation, which will be discussed later.

5.4.1 Discriminant analysis

The classical statistical approach to discrimination and class prediction was introduced by R. A. Fisher as early as 1936 and is known simply as discriminant analysis (DA). Fisher's original idea was to choose a linear combination of predictors to maximize the difference of that linear sum between classes relative to its variance within classes. Later he and others developed a theory that justified his original idea, and incidentally gives us some guidance about when this approach works well and when it does not.

For variables $x_1, x_2, \ldots, x_M$, a discriminant function L has the form

$$L = a_1 x_1 + a_2 x_2 + \cdots + a_M x_M + c. \tag{5.1}$$

Where $a_1, a_2, \ldots, a_M$ are the weights for the variables $x_1, x_2, \ldots, x_M$. The aim of DA is to pick the values for the a_i, and for c so that the values of L are negative for one class and positive for another class, as much as this is possible.

In practice, DA does well when the following assumptions are true: for both classes, the joint distribution of the values of the variables x_i is a multivariate Normal distribution, and within both classes separately, these variables have the same covariance matrix. For such a situation one can show that DA is the most efficient method for discrimination. However, genomic data rarely satisfy these conditions in practice. Most commonly, the covariances of the observed measures differ between the two classes, especially when comparing diseased to normal samples. Furthermore, genomic data often depart from the Normal distribution in ways that compromise multivariate theory. For example, among cancer samples the variances of individual genes are much larger than among normal samples, and the distribution of the values is skewed to the right (i.e., large outliers).

In principle the issue of different covariance matrices among the different classes could be addressed by estimating the two covariance matrices separately, employing a so-called Quadratic Discriminant Analysis (QDA). However, in the 'small N large P' situation, the estimates of covariance matrices are unstable, and in practice this technique doesn't perform well on genomic data, and is therefore rarely used.

5.4.2 Modern procedures

The so-called 'Naïve-Bayes' method is one of the oldest and simplest forms of ML, and often outperforms more sophisticated methods on complex problems. The key idea is to compute the distributions of all variables for each class, then to multiply the densities of these distributions, as if the variables were independent, to come up with an estimated joint distribution of all variables, for each class. Then if the values of the variables are known, the probabilities of classes can be computed by applying Bayes' rule. In practice the measured variables are usually not independent, but this classifier does surprisingly well nevertheless.

If all the variables are distributed Normally, the Naïve Bayes approach to classification turns out to be equivalent to the simplest form of discriminant analysis, in which each variable is independent. Why does this simple approach work? In the small N large P situation, in high dimensions, the estimates of covariance are very inaccurate; it seems that with small sample sizes the typically large errors in covariance estimates render these estimates more misleading than helpful. A helpful compromise, which has become more popular recently, is to address this problem by blending the estimates of the covariance matrix with the identity, weighting the estimated covariance more if there are a large number of samples (Hastie *et al.*, 2001).

5.4.2.1   Support vector machines

One new ML method now in the spotlight is Support Vector Machines (SVMs; Cristiani and Shawe-Taylor, 2000); the word 'machine' in the name might be better rendered 'mechanism' or 'algorithm.' This approach has the potential to improve on the classical technique of discriminant analysis by using nonlinear combinations of measured predictive variables, hence giving a more complex and flexible repertoire of ways to separate the two groups. However, this very flexibility raises the possibility of over-fitting. Nevertheless, anecdotal reports suggest that the SVM approach can be made quite competitive with modern statistical methods in many situations, and seems to work better than classical methods in situations where the predictive data are quite sparse (most observations are 0).

It is not easy to figure out the theory behind the SVM approach; the arguments embrace finiteness and eschew the idealization and asymptotic analysis that characterizes so much of classical statistics. The theory for SVMs is not much use in practice as the theoretical bounds are too big to be useful (Cristiani and Shawe-Taylor, 2000). One sometimes hears that SVMs beat the 'curse of dimensionality', but this is not really true (Hastie *et al.*, 2001). That is, SVMs allow convenient mechanisms, in the form of kernel functions, for picking representations in higher dimensions, which are indifferent to dimensionality. However these kernel functions don't provide a magic bullet to select distinct appropriate scales for individual variables, which is where the curse of dimensionality really bites. In my opinion the value of SVMs comes when most variables can affect the outcome, but only a few do in any one instance, which often is the situation with sparse data.

Most modern ML procedures are very flexible, as are SVMs, and they can produce plausible models for pure noise. A conscientious researcher must be concerned with how to avoid exaggerated claims. A practical method for assessing the reliability of any predictive machine learning algorithm is cross-validation.

Cross-validation proceeds as follows:

(1) Hold back some subset of the data (say 10% or 20%).

(2) Perform the machine learning or statistical procedure on the remaining data.

(3) Predict the outcomes for the data that are left out.

(4) Compare the predictions of (3) with the actual (known) outcomes.

(5) Repeat steps 1–4 for many subsets left out, recording the correct predictions at each iteration.

(6) Characterize the overall predictive accuracy.

Many ML methods have tuneable parameters; the cross-validation procedure enables one to make a plausible choice of these parameters, by cross-validating with many choices of these parameters, and then picking the parameters with the best cross-validation score.

It is sometimes misleading to compare the accuracy of a ML procedure on two different data sets with different proportions of categories or outcomes, using as a measure the proportion correct. For example, if one data set is split 50 : 50 between two categories and another is split 90 : 10, it will be easy to obtain 90% accuracy on the second data set, simply by guessing the first category every time. A better measure of how much the ML procedure does for the problem is to compare the accuracy to the best accuracy achievable by chance. A good statistic for this is Cohen's Kappa.

Statisticians are enthusiastically addressing these issues right now, and I anticipate a convergence of machine learning with statistics, embracing classical statistical concerns with random fluctuations. Many modern statistical procedures use regularization, by imposing a cost for complex predictors to limit the range of possible predictor functions. A very clear introduction to these issues is provided in (Hastie *et al.*, 2001).

5.5 References

Benjamini, Y. and Hochberg, Y. (1995) Controlling the false discovery rate: a practical and powerful approach to multiple testing. *J Roy Stat Soc B*, **57**, 289–300.

Cristiani, N. and Shawe-Taylor, J. (2000) *An Introduction to Support Vector Machines*, Cambridge University Press, Cambridge.

Croux, C., Filzmoser, P., and Oliveira, M.R. (2006) Algorithms for projection–pursuit robust principal component analysis. KU Leuven Working Paper No. KBI 0624, Catholic University of Leuven.

Do, J.H. and Choi, D.K. (2008) Clustering approaches to identifying gene expression patterns from DNA microarray data. *Mol. Cells*, **25**, 279–88.

Garge, N.R., Page, G.P., Sprague, A.P., *et al.* (2005) Reproducible clusters from microarray research: whither? *BMC Bioinformatics*, **6**(Suppl 2), S10.

Hastie, T., Tibshirani, R., and Friedman, J. (2001) *The Elements of Statistical Learning*, Springer, New York.

Hyvärinen, A., Karhunen, J., and Oja, E. (2001) *Independent Component Analysis*, John Wiley & Sons, Inc., New York.

Johnson, R.A. and Wichern, D.W. (2007) *Applied Multivariate Statistical Analysis*, Prentice-Hall, Upper Saddle River, NJ.

Kerr, G., Ruskin, H.J., Crane, M., and Doolan, P. (2008) Techniques for clustering gene expression data. *Comput. Biol. Med.*, **38**, 283–93.

Lee, S.I. and Batzoglou, S. (2003) Application of independent component analysis to microarrays. *Genome Biol.*, **4**, R76.

Mardia, K.V., Kent, J.T., and Bibby, M. (1979) *Multivariate Analysis*, Academic Press, London.

Perou, C.M., Sorlie, T., Eisen, M.B., *et al.* (2000) Molecular portraits of human breast tumours. *Nature*, **406**, 747–52.

Reiner, A., Yekutieli, D., and Benjamini, Y. (2003) Identifying differentially expressed genes using false discovery rate controlling procedures. *Bioinformatics*, **19**, 368–75.

Spellman, P.T., Sherlock, G., Zhang, M.Q., *et al.* (1998) Comprehensive identification of cell cycle-regulated genes of the yeast Saccharomyces cerevisiae by microarray hybridization. *Mol. Biol. Cell*, **9**, 3273–97.

Storey, J.D. and Tibshirani, R. (2003a) Statistical methods for identifying differentially expressed genes in DNA microarrays. *Methods Mol. Biol.*, **224**, 149–57.

Storey, J.D. and Tibshirani, R. (2003b) Statistical significance for genomewide studies. *Proc. Natl. Acad. Sci. U.S.A.*, **100**, 9440–5.

6

Bayesian methods in genomics and proteomics studies

Ning Sun and Hongyu Zhao

6.1 Introduction

Bayesian methods have become widely adopted in bioinformatics in recent years, thanks to their many advantages over frequentist-based methods. First, they offer an intuitive approach to incorporating prior biological knowledge in data analysis and interpretation. Second, different types of genomics and proteomics data can be integrated in a principled fashion under a consistent modeling framework. Third, many Bayesian computational tools that have been developed over the past 30 years are readily available to complex models for genomics and proteomics data that are difficult to deal with from a frequentist perspective. In addition to numerous journal articles published on Bayesian methods for computational biology problems, several books dedicated to Bayesian models in bioinformatics have appeared, for instance Dey *et al.* (2010), Do *et al.* (2006), and Mallick *et al.* (2009), as well as tutorials (Wilkinson, 2007). In this chapter, we start with a discussion of the fundamental Bayes theorem and several simple examples of how it can be applied to study some biological problems. We then cover several more complex problems that have benefited from Bayesian modeling and analysis. We conclude this chapter with references to other problems that Bayesian methods have been applied to, potential issues in Bayesian methods, and books that are commonly used in the teaching of Bayesian methods.

Knowledge-Based Bioinformatics: From Analysis to Interpretation Edited by Gil Alterovitz and Marco Ramoni
© 2010 John Wiley & Sons, Ltd

6.2   Bayes theorem and some simple applications

The Bayes theorem underlies all Bayesian methods. It is generally introduced in most probability and statistics textbooks in the following form: $P(B|A) = P(A$ and $B)/P(A)$, where A and B represent two events, $P(B|A)$ is the conditional probability that B occurs given that A has already happened, $P(A$ and $B)$ is the joint probability that both A and B occur, and $P(A)$ is the unconditional (marginal) probability that A occurs. When events A and B are independent, that is, $P(A$ and $B) = P(A)P(B)$, then $P(B|A) = P(B)$. The value of $P(B|A)/P(B)$ characterizes the degree of dependency of event B on event A. Let us first consider a simple example using the Bayes theorem. If we are interested in the dependency of the neighboring nucleotides in a genome, event A may correspond to a base being a specific nucleotide, say 'C,' and event B may correspond to the next base being a specific nucleotide, say also 'C.' Suppose that the proportion of bases in this genome being 'C' is 0.25, that is, $P(A) = 0.25$, and the proportion of two consecutive bases being 'CC' is 0.10, that is, $P(A$ and $B) = 0.10$. Then $P(B|A) = P(A$ and $B)/P(A) = 0.10/0.25 = 0.4$. Therefore, $P(B|A)/P(B) = 0.4/0.25 = 1.6$; that is, having a base being 'C' increases the probability of the next base being 'C' by 60%. In this case, knowing that the previous base is 'C' (prior information) affects the chance that the current base is 'C.'

For a second example, we consider the inference of the ethnic origin of an individual based on genetic marker information, which is commonly encountered in genetics or forensic studies. For example, we may collect genetic markers at a number of single nucleotide polymorphisms from an individual and use these markers to infer whether this person is a Caucasian or an African. In this case, event B is the ethnic origin of this individual, and event A is the genetic marker data for this person. To solve this problem, we can apply the Bayes theorem as follows:

P(an individual is a Caucasian|marker data)

= P(an individual is a Caucasian and marker data)|P(marker data)

= P(an individual is a Caucasian)P(marker data|an individual is

a Caucasian)|P(marker data).

We note that we have used the following results:

$$P(B|A) = P(A \text{ and } B)|P(A) = P(B)P(A|B)/P(A)$$

above; that is, we used the Bayesian theorem twice with one conditional on A and the other conditional on B. There are three quantities that need to be evaluated in this formulation: P(an individual is a Caucasian), P(marker data | an individual is a Caucasian), and P(marker data). The information about the value of P(an individual is a Caucasian) often comes from the knowledge on the general population where this individual is sampled from, therefore some injection of prior information in this calculation. To calculate P(marker data|an individual is a Caucasian), we need information on the properties of the markers in a Caucasian population, such as allele and genotype frequencies. Again, some prior knowledge about the markers is needed in this calculation. Lastly, to calculate P(marker data), we need to sum over all possible ethnic origins for this individual. In the case of two origins, for instance Caucasian and African, we can calculate P(marker data) as

P(marker data)

$= $ P(an individual is a Caucasian and marker data)

$+ $ P(an individual is an African and marker data)

$= $ P(an individual is a Caucasian)P(maker data | an individual

is a Caucasian) $+ $ P(an individual is an African)P(marker data|

an individual is an African).

We may consider P(an individual is a Caucasian) as the prior information about the ethnicity of a sampled individual in the population being studied, and P(an individual is a Caucasian|marker data) as the posterior probability for this person's ethnicity after we collect the relevant information about this individual, that is, genetic marker data. Therefore, the Bayes theorem offers an intuitive way to combine prior information and data to achieve our inferential objective.

Although A and B represent events in the above notion, the equation still holds if they are replaced by more general entities, such as data sets or parameters. For example, consider A as the probability that a coin lands on heads; therefore we can replace A by a parameter h, the chance of seeing a head. Now consider B as the results from tossing this coin N times; therefore we can replace B by a set of observations $Y = (Y_1, \ldots, Y_N)$, where $Y_i = 1$ or 0 corresponds to a head or tail from the ith tossing. Then $P(B|A) = P(Y|h) = h^H(1 - h)^T$, where H is the total number of heads and T is the total number of tails observed from N experiments. We can use the Bayes theorem to infer the probability that this coin lands on heads based on the observations Y, $P(h|Y)$, through the Bayes theorem as follows. Note that we can exchange the roles of A and B so that $P(A|B) = P(A \text{ and } B)/P(B)$. Therefore, $P(h|Y) = P(h \text{ and } Y)/P(Y) = P(h)P(Y|h)/P(Y)$. It is easy to see that the posterior distribution of h depends on three quantities: the prior distribution for h, $P(h)$; the probability distribution for the observations Y conditional on the parameter h, $P(Y|h)$; and the probability for Y integrated over all possible values of h, $P(Y) = \int P(Y|h)P(h)dh$, which is independent of h. Depending on the specific choices of the prior distribution for h, we will have different posterior inference conditional on the observed data Y. Although simple, this example illustrates the general approach for inferring parameters of interest, denoted by θ, from a prior distribution $\pi(\theta)$ and observations Y, through the following equation:

$$\pi(\theta|Y) = \pi(\theta)P(Y|\theta)/P(Y) \propto \pi(\theta)P(Y|\theta).$$

Consider a population genetics problem similar to the one discussed above. Instead of inferring one individual's ethnic background, our goal is to infer the allele frequencies of a set of genetic markers from the collection of a set of individual samples. In the case of a single marker, the observed data are the genotypes from a set of individuals at this marker, and the parameter is the allele frequency of this marker. Then we have P(allele frequency|marker data) = P(marker data|allele frequency) P (allele frequency)|P(marker data). P (marker data|allele frequency) can be calculated based on the Hardy–Weinberg equilibrium. The prior information comes in the form of P(allele frequency), which is usually assumed to have the Dirichlet distribution, and an integration is needed to calculate P(marker data) where the integration is over the prior distribution for the allele frequencies.

From these three examples, we can see that the simple Bayes theorem can be applied to a variety of problems where the events A and B can represent different entities. For most Bayesian methods, A and B usually correspond to the observed data and parameters of interest, respectively, which can be generically denoted by Y and θ, and the goal is to infer the posterior distribution of $\pi(\theta|Y) = \pi(\theta)P(Y|\theta)/P(Y)$. There are three key elements in applying the Bayesian theorem to a specific problem: model specifications, which are needed to evaluate $P(Y|\theta)$; prior specifications, which are needed to define $\pi(\theta)$; and computational methods needed to infer the posterior distributions, because it is usually not easy to directly evaluate $P(Y)$. Statistical inference is usually achieved through sampling from the posterior distribution $\pi(\theta|Y)$. It is important to specify model forms and prior distributions such that (1) the models are comprehensive enough to appropriately model the observed data; (2) the prior distributions can reflect the degree of knowledge about the model parameters; and (3) it is feasible to infer the posterior distributions through appropriate computational methods. One major driving force that has made Bayesian methods widely employed in recent years is the development and applications of Markov chain Monte Carlo methods for posterior inferences (Gilks *et al.*, 1995). Among different Markov chain Monte Carlo methods, the Gibbs sampler is the most commonly used one when there are a large number of parameters involved in a model. Let θ denote the collection of all the model parameters. It may be difficult to sample from the joint posterior distribution of $\pi(\theta|Y)$ when θ has a large dimension. On the other hand, if θ can be partitioned into C subsets in the form of $\theta = (\theta_1, \theta_2, \ldots, \theta_C)$, and if it is relatively easy to sample from $\pi(\theta_1|\{Y, \theta_2, \theta_3, \ldots, \theta_C\})$, $\pi(\theta_2|\{Y, \theta_1, \theta_3, \ldots, \theta_C\}), \ldots, \pi(\theta_C|\{Y, \theta_1, \theta_2, \ldots, \theta_{C-1}\})$, the Gibbs sampler proceeds by repeatedly sampling $\theta_1, \theta_2, \ldots$, and θ_C from these distributions to arrive at an empirical distribution of the joint posterior distribution of $\pi(\theta|Y)$. In the following, we discuss several Bayesian methods that have proven useful in genomics and proteomics studies.

6.3   Inference of population structure from genetic marker data

Genome-wide association studies have led to many discoveries of genes associated with common diseases in recent years. Such studies usually involve hundreds or thousands of individuals, each genotyped at hundreds of thousands of genetic markers. In the case–control association study setting, each individual either has the disease of interest or is normal. Putative markers associated with disease are often identified from comparing the genotype distributions between the cases and controls. In such analysis, the presence of sample heterogeneity in their genetic background, often called population stratification, may complicate association analysis, leading to false positive results. For example, if the cases and controls differ in the proportions of people from a specific ethnic background, any marker having ethnic differences may be incorrectly inferred to be associated with disease phenotype. One strategy to deal with sample heterogeneity is to first infer population structure from the collected genotype data, which can be represented as

$$G = \{G_1, G_2, \ldots, G_N\}, G_1 = \{g_{11}g_{12}\cdots g_{1M}\}, \ldots, G_N = \{g_{N1}g_{N2}\cdots g_{NM}\},$$

where N is the number of sampled individuals, M is the number of markers, and g_{ij} is the genotype for the ith individual at the jth marker, where $i = 1, 2, \ldots, N$, and $j = 1, 2, \ldots, M$. For simplicity, we only consider biallelic markers. Let the two alleles for a given marker be labeled by A and a; then the genotype at each marker can be coded as 0, 1, or 2 according to the number of allele A that an individual carries. Let us assume that there are K subpopulations in the overall sample. The objective is to infer the subpopulation membership for each individual, $T = \{t_1, t_2, \ldots, t_N\}$, where $t_i = 1, 2, \ldots,$ or $K, i = 1, 2, \ldots, N$. T can be considered the primary set of parameters to be inferred from the observed data, G. To facilitate the inference of T, we introduce other parameters relevant to our inference; that is, the allele frequencies $F = \{F_1, F_2, \ldots, F_M\}$, $F_1 = \{f_{11}, f_{12}, \ldots, f_{1M}\}, \ldots, F_K = \{f_{K1}, f_{K2}, \ldots, f_{KM}\}$, where f_{kj} is the frequency of allele A of the jth marker in the kth population, $j = 1, 2, \ldots, M, k = 1, 2, \ldots, K$. Based on the Bayes theorem, we can infer T and F as follows:

$$P(\{T, F\}|G) = P(\{T, F\})P(G|\{T, F\})/P(G) = P(T)P(F)P(G|\{T, F\})/P(G),$$

where we assume that the priors for T and F are independent. Although this formulation is straightforward, simultaneously inferring the joint posterior distribution of T and F is nontrivial. On the other hand, it is much easier to infer the posterior distribution of T conditional on F and G and infer the posterior distribution of F conditional on T and G. First, conditional on the allele frequency information F for each marker and the genotypes across all the M markers $\boldsymbol{G}$ for each individual, we can derive the posterior probability that an individual is from the kth subpopulation as

$$P(t_i = k|\{\boldsymbol{F}, \boldsymbol{G}\}) = P(t_i = k)P(\{\boldsymbol{F}, \boldsymbol{G}\}|t_i = k)/P(\{\boldsymbol{F}, \boldsymbol{G}\})$$

$$= P(t_i = k)P(\boldsymbol{F}|t_i = k)P(\boldsymbol{G}|\{\boldsymbol{F}, t_i\})/P(\{\boldsymbol{F}, \boldsymbol{G}\})$$

$$\propto P(t_i = k)P(\boldsymbol{G}|\{\boldsymbol{F}, t_i\}),$$

where $P(\boldsymbol{G}|\{\boldsymbol{F}, t_i\})$ can be obtained through information on allele frequencies and Hardy–Weinberg equilibrium for each marker within each subpopulation. Second, conditional on the subpopulation membership for each individual $\boldsymbol{T}$ and the genotype information $\boldsymbol{G}$, it is easy to infer the posterior distribution of the allele frequencies for the markers for each subpopulation by considering the marker genotypes for all those individuals inferred to belong to this subpopulation. We omit the details on the exact forms of these distributions, which can be found in the paper by Pritchard and colleagues (Pritchard *et al.*, 2000). Through the Gibbs sampler, we can alternate between inferring the posterior distributions of $\boldsymbol{T}$ and the posterior distributions of $\boldsymbol{F}$ by iteratively sampling $\{t_1, t_2, \ldots, t_N\}$ from $P(\boldsymbol{T}|\{\boldsymbol{F}, \boldsymbol{G}\})$ for a realization of $\boldsymbol{T}$ and then sampling $\{\boldsymbol{F}_1, \boldsymbol{F}_2, \ldots, \boldsymbol{F}_M\}$ from $P(\boldsymbol{F}|\{\boldsymbol{T}, \boldsymbol{G}\})$ for a realization of $\boldsymbol{F}$ until convergence. The key for the inference of $\boldsymbol{T}$, the main objective for this problem, is the introduction of another set of unknown parameters $\boldsymbol{F}$ that characterize the allele frequency distribution properties of each subpopulation. With this new set of parameters, the conditional distributions of $P(\boldsymbol{T}|\{\boldsymbol{F}, \boldsymbol{G}\})$ and $P(\boldsymbol{F}|\{\boldsymbol{T}, \boldsymbol{G}\})$ have relatively simple forms and allow us to easily sample from their conditional distributions. More sophisticated models have been developed along this line of reasoning (Falush *et al.*, 2003), and this modeling approach has led to many significant advances in characterizing the general structure of different human populations (Rosenberg *et al.*, 2002).

6.4   Inference of protein binding motifs from sequence data

It is well known that transcription factors recognize their DNA binding targets through factor-dependent motifs, represented either by a DNA sequence, for instance CACGTG, or a position weight matrix (PWM), for example

	1	2	3	4	5
A	1	0	0	0.9	0
C	0	0.8	0.3	0.1	0
G	0	0.2	0.7	0	0
T	0	0	0	0	1

where this matrix represents a motif of length five and each entry in this matrix represents the probability that a specific base in this motif has a given nucleotide.

Therefore, the identification of motifs may lead to better understanding of how different genes are regulated by transcription factors. One way to infer binding motifs is to search for common patterns from a set of sequences that are known to be bound by a given transcription factor. We introduce the following notations to formulate the statistical problem. Let the K observed sequences be denoted by $S = \{S_1, S_2, \ldots, S_K\}$, where $S_1 = \{s_{11}s_{12}\cdots s_{1L}\}, \ldots, S_K = \{s_{K1}s_{K2}\cdots s_{KL}\}$, and where, for simplicity, we assume that these sequences have the same length, L, and each base s_{ij} is of one of four possible nucleotides. The objective is the inference of the PWM $W = \{w_{Dj}\}$, where w_{Dj} is the probability that the jth base is of nucleotide D, where D = 'A,' 'C,' 'G' or 'T,' $j = 1, 2, \ldots, M$, and we assume the motif length, M, is known. That is, we would like to infer the PWM conditional on the observed sequence data S. Now we introduce a new set of unknown parameters to facilitate statistical inference. This new set of parameters is the starting positions of the motif in each sequence $T = \{t_1, t_2, \ldots, t_K\}$, where $t_i = 1, 2, \ldots$, or $L - M + 1$. Although it is not easy to sample from the joint posterior distribution of $\{W, T\}$, we can infer their distributions through the Gibbs sampler by iteratively sampling from the posterior distributions of $P(W|\{T, S\})$ and $P(T|\{W, S\})$, respectively, which have relatively simple forms. First, conditional on the starting positions of the motif for each sequence, it is easy to tabulate all the motifs across all the sequences and sample from the posterior distribution of $P(W|\{T, S\})$. Second, conditional on the PWM W, we can calculate the conditional probability that the motif starts at each position along the sequence and then sample the starting position based on these conditional probabilities. We can iterate between these two steps until convergence. The basic idea of this algorithm was first described by Lawrence and colleagues (Lawrence *et al.*, 1993) in a somewhat different context, and then widely adopted to much more diverse and complex models (Zhou and Wong, 2004). The first two examples involve the use of one type of data, and the key was the introduction of a new set of parameters that facilitate statistical modeling and inference.

6.5 Inference of transcriptional regulatory networks from joint analysis of protein–DNA binding data and gene expression data

In this section, we discuss the use of Bayesian methods to integrate data from multiple sources. The objective is to infer the regulatory targets of transcription factors from joint analysis of gene expression data and protein–DNA binding data. Gene expression data can be gathered from microarrays having probes targeting transcripts, whereas protein–DNA interaction data can be obtained through chromatin immunoprecipitation experiments coupled with microarrays with probes targeting regulatory regions. These two data types reveal different aspects of the gene regulation process, with the protein–DNA binding data suggesting the potential targets of transcription factors, and the gene expression data showing the results of the complex regulation process. If the protein–DNA interaction data were perfect, that is, there is no experimental noise and all binding targets are functional, such data by itself would be sufficient to deduce the regulatory network. However, as with any high-throughout data, there can be substantial noises in protein interaction measurements, and more importantly, the observed physical binding between a transcription factor and the regulatory region of a gene does not necessarily imply that the binding is functional. Therefore, the incorporation of gene expression data can help resolve some of the ambiguities and errors in the protein–DNA interaction data. For example, if a set of five genes have very similar expression profiles across a large number of experiments, and four of them are bound by the same transcription factor with the fifth one showing marginal evidence of binding, then under the assumption that these five genes are similarly regulated, we may infer that the fifth gene is also regulated by the same transcription factor despite the fact that the binding evidence is marginal. We can formulate this rationale with the following statistical model $Y = A\beta + e$, where $Y = (y_1, y_2, \ldots, y_N)^T$ denotes the gene expression level of the N observed genes, $\beta = (\beta_1, \beta_2, \ldots, \beta_J)^T$ denotes the activities (which are unobservable) of the J transcription factors, $A = \{a_{ij}, i = 1, \ldots, N, j = 1, \ldots, J\}$ denotes the functional binding intensity between the ith gene and the jth transcription factor, and $e = (e_1, e_2, \ldots, e_N)^T$ denotes the noises not explained by the model. As for a_{ij}, we assume that $a_{ij} = b_{ij} \times r_{ij}$, where b_{ij} is the observed binding level between the ith gene and the jth transcription factor and r_{ij} is an indicator (unobserved) with value being either 1 or 0 corresponding to whether the jth transcription factor does (1) or does not (0) regulate the ith gene. See (Sun et al., 2006) for the justification of this model. The linear regression model $Y = A\beta + e$ integrates both gene expression data (Y) and protein–DNA interaction data (b_{ij}) through the unobserved regulatory network defined by $\{r_{ij}\}$ and the unobserved transcription factor activities (β). Although seemingly complex, the Gibbs sampler can be applied for statistical inference as follows. First, conditional on a known regulatory network $\{r_{ij}\}$, then the regression model becomes a standard regression problem where the goal is to infer the transcription factor activities β. Under the Bayesian setting, the posterior distribution of β can be easily derived and a set of activity levels can sampled from this posterior distribution. Second, conditional on the sampled transcription factor activity levels, we can derive the conditional probability for any specific regulatory pattern given the observed expression data and the activity levels. Then we can sample a specific regulatory pattern from the conditional distribution. We then iterate between these two sampling steps to infer the regulatory networks from the sampled parameter values.

6.6   Inference of protein and domain interactions from yeast two-hybrid data

Protein interactions play a central role in many cellular processes, such as signal transduction, gene regulation, and cell cycle control. Alterations in protein interactions perturb the normal cellular processes and contribute to many diseases. The correct identification of protein interactions can help us assign the cellular functions of novel proteins, investigate the mechanisms of intracellular biochemical pathways, and understand the underlying causes of diseases. Different methods have been developed to gather high-throughput protein interaction data, with the yeast two-hybrid (Y2H) experimental technique providing the most direct evidence of physical protein interactions. Although extensive data have been generated through the Y2H, this experimental approach suffers from high error rates, with the estimated false negative rate above 0.5. Due to the large number of possible non-interacting protein pairs, although the false positive rate, defined as the ratio of the number of incorrect interactions observed over the total number of non-interacting proteins, is small (1×10^{-3} or less), the false discovery rate, defined as the ratio of the number of incorrect interactions observed over the total number of observed interactions, is much greater and is estimated to be 0.2 to 0.5, indicating that a large portion of the observations from the Y2H technique are incorrect. Therefore, it is desirable to reduce the errors in Y2H data, and one approach is through the integration of data from a number of model organisms, such as yeast, worm, fruit fly, and humans. An appropriate statistical model is needed to pool data from different organisms together. Noting that domains are structural and functional units of proteins and are conserved during evolution, and protein interactions are mediated through domain pairs, one possible strategy is to utilize domain information as the evolutionary connection among these organisms. In this setting, there are two sources of information used: the observed protein interaction data from Y2H experiments, and the annotated domain information for each protein. For a total of N proteins and Y2H data from K different organisms, the Y2H data can be represented as $\{O_{ijk}, i, j = 1, \ldots, N, k = 1, \ldots, K\}$, where $O_{ijk} = 1$ if proteins i and j are observed to interact with each other in the kth organism. Based on the domain annotation information, we use D_i to represent the collection of domains in protein i. We then make the following assumptions to develop a statistical model for data integration: (1) Domain interactions are independent, so whether two domains interact or not does not depend on the interactions among other domains. (2) The probability that two domains m and n interact is the same among all the organisms. (3) Two proteins i and j interact if and only if at least one pair of domains from the two proteins interact. Let λ_{mn} denote the probability that domains m and n interact with each other, and $P_{ijk} = 1$ or 0 denote whether proteins i and j interact with each other (1) or not (0). With these assumptions and notations, we have $P(P_{ijk} = 1) = 1 - \Pi_{D(m,n)}(1 - \lambda_{mn})$, where the product is over all $D(m, n)$ domain pairs from protein pair i and j in organism k. Due to experimental errors, the observed interaction data $\{O_{ijk}\}$ may differ from the true interaction data $\{P_{ijk}\}$. Let fn and fp represent the false negative rate and false positive rate of the protein interaction data. We then have $P(O_{ijk} = 1) = P(P_{ijk} = 1)(1 - fn) + (1 - P(P_{ijk} = 1))fp$. Although a likelihood-based method can be used to derive frequentist-based solutions to this problem (Liu *et al.*, 2005), the false negative and false positive rates of the observed protein interaction data have to be treated as known, which is rarely the case in practice. With proper specifications of priors for the model parameters, such as *fn*, *fp*, and λ_{mn}, a Bayesian approach was developed so that the domain interaction probabilities, the false positive rate and the false negative rates of the observed data can be estimated simultaneously (Kim *et al.*, 2007). Compared to the likelihood-based methods, the Bayesian-based methods may be more efficient in dealing with a large number of parameters and more effective in allowing for different error rates across different data sets. Moreover, assuming that the majority of the domain pairs do not interact, that is, by imposing a sparse prior distribution, the prediction accuracy can be further improved (Kim *et al.*, 2010).

6.7 Conclusions

The general Bayesian framework coupled with computational tools offer a powerful approach for modeling and analyzing complex genomics and proteomics data. Bayesian methods have been applied to many more problems than we have discussed in this chapter (e.g., Alterovitz *et al.*, 2007; Ding, 2006; Spyrou *et al.*, 2009; Zhang et al., 2010; Zhang and Liu, 2007). As noted in the introduction section, several books have been published focusing on Bayesian methods in computational biology, showcasing the active developments in this area.

As shown in the examples, statistical inference for Bayesian models is usually based on the Gibbs sampling scheme where the posterior distributions are inferred from iteratively sampling from a set of conditional distributions. Aside from significant computational demands, it is critical to ensure that the samples thus obtained can be used to represent the correct posterior distributions. Therefore, we need to monitor the convergence of the samples, and multiple runs are sometimes needed to check for the consistency across many runs.

As for any Bayesian method, the choices of prior distributions may impact the conclusions drawn based on the posterior distributions. Although prior distributions may be dominated by the observed data when there are sufficient data, there is no guarantee this would always be the case, especially when the number of parameters is large, which is often the case in the analysis of genomics and proteomics data. Some types of sensitivity analysis may be needed to evaluate the impacts of the changes in the prior distributions.

Despite these potential caveats, we believe that the Bayesian approach can offer a consistent framework for knowledge and data integration in the analysis of genomics and proteomics data, which is critical when many types of data need to be jointly analyzed to extract the most information from these data. For example, if we know the regulatory relationship among a set of transcription factors and genes, this knowledge can be easily brought into the analysis by fixing these relationships. However, such prior knowledge may be difficult to incorporate under a frequentist approach. Computationally, Bayesian methods are well suited for complex models where many parameters, sometimes numbered in hundreds or thousands, are involve in the model. Although it may be possible to infer these parameters in a frequentist setting, such as through the expectation-maximization algorithm to maximize the likelihood, the likelihood surface may have many modes, making the inference unstable and difficult. On the other hand, Bayesian analysis through sampling from the posterior distributions may lead to more stable results. Sparsity constraints can also be easily incorporated through sparse priors. Because we would like to convey the basic ideas and approaches of Bayesian methods in this chapter, we have not attempted to provide the detailed implementations of these models. Interested readers can refer to the original publications. In addition, several excellent books are available to provide the foundations on Bayesian analysis (Box and Tiao, 1992; Carlin and Louis, 2009; Gelman *et al.*, 2004; Gilks *et al.*, 1995) as well as computational implementations (Albert, 2009).

6.8   Acknowledgements

Supported in part by NIH grants R21 GM 84008 and R01 GM59507.

6.9   References

Albert, J. (2009) *Bayesian Computation with R*, Springer-Verlag, New York.

Alterovitz, G., Liu, J., Afkhami, E., and Ramoni, M.F. (2007) Bayesian methods for proteomics. *Proteomics*, **7**(16), 2843–55.

Box, G.E.P. and Tiao, G.C., (1992) *Bayesian Inference in Statistical Analysis*, Wiley, New York.

Carlin, B.P. and Louis, T.A. (2009) *Bayesian Methods for Data Analysis*, CRC Press, Boca Raton.

Dey, D.K., Ghosh, S., and Mallick, B.K. (2010) *Bayesian Modeling in Bioinformatics*, Chapman & Hall/CRC, Boca Raton.

Ding, Y. (2006) Statistical and Bayesian approaches to RNA secondary structure prediction. *RNA*, **12**(3), 323–31.

Do, K.-A., Müller, P., and Vannucci, M. (2006) *Bayesian Inference for Gene Expression and Proteomics*, Cambridge University Press, New York.

Falush, D., Stephens, M., and Pritchard, J.K. (2003) Inference of population structure using multilocus genotype data: linked loci and correlated allele frequencies. *Genetics*, **164**(4), 1567–87.

Gelman, A., Carlin, J.B., Stern, H.S., and Rubin, D.B. (2004) *Bayesian Data Analysis*. Chapman & Hall/CRC, Boca Raton.

Gilks, W.R., Richardson, S., and Spiegelhalter, D. (1995) *Markov Chain Monte Carlo in Practice: Interdisciplinary Statistics*. Chapman & Hall/CRC, Boca Raton.

Kim, I., Liu, Y., and Zhao, H. (2007) Bayesian methods for predicting interacting protein pairs using domain information. *Biometrics*, **63**(3), 824–33.

Kim, I., Liu, Y., and Zhao, H. (2010) Sparsity priors for protein-protein interaction predictions, in *Bayesian Modeling in Bioinformatics* (eds D. Dey, S. Ghosh, and B. Mallick), Chapman & Hall, in press.

Lawrence, C.E., Altschul, S.F., Boguski, M.S., *et al.* (1993) Detecting subtle sequence signals: a Gibbs sampling strategy for multiple alignment. *Science*, **262**(5131), 208–14.

Liu, Y., Liu, N., and Zhao, H. (2005) Inferring protein-protein interactions through high-throughput interaction data from diverse organisms. *Bioinformatics*, **21**(15), 3279–85.

Mallick, B.K., Gold, D., and Baladandayuthapani, V. (2009) *Bayesian Analysis of Gene Expression Data*, John Wiley & Sons, Inc., Hoboken.

Pritchard, J.K., Stephens, M., and Donnelly, P. (2000) Inference of population structure using multilocus genotype data. *Genetics*, **155**(2), 945–59.

Rosenberg, N.A., Pritchard, J.K., Weber, J.L., *et al.* (2002) Genetic structure of human populations. *Science*, **298**(5602), 2381–5.

Spyrou, C., Stark, R., Lynch, A.G., and Tavare, S. (2009) BayesPeak: Bayesian analysis of ChIP-seq data. *BMC Bioinformatics*, **10**, 299.

Sun, N., Carroll, R.J., and Zhao, H. (2006) Bayesian error analysis model for reconstructing transcriptional regulatory networks. *Proc. Natl. Acad. Sci. U.S.A.*, **103**(21), 7988–93.

Wilkinson, D.J. (2007) Bayesian methods in bioinformatics and computational systems biology. *Brief. Bioinformatics*, **8**(2), 109–16.

Zhang, Y. and Liu, J.S. (2007) Bayesian inference of epistatic interactions in case-control studies. *Nat. Genet.*, **39**(9), 1167–73.

Zhang, W., Zhu, J., Schadt, E.E., and Liu, J.S. (2010) A Bayesian partition method for detecting pleiotropic and epistatic eQTL modules. *PLoS Comput. Biol.*, **6**(1), e1000642.

Zhou, Q. and Wong, W.H. (2004) CisModule: de novo discovery of cis-regulatory modules by hierarchical mixture modeling. *Proc. Natl. Acad. Sci. U.S.A.*, **101**(33), 12114–9.

7

Automatic text analysis for bioinformatics knowledge discovery

Dietrich Rebholz-Schuhmann and Jung-jae Kim

7.1 Introduction

Automatic literature analysis has been successfully integrated into bioinformatics research and services to promote efficient knowledge discovery. *Biomedical text mining* is an interdisciplinary domain from computational linguistics for text analysis and from computational biology in biomedical research. The two original domains share well-established IT solutions and one important data resource, that is, the biomedical scientific literature. In this chapter, we explain the integration of automatic text analysis into the knowledge discovery efforts in biomedical research.

Biomedical research becomes driven by the results from high-throughput experiments and by the exploitation of large-scale electronic data resources. The former produces new results, while the latter gives access to standardized knowledge. The biomedical scientific literature, consisting of scientific papers reporting on the 'new' results, is being integrated into the electronic data resources by database curators. The literature analysis is thus crucial for the knowledge discovery in biomedical research.

The scientific literature in biomedicine (i.e., MEDLINE abstracts and full-text documents) is a rich repository of up-to-date public knowledge that has been

Knowledge-Based Bioinformatics: From Analysis to Interpretation   Edited by Gil Alterovitz and Marco Ramoni
© 2010 John Wiley & Sons, Ltd

verified through peer review. Individual researchers formulate their novel hypotheses by consulting related papers from the literature and based on the results from their lab experiments. The novelty of the hypotheses has to withstand a thorough literature analysis and an assessment against existing data resources. The researchers exploit interactive literature analysis such as PubMed searches as well as automatic text processing solutions such as protein interaction networks from the scientific literature to increase the accuracy and efficiency of the literature-related works. In principle, text mining solutions support knowledge discovery by providing fast access to documents and thus to the information contained in the documents.

7.1.1   Knowledge discovery through text mining

Knowledge discovery is, in a specific sense, defined as a domain of computer science seeking meaningful patterns from data with automated techniques, in particular employing data mining techniques (Fayyad et al., 1996). Similar techniques are applied to data retrieved from the scientific literature. In addition to this pattern recognition approach, text mining targets the extraction of information that is explicitly expressed in the syntax and semantics of natural languages (e.g., English). The explicit information has of course been expressed by the authors of the text and is therefore known to them, but it is potentially unknown to a large number of researchers, who would benefit from automated text mining solutions that extract and deliver the contained information in a structured form. Altogether, text mining can untangle, normalize and deliver facts from natural language text to improve the background knowledge and scientific performance of biomedical researchers.

A number of tasks for knowledge discovery, which are to disclose valuable information from the literature, have been addressed by, or with the help of, text mining solutions, providing new insights for biomedical research questions. For example, researchers have analyzed the scientific literature to identify the sub-cellular locations of proteins at high accuracy (Brady and Shatkay, 2008). Genes from microarray experiments have been clustered to identify genes with similar functions, based on both the similarity of their expression profiles and the similarity of their contextual features from the scientific literature (Blaschke et al., 2001; Küffner et al., 2005). Other bioinformatics tasks, which have profited from the contextual features from the literature, include the prediction of protein functions based on protein interaction networks (Jaeger et al., 2008), the annotation of mutations and residues of genes and proteins (Nagel et al., 2009), and the prediction of gene–disease associations based on gene annotations from biomedical data resources (Lage et al., 2007). Last, but not least, the scientific literature has been analyzed to identify protein–protein interactions and protein interaction networks (Blaschke et al., 1999; Jaeger et al., 2008).

In addition to the fully automated text mining approaches, solutions for interactive search driven by user queries have been developed to enable efficient retrieval and exploitation of the scientific literature. Also, we can expand and annotate existing biomedical data resources by extracting relevant information from the literature. We can interpret the results of high-throughput experiments by exploiting the literature with data mining techniques. All these applications are due to the presence of experiment details, findings, and hypotheses in the scientific papers of the literature.

7.1.2   Need for processing biomedical texts

The need for text mining in the biomedical domain primarily comes from the huge size of the literature (currently over 19 million abstracts in MEDLINE), where each piece of text in the literature encodes valuable information. Though the literature is the main source of biomedical text mining, other sources such as patents and electronic medical records (EMRs) have been considered as well.

Text processing starts with the decomposition of text. A text is specified by the sequence of its tokens, or words, apart from formatting details, tables, and figures. Tokens can be words or symbols such as punctuation marks, quotation marks, and slashes. The combination of subsequent tokens may form a composite term (e.g., '*Bcl-2*', 'cancer cell growth'). The tokens and terms are indexed by search engines and used as features in data mining approaches, for example, for clustering and classification of texts. More complex structures of texts are based on the language syntax that combines tokens into phrases and sentences. These compositional structures denote relationships between terms, facts, and events and can thus be used as bases for logical reasoning and discourse analysis. Figure 7.1 depicts part-of relations between the textual units, examples of the units, and a relation between terms.

A search engine has its own index of tokens and terms from texts and, given a query from a user, returns relevant documents by matching the query to the index. Even after the filtering through search engines, it is a time-consuming job for researchers to read a large number of filtered documents to locate the information of their interest. The user may devise a very specific query to grasp only a handful of documents, but will then miss many other related documents. Text mining can assist the user in locating interesting information from loosely filtered documents.

Since the scientific papers are still very long to read, text mining can help users by locating and highlighting the most relevant terms, sentences, and paragraphs from the retrieved texts. Text mining can also help users by classifying texts into semantic categories among which the users can choose the most relevant topic.

More sophisticated solutions locate named entities (e.g., a protein name, a disease name) in text and, if required, link them to the identifiers of corresponding entries in reference databases (e.g., UniProt, OMIM). They then recognize binary relations between the entities (e.g., protein–protein interactions) and more complex events of the entities. They may transform the facts into an ontological representation (Daraselia *et al.*, 2004; Cimiano *et al.*, 2005).

The potential users of these text mining solutions can be roughly grouped as follows: (1) researchers who have the job of populating a scientific database

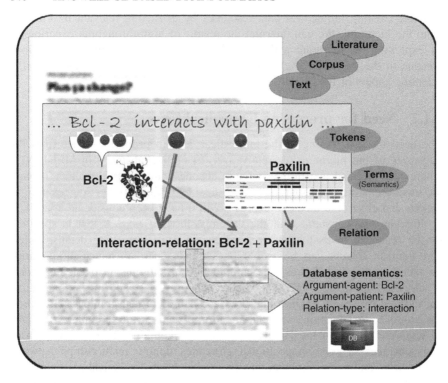

Figure 7.1   A text is decomposed into tokens. Tokens form a term that has a semantic type. Sentences convey facts, for example, represented as a relation between terms. Facts from texts can be transformed into database semantics.

with facts (called biocurators); (2) researchers who mine scientific literature for automatic integration of the literature into other data resources (bioinformaticians); and (3) researchers who seek background information on specific questions of their research (biomedical researchers). The first two research groups might employ large-scale means for the sophisticated text processing tasks, whereas the last group might submit queries of the specific questions to interactive Web interfaces of text mining solutions.

Those biomedical texts are not the only data sources scrutinized by text mining techniques. Although the techniques for text mining originate in general from the field of natural language processing (NLP), NLP techniques have also been applied to genomic and proteomic sequences since these are based on alphabets as well (e.g., A-C-G-T, Ala-Cys-Glu). They have adapted the algorithms of pattern matching and parsing to such bioinformatics tasks as sequence alignment, motif identification, and secondary structure prediction (Smith and Waterman, 1981; Knudsen and Hein, 2003; Bradley *et al.*, 2008). However, we will not discuss this research topic in this chapter.

7.1.3   Developing text mining solutions

The successful application of text mining for knowledge discovery depends on (1) the information needs of users; (2) the content of available literature; and (3) the selection of appropriate methods to deliver the content with good quality to the users. Developers of text mining solutions should thus understand the needs of the users, the NLP techniques required for meeting the needs, and the data resources needed for the text processing. Text mining in the biomedical domain increasingly profits from semantic resources like ontologies. This chapter provides the basic information for the developers in Sections 7.3 and 7.4.

In principle, any research question that is related to the expression of facts in text could profit from text mining technologies. In reality, a number of constraints have to be kept in mind to perform a requirement analysis. First, the literature must be accessible in an electronic form where the text is easily discernible (cf. PDF format, MS Word format). This is often not the case, when a document from ancient archives has to be analyzed.

Second, the information in the documents has to be represented in a way that meets the standardization criteria of the scientific domain (e.g., the terminology) and the expressions for the information have to be grammatically appropriate. In Poems and fictions, the authors do not have such constraints except the language rules. In the biomedical domain, international collaborations have been set up to ensure the standardization of the terminology (Gerstein and Krebs, 1998; Leitner and Valencia, 2008).

Third, it has to be clear how the representation of the information in text can be identified with automatic means. It is not necessary that the information contained in the literature should be of large quantity, but typically text processing techniques are required only if a large amount of documents have to be processed.

Last but not least, the facts from the literature should not be confused with raw data obtained from experiments, where the former are the results from the interpretation of the latter by the authors. On the other hand, the scientific literature is the first-hand delivery of the author's interpretations that have gone through the verification process instantiated by peer review.

A typical type of information expected from the biomedical literature is protein–protein interactions. They can be used, for example, for constructing protein networks of biological pathways, for verifying relations between co-expressed genes in microarrays, and for collecting candidate genes for a disease pathway. In general, the information automatically obtained from the literature can be used for the curation of data resources, the interpretation of high-throughput experiment data, and the design of laboratory experiments.

We emphasize that some of the text mining tasks are easy even for beginners to tackle with simple methods like co-occurrence analysis and pattern matching. It depends on the complexity of the targeted information. We provide several success stories of biomedical text mining in Section 7.5.

7.2   Information needs for biomedical text mining

Text mining analyses facts and features reported by authors in the literature. The researchers of a research domain in biomedicine contribute to the biomedical literature, require up-to-date information of their domain, and also seek recent discoveries in other related domains. These practices upon the literature apply in the same way to patents and EMRs.

The needs for information from the literature can be classified into the following two groups: (1) needs for information that can be identified in text and then transformed into a normalized format such as protein–protein interactions and gene regulatory events; and (2) needs for sets of text documents or text passages matching user queries. For text mining we define *normalization* as the process of semantically characterizing the information extracted from text. For instance, the protein–protein interactions extracted from text are characterized as an interaction event involving the two participating proteins.

Table 7.1 summarizes the characteristics of the two groups that will be discussed below. The two groups are complementary to each other in the sense that a solution for one group can be used to serve the benefits for the other group. They are not completely distinctive such that there are applications that do not clearly fall into either group. For example, iHOP which is dedicated to retrieve the sentences that express protein-protein interactions produces a set of sentences without any explicit user query, helping curators populate the databases of protein-protein interactions.

7.2.1   Efficient analysis of normalized information

Text mining can extract the terms of entities and concepts, and facts about the terms, typically binary relations between the terms. We discuss information needs for two objectives: integration of bioinformatics data and knowledge discovery.

7.2.1.1   Term identification for bioinformatics data integration

Understanding a scientific text requires identification of domain-specific terms, for example, of genes/proteins, diseases, sequence variation of genes and proteins, protein residues, chemical entities, and biomedical concepts such as Gene Ontology concepts.

The identification of gene and protein names has progressed more than that of the other types of terms in recent years, and the state-of-the-art solutions show around 85% balanced F-score for human protein names (Wermter *et al.*, 2009). Solutions taking different species into consideration have to include the species identification in the analysis and possibly make use of sequence variability to produce optimal results (Divoli *et al.*, 2008; Hakenberg *et al.*, 2008; Winnenburg *et al.*, 2008).

The term recognition for diseases and chemical entities has progressed less, and produces poorer performance than the gene/protein name identification

Table 7.1   Comparision between the two types of information needs.

	Information type	
	Text documents and text passages	Normalized information
Research area	**Information retrieval**	**Information extraction**
Examples of output	A set of text passages, a collection of documents, a list of sentences	Protein–protein interaction, gene–disease relation
Specification of information request	User queries written in a query language	Relational database schema
Anticipated general purpose	Seeking relevant information from a large collection of documents (e.g., the Web, MEDLINE)	Database population by gathering information that matches the database schema
Applications serving the needs	Search engine, solutions for document classification	Solutions for database population
Application domains	Any domain	Domains of the databases
Approaches	Statistical approaches to decomposing documents into features and analyzing the distribution of the features	Statistical, pattern-based, machine learning approaches to attributing semantic roles to textual features and relating them to each other
Requirements that constrain the approaches	Users still have to read the retrieved text to find the information. Users repeat the search until they find the information.	Fine-grained methods of natural language processing are required. Normalization with standard databases and ontologies is often required.

(Wilbur *et al.*, 1999; Jimeno-Yepes *et al.*, 2008; Klinger *et al.*, 2008). The identification of mutations in the literature can be achieved at high precision levels since the representation of mutation follows a well-defined nomenclature (Rebholz-Schuhmann *et al.*, 2004; Yeniterzi and Sezerman, 2009).

There are only a few of standard representation schemes for biomedical named entities such as gene name and chemical name (Howe and Rhee, 2008). On the other hand, there are many potentially important types of terms to be annotated in the literature (Cohen and Hunter, 2008). This fact shows us the need for automatic annotation of the literature with so-called standard, or reference, databases (e.g., GenBank, Swiss-Prot, ChEBI, OMIM). We have proofs that such annotation can efficiently support database curation (Alex et al., 2008; Karamanis et al., 2008).

The identification of conceptual terms such as ontology terms provides text mining with a different perspective. Conceptual terms represent biological knowledge. The association of named entities with the conceptual terms thus helps us understand how the biological entities work for the biological world. One well-known example is the annotation of gene products with Gene Ontology (GO) concepts (Gene Ontology Consortium, 2000), which describe the properties and functions of gene products in an abstract level (Hirschman et al., 2005; Cakmak and Ozsoyoglu, 2007).

The identification methods introduced above can be combined with each other to deliver more complex information extraction solutions. For example, the identification of protein name, protein residue, species, and GO concepts can be used to annotate mutations with functional information (Nagel et al., 2009). Similar approaches have recently identified mutations of protein kinases, changes in the stability of G-protein-coupled receptors and changes in the activity of lipases and amylases (Krallinger et al., 2009; Winnenburg et al., 2009; Yeniterzi and Sezerman, 2009).

The term recognition methods aforementioned can be applied to the identification of the terms related to a given set of seed terms (Spasic et al., 2008; Waagmeester et al., 2009). This application is effective for knowledge discovery because of the representative role of terminology in the biomedical domain. They use the seed terms to retrieve all relevant documents from the literature, analyze the frequencies of terms in the documents, and manually extract relevant terms among the most frequently occurring.

In summary, we need to recognize terms in text and to associate them with standard databases and ontologies. This is the basic step for integrating the literature with bioinformatics data, where the bioinformatics data in various resources are also being integrated into the standard resources.

7.2.1.2   Fact extraction for knowledge discovery

The terms themselves are not the main concern of biomedical researchers, but the facts about the terms contained in the scientific literature are. For a domain expert, an ideal application of text mining would be to periodically report recent novel knowledge from the literature. However, this is not yet achievable, since no standard representation of knowledge demands is available and also since many more solutions have yet to be developed for knowledge discovery.

A lot of research has been invested into the extraction of facts from the biomedical literature. All solutions have been geared with the information needs of biomedical researchers to extract a wide range of types of facts from the literature, including protein–protein interactions (Hunter et al., 2008; Chowdhary et al., 2009; Rajagopala et al., 2008), phosphorylation events for proteins (Narayanaswamy *et al.*, 2005; Yuan *et al.*, 2006), sub-cellular locations of proteins (Brady and Shatkay, 2008; Fyshe *et al.*, 2008), causal relations between genes and diseases (Rindflesch *et al.*, 2003; Seki and Mostafa, 2007; Ahlers *et al.*, 2007; Cheng *et al.*, 2008; Ozgür *et al.*, 2008), association of phenotypes with genes (Korbel *et al.*, 2005; Chun *et al.*, 2006; Lussier *et al.*, 2006), association of diseases with drugs (Chen *et al.*, 2008), contrastive relationships between proteins (Kim *et al.*, 2006), negation of protein–protein interactions (Sanchez-Graillet and Poesio, 2007), information about ubiquitin-protein ligases (Lee *et al.*, 2008), and gene regulatory events (Saric *et al.*, 2005; Rodríguez-Penagos *et al.*, 2007).

These target information are mainly binary relations. It is still an open issue to integrate them into biomedical pathways and networks (Oda *et al.*, 2008). Furthermore, it is expected to extract more types of facts, including all the event types of Gene Ontology (Gene Ontology Consortium, 2000). Concerning these goals, we should address such issues as applicability, robustness, and interoperability of text mining systems.

Text mining techniques can also be used to enhance bioinformatics applications. For instance, it has been attempted to identify microarray data which contain both disease and normal control states by analyzing free-text descriptions of the data (Dudley and Butte, 2008), understand potential relations between co-expressed genes by utilizing co-occurrences in the literature (Frijters *et al.*, 2008), and select genes that are highly likely to be involved in a disease pathway (Yu *et al.*, 2008).

Text mining can also be used to recognize scientific trends in the literature. The scientific literature has been analyzed to understand which scientific trends are shared between the medical informatics and the bioinformatics research domain (Rebholz-Schuhmann *et al.*, 2007a), identify the journals that tend to set the trends by following the chain of arguments through different scientific journals (Rzhetsky *et al.*, 2006), and identify paradigm shifts in the analysis of neurodegenerative diseases (Lisacek, 2005).

In summary, text mining can extract both known and unknown facts from the literature, link them into structures, and utilize them to interpret bioinformatics data.

7.2.2 Interactive seeking of textual information

Nowadays all biomedical researchers and bioinformaticians have experience in retrieving information from the scientific literature through search engines (e.g., PubMed, Google Scholar, ScienceDirect, Web of Science, CiteXplore) (Zhou *et al.*, 2006; Pezik *et al.*, 2009). The search engines make use of the tokens contained in documents to produce a list of documents that are relevant to a query given by a user. However, users often find that relevant texts or text passages are not always placed at the top of the search results. The successful placement of relevant texts should consider issues below, concerning both search engines and user queries.

It is intuitively clear and scientifically proven that an underspecified query (e.g., 'cancer') leads to the retrieval of a large number of documents with a lot of false positive results. For example, the query term 'cancer' not only refers to a disease but also to a sign of the zodiac. Moreover, the query term can be matched to author affiliations (e.g., 'National Cancer Institute'), while users might expect to find the query term in titles or abstracts. Even if the query term has the intended meaning in the document, users may still have narrower foci.

This fact leads us to the issue of *query refinement*, which is to provide an alternative query. A general query makes search engines retrieve most of the relevant documents, but the search engines could still fail to push them up to the top of the search results due to the under-specification of user interests. A more specific query may place some of the relevant documents in higher positions, but will probably lose other relevant documents. Another problem is that it is not easy to make up specific queries that represent the user interests precisely and produce good results at the same time. In practice, a user would start with a general query that refers to the whole domain of interest and add to the query the terms found in the top-ranked documents to focus on sub-domains or vice versa (Ellis, 2005). These observations indicate the need for interactive interfaces toward query refinement which guide users to effectively refine their queries. Advanced search engines may also consider distinctive behavior patterns of users to cope with the information seeking needs (Kim and Rebholz-Schuhmann, 2008).

Query expansion is a technique which adds related terms into a given query, concatenated with the Boolean OR operator, to raise the coverage of the search results. For instance, the single-term query 'cancer' may miss relevant documents that do not contain the term, but contain related terms like 'tumor' and 'oncogene.' Since it is impossible for users to list all the related terms in their queries, many search engines (e.g., CiteXplore, HubMed, PubMed) provide ways for automatic query expansion, for example including the synonyms, hypernyms, and hyponyms of given query terms and the terms that frequently co-occur or have similar distributions with the query terms. It is still an open issue whether semantic resources such as MeSH and Gene Ontology can improve document retrieval through automatic query expansion or query refinement (Tsuruoka *et al.*, 2008; Jimeno-Yepes *et al.*, 2009; Trieschnigg *et al.*, 2009).

Even when we get a reasonably small number of relevant biomedical papers from search engines, we may not have enough time to read all the papers. If we want to scan them through, we need automated aids for highlighting keywords and key sentences. This *highlighting* has been demonstrated to improve the curation work of biocurators (Mueller *et al.*, 2004; Karamanis *et al.*, 2008).

Another way of filtering relevant documents is via the categorization of documents. *Document classification* or *document clustering* is to assign independent classes to documents, where each document can have multiple assignments and the classes are not necessarily predefined. Once users locate an interesting document, they can follow the categories of the document to locate the other documents of the categories, which are expected to be related to the first chosen document. Users can also find relevant documents by selecting semantic categories of interest when the document classes are tagged with semantic labels (Doms and Schroeder, 2005; Shatkay *et al.*, 2008; Theodosiou *et al.*, 2008). We can also use the semantic labels to restrict our queries for search engines (Mueller *et al.*, 2004). The semantic labels for this usage are generally from ontologies such as Gene Ontology.

Question answering is another useful approach to interactive information seeking. The basic assumption of conventional search engines is that, given a query, the user wants to receive documents that are similar to the query. In question answering, the users can state their demands as interrogative sentences (or questions) using interrogative words (e.g., 'what,' 'how'). Given a question, the question answering system tries to locate the text passages that have the answers of the question. However, it is not yet achievable to understand the semantics of arbitrary natural language questions. For practical reasons, the ongoing projects of question answering have thus focused on definitional questions (Yu and Kaufman, 2007) or treated the questions as if they are the queries for conventional search engines, focusing on such issues as query expansion (Hersh *et al.*, 2009).

In summary, all approaches introduced above support individual researchers in their knowledge discovery from the literature. They assume that a piece of interesting information is there and that a user forms a query to retrieve the information, ranging from a general term to a semantic label and to a specifically described question. A text mining solution converts the query into a computable format and filters documents or text passages that match the query. It is thus essential for the success of the solutions to assist users in specifying their demands clearly, to understand the underlying meaning of user specifications, and to generate the best computable form of the query.

7.3   Principles of text mining

Text mining is the task of deriving from text those information that meet user requests. It requires the understanding of natural language to some extent and is thus closely related to natural language processing (NLP). We give an overview on the components of NLP solutions and describe existing approaches to text mining applications using the terminology of NLP, focusing on the biomedical domain.

7.3.1   Components

We provide the definitions of NLP components together with references that provide more detailed information. Review papers on biomedical text mining would be also helpful for readers to understand underlying techniques (Cohen and Hersh, 2005; Scherf *et al.*, 2005; Spasic *et al.*, 2005; Erhardt *et al.*, 2006; Jensen *et al.*, 2006; McNaught and Black, 2006; Roberts, 2006; Thomas *et al.*, 2007; Zweigenbaum *et al.*, 2007; Cohen and Hunter, 2008; Kim and Rebholz-Schuhmann, 2008; Rzhetsky *et al.*, 2008; Winnenburg *et al.*, 2008; Tuncbag *et al.*, 2009).

7.3.1.1 Software components

Tokenization is the process of splitting a text into tokens. A *token* is a word or a symbol and serves as the basic unit for further processes of NLP.

Part-of-speech (POS) is a linguistic category of words such as Noun and Verb. *POS tagging* is the process of resolving the POS tags of tokens. A well-known set of POS tags in English has been proposed by the Penn Treebank Project[1].

Named entity recognition is the task of recognizing *named entities* such as person names and city names in text and associating them with appropriate semantic categories such as Person and City (Park and Kim, 2006). It may be further required to ground the named entities from text with database entries (e.g., UniProt entries for protein names). We call the grounding task *term grounding* or *term normalization*. *Term recognition* is a more general task than named entity recognition: it includes the identification of conceptual terms like ontology terms.

Parsing is to identify syntactic relations between tokens in a sentence, for example those between a noun and its modifiers and between a verb and its subject and object. The syntactic relations of a grammatical sentence can be merged into a connected graph representation. Several models of syntactic structures, including phrase structure, predicate argument structure, and dependency structure, have been proposed and automatically converted into each other (Clegg and Shepherd, 2007; Miyao *et al.*, 2009). The state-of-the-art parsers are based on such grammar formalisms as context-free grammar (CFG) (Klein and Manning, 2003), dependency grammar (Briscoe *et al.*, 2006), head-driven phrase structure grammar (HPSG) (Sagae *et al.*, 2007), and combinatory categorical grammar (CCG) (Clark and Curran, 2007).

Full parsing is not always required for NLP applications. *Partial parsing* is to identify only parts of the syntactic structure of a sentence (Abney, 1996). Shallow parsing is often confused with partial parsing. We may define *shallow parsing* as a specific approach to partial parsing that identifies sentential syntactic structures only up to a predefined level or type of syntactic structure, for example, noun phrases and verb phrases (Kim, 2006). *Chunking* is the task of breaking a sentence into *chunks*, which are non-recursive intra-clausal constituents such as basic noun phrases (Abney, 1996).

An anaphoric expression (e.g., 'it', 'the person') is a word or a phrase that refers to a preceding expression in the same text, called the *antecedent* (Mitkov, 2002). *Anaphora resolution* is the task of identifying the antecedent of an anaphoric expression. A related task, called *abbreviation resolution*, is to resolve the full names of abbreviations, which are often expressed in the form of acronyms (e.g., 'natural language processing (NLP)').

Document retrieval or *information retrieval* is to retrieve documents that are relevant to given user queries, mostly Boolean queries (Pezik *et al.*, 2009). In particular, we call the retrieval of unstructured texts *text retrieval*. When the retrieved text is too long for users to read through, it is necessary to select a

---

[1] www.cis.upenn.edu/~treebank/

passage (e.g., sentence, paragraph) in the text that is most relevant to the user query. We call the task *passage retrieval*.

Document classification is the task of classifying documents into pre-defined categories, while *document clustering* is to group similar documents into yet unspecified clusters. It is possible to classify or cluster documents into a hierarchical structure.

Information extraction is the task of extracting information of predefined types from text (McNaught and Black, 2006). The target information is typically relations between named entities such as protein–protein interactions and gene–disease relations. Information extraction is mostly performed in an offline manner, while the query response of information retrieval is performed online.

Question answering is, given a Wh-question from a user, to find the passage that answers the question (Hersh and Voorhees, 2009). It is different from document retrieval since it does not deliver a whole document, but the information found in the document. It is also distinguished from information extraction in that it deals with variable user queries, not only with fixed user requests.

7.3.1.2 Resources

A *lexicon* contains lexical information of words (Hirst, 2004). It may have definitions, POS information, derivative forms, sub-categorization frames, synonyms, semantic categories, and lexical semantics. Recent work has produced lexicons specialized in the biomedical domain, including UMLS Specialist Lexicon, BioLexicon (Sasaki *et al.*, 2008) and Geno (Wermter *et al.*, 2009).

An *ontology* is a formal representation of concepts, providing a controlled vocabulary. Ontologies are well-known forms for knowledge representation and knowledge sharing. A lexicon is not an ontology due to the lack of the formality such as the representation of equivalence, disjointness, and union (Hirst, 2004). Ontology terms can be limitedly used for representing textual semantics (Spasic *et al.*, 2005), where the limitations of ontologies for the usage have been discussed (Tsujii and Ananiadou, 2005). OBO Foundry has a collection of referential ontologies in the biomedical domain (Smith *et al.*, 2007).

A *corpus* is a collection of texts. The texts can be annotated with the information to be extracted by text mining solutions, called *corpus annotation*. The annotated corpora can be used for training and testing text mining systems and NLP components such as POS tagger and parser.

7.3.1.3 Evaluation metrics

Precision and recall are the most commonly used measures for evaluating text mining systems against annotated corpora. *Precision* is the ratio of the facts that are correctly extracted by the system (namely, *true positive (TP)*) over all the facts extracted by the system. The facts that are incorrectly extracted are called *false positive (FP)*. *Recall* is the ratio of TP facts over the facts that are annotated on the corpus as true examples. The annotated facts which are not extracted by the system are called *false negative (FN)*. *F-score* is the harmonic mean of precision and recall.

$$\text{Precision} = \frac{TP}{TP + FP} \qquad \text{Recall} = \frac{TP}{TP + FN} \qquad (7.1)$$

$$\text{F-score} = \frac{2 \times \text{Precision} \times \text{Recall}}{\text{Precision} + \text{Recall}}.$$

Equation (7.1): Definitions of evaluation measures.

7.3.2   Methods

7.3.2.1   Text retrieval

In general, text retrieval systems work in two modes (Baeza-Yates and Ribeiro-Neto, 1999). In the offline mode, they identify tokens and terms in a set of documents and create an index for them. In the online mode, they analyze user queries, match them to the index to retrieve relevant documents, and rank the retrieved documents according to their scores of relevancy to the queries. Search engines are the typical applications that exploit text retrieval.

Existing text retrieval solutions have further employed in-depth NLP techniques to narrow down the search results (Kim and Rebholz-Schuhmann, 2008). For instance, passage retrieval extracts short passages of the documents which users will scan to check the relevance of the documents to their queries. Document classification enables users to focus on a smaller collection of texts.

7.3.2.2   Term recognition

There are many open issues in term recognition (Ananiadou and Nenadic, 2006; Park and Kim, 2006). For instance, we should deal with the high degree of term variation (e.g., 'D(2)'–'D2', 'IGA'–'IG alpha') and term ambiguity (e.g., 'orb' either as a gene name or as a common English word, 'CD4' either as a gene name or as a cell name) (Howe and Rhee, 2008). In some cases, we need to recognize terms that are not yet registered in term repositories, bearing in mind that biomedical knowledge is ever and rapidly growing.

A typical term recognition system may work as follows: It first locates all noun phrases in text, on the assumption that all terms are noun phrases, either standalone or embedded (e.g., 'CD4' in 'CD4 activation'); the system then determines whether a noun phrase is, or contains, a term; and, if required, it grounds the recognized terms with corresponding entries of standard databases.

The existing approaches to term recognition can be roughly classified into the following four groups (Ananiadou and Nenadic, 2006; Park and Kim, 2006): (1) dictionary-based approaches that match terms of standard repositories to text; (2) rule-based approaches that learn rules and patterns for term recognition from annotated corpora and recognize terms by using the rules; (3) machine learning-based approaches that implement machine learning models for term recognition;

and (4) hybrid approaches that combine two or more of the above methodologies, mostly in a sequential way, to achieve the optimal result.

Previous solutions to term recognition have been developed mainly for unstructured text. It is a recent achievement to develop online solutions for formatted documents such as MS Word documents and PDF files, where most of scientific papers are available in such formats (Pavlopoulos *et al.*, 2009; Rebholz-Schuhmann *et al.*, 2010).

The recognition of ontology terms in text is much more difficult than other tasks of term recognition, because ontology terms are generally so artificial that they do not literally appear in the literature or are confused with other terms in the absence of contextual evidence (Tsujii and Ananiadou, 2005; Winnenburg *et al.*, 2008). To address this issue, researchers have located the component words of ontology terms in text and then estimated the probability that the words together represent the ontology concepts (Gaudan *et al.*, 2008) or have automatically collected language patterns that might represent the ontology concepts (Cakmak and Ozsoyoglu, 2007).

7.3.2.3 Relation identification

The linguistic expression of a relation involves a keyword and its semantic arguments, where the keyword indicates the semantic type of the relation and the arguments indicate the related concepts. Such an expression often shows two types of locality: the arguments are located in a local context of the keyword, namely *spatial locality*, and have syntactic relations with the keyword, namely *structural locality* (Kim, 2006).

The simplest approach to dealing with the spatial locality is to identify co-occurrences between terms, based on the assumption that if two terms co-occur frequently, then they are likely to be related to each other (Jensen *et al.*, 2006). But, this method is unable to recognize the semantic type of the relation between the co-occurring terms. To address this issue, many have used language patterns that represent relations of pre-defined types (Cohen and Hersh, 2005). For instance, the pattern 'A interact with B' has been frequently used for extracting protein–protein interactions. It is still an ongoing work to automatically learn such language patterns for relation identification (Huang *et al.*, 2004; Hao *et al.*, 2005).

Language patterns are effective in dealing with the spatial locality, but not with the structural locality. To address the latter, some approaches utilize the results of either full parsing or partial parsing (Kim *et al.*, 2009). For matching the language patterns to the parse results, they annotate the patterns with syntactic information, for example, that the variable A in the pattern 'A interact with B' is the subject of the verb 'interact.' Other approaches deal with the structural locality not through syntactic information, but through statistical information on the contexts of the relation keywords (Chowdhary *et al.*, 2009).

Many recent approaches to relation identification have employed machine learning (ML) algorithms such as Maximum Entropy, Support Vector Machine, and Conditional Random Field (Kim *et al.*, 2009). A typical ML system would first locate a trigger word in text, which generally refers to an event of interest. It then recognizes potential arguments of the event in the same sentence as that of the trigger word and fills in the slots of an event template with the arguments. ML systems can also consider the relation identification as a classification task such that it is to decide whether a set of a trigger word and potential participants together are formed into a relation or not (Hahn *et al.*, 2009). They have utilized various features for learning models, including lexical, chunking, parse, semantic, proximity, and directionality features. These features address the two types of locality at the same time in a single ML model.

Some text mining solutions seek to extract not the information of interest itself, but the sentences that are likely to contain such information. They are practical since, before adding extracted information into databases, the curators of the databases always read source sentences to check the validity of the information. The relevant sentence detection may demand less computing than the information extraction (Krallinger *et al.*, 2008).

The task of database population may require the extraction of not only the main relation information (e.g., protein–protein interaction), but also the other information of the target database (e.g., interacting protein domain, related disease) (Lee *et al.*, 2008).

We may represent text mining results by using ontologies for knowledge sharing (Daraselia *et al.*, 2004; Cimiano *et al.*, 2005; Saric *et al.*, 2005). Owing to the formality of the ontologies, we can then perform logical reasoning over the results (Donnelly *et al.*, 2006).

7.3.2.4   Bioinformatics tasks

Ng (2006) describes previous approaches to enhancing bioinformatics systems by examining the literature. For instance, it has been proven that we can improve sequence homology search by finding textual support for homologies, and improve functional classification of proteins by combining sequence information with textual information. We can also gather keywords for a gene cluster, which might explain the common biological function of the member genes, and enhance expression data analysis with literature knowledge.

Text mining techniques that have been employed for bioinformatics tasks so far are mainly basic analyses of term frequencies and term co-occurrences. For instance, the identification of keywords that are representative for a gene cluster is based on the exclusive frequencies of the keywords (Frijters *et al.*, 2008). While these analyses would not show excellent performance for general purpose tasks, they have been proven to be effective for highly focused tasks like those above.

7.4   Development issues

The development of a text mining system includes such phases as design, construction, integration, and evaluation. During the design phase, we gather the information needs from users. During the construction phase, we develop corpora, annotate them with the information to be extracted if required, and assess parts of them to specify the use of language for the target information. We then collect or develop NLP components that can analyze the language usages. During the integration phase, we integrate the NLP components and required resources into a system. During the evaluation phase, we evaluate the system against the corpora. In this section, we only discuss issues that are particularly related to text mining. Table 7.2 summarizes the development phases and their specific tasks for text mining systems.

Table 7.2   Development phases for text mining systems.

Development phase	Tasks
Design	Specification of information needs
Construction	Corpus construction and annotation
	Analysis of language usages
	Collecting and developing NLP modules
Integration	Implementing interfaces for NLP modules and resources
	Integrating NLP modules and resources into a system
Evaluation	Evaluation of the system against annotated corpora
	Evaluation of the system against all corpora

7.4.1   Information needs

It should be the first step to explicitly state the requirements for the text mining task (Cohen and Hunter, 2008). One of the unusual decisions to be made for biomedical text mining systems is whether the system should associate its results with existing databases. This grounding is often expected since the biomedical domain has many databases that are maintained by expert curators and cited by researchers working in the domain. In short, this step should include enough time to understand the needs from potential users (Karamanis et al., 2008).

7.4.2   Corpus construction

The next step is to construct corpora from which the text mining system will extract the information of interest. We should carefully consider the issues below.

First, we should determine the selection criteria of input texts for the corpus construction. In the biomedical domain, MEDLINE abstracts are freely available, while full-text articles often produce overheads due to copyright issues and the diversity of file formats. On the other hand, the full-texts have more information than the abstracts (Corney et al., 2004). Other types of text have also been considered, including electronic medical records (Heinze et al., 2001), patents (Rhodes et al., 2007), figure and table captions (Hearst et al., 2007), MeSH terms of MEDLINE citations (Neveol et al., 2007), and natural language comments in databases (e.g., UniProt comments) (Dudley and Butte, 2008; Mottaz et al., 2008).

Second, we should carefully determine the scope and size of corpora. It is not always best to have the largest corpus. It has been reported that the performance of text mining systems, in terms of precision and recall, is dependent upon the selection of corpora (Rodriguez-Penagos *et al.*, 2007). Typically, a text mining system shows better precision from a corpus focused on a specific domain and better recall from a corpus with a wide scope. It is thus recommended to have multiple corpora with various foci and scopes.

Third, we need gold standard annotations of target information on the corpus for the purpose of training and evaluating the system. We can develop the system based on a subset of the annotations, named *training corpus*, and test the system against the rest of the annotations, named *test corpus*. We do not have to annotate all the corpora, but can choose, in general by random selection, a small portion of the corpus for the corpus annotation.

Significant efforts have been exerted on the corpus annotation. The following criteria have been proposed for the corpus annotation (Wilbur *et al.*, 2006): focus (e.g., scientific vs. general), polarity (positive vs. negative statement), level of certainty, strength of evidence, and direction/trend (increase or decrease in certain measurement). It is suggested that we should use standard formats for sharing the annotations (Rebholz-Schuhmann *et al.*, 2006; Johnson *et al.*, 2007) and include structural and linguistic information into the annotations (Cohen *et al.*, 2005). Different annotated corpora may result in different performances of the same text mining system, which leads us to the need for the standardization of corpus annotation (Pyysalo *et al.*, 2008).

7.4.3   Language analysis

A step that is often neglected is to analyze the linguistic characteristics of the information to be extracted (Cohen and Hunter, 2008). This analysis may lead us to the realization of the depths of the problems before us (Netzel *et al.*, 2003), help us to understand whether the problems are solvable with the available NLP techniques (Light *et al.*, 2004), and give us the insight on how to tackle the problems both effectively and efficiently (Kim *et al.*, 2006). Of course, we should be cautious neither to over-generalize nor to over-simplify the problems.

As the result of this step, we will understand what kinds of NLP modules are required for a specific text mining task. A comprehensive number of NLP modules are publicly available at the moment[2]. If there is any missing module, we would develop one, consulting the existing methods introduced in Section 7.3.2.

7.4.4   Integration framework

Once we acquire the required NLP modules in hand, we should integrate them into a system. If some modules are complementary to each other, they should be arranged in parallel (Leroy and Chen, 2005). But, in most cases, we need a sequential model for the integration.

---

[2] http://zope.bioinfo.cnio.es/bionlp_tools/

We here introduce two integration frameworks for biomedical text mining systems. Kirsch and colleagues (Kirsch *et al.*, 2006) have proposed such a model for Web services, where each service encompasses a NLP module with a socket wrapper and interacts with clients by the exchange of XML annotated texts. Kano and colleagues (Kano *et al.*, 2009) have presented U-Compare, an integration framework based on UIMA (Unstructured Information Management Architecture). It provides a type system that defines data types used by NLP tools, ranging from sentenciser to parser. Please note that both integration frameworks are Web-based services and support XML.

In addition, the resources for text mining such as lexicon, language patterns, knowledge base, and queries should also be integrated in a semantic level. The information selected from a resource by a module should be used as input to another module and be semantically associated with those from another resource. Ontologies are the common means for the semantic integration (Daraselia *et al.*, 2004; Cimiano *et al.*, 2005). Typically, the resources share the concepts and properties of an ontology for knowledge representation.

7.4.5 Evaluation

Last but not least, we should consider how to evaluate our systems against the common evaluation measures such as precision, recall, and F-score (Cohen and Hunter, 2008). Text mining systems would never be free from defects due to the universal ambiguity of natural languages. The trade-offs between evaluation measures are also well known. For example, if we tune our text mining system to increase the precision, the recall of the system generally is lowered, and vice versa. It is thus necessary to estimate the performance of text mining systems in terms of a combined measure like the F-score. To obtain a better average measure, it is often the practice to divide the whole corpus into several sets and to change the roles of training and test corpora among them, instead of a fixed division between the two corpora. The evaluation based on these alternated training/test corpora is called *cross-validation*.

We should bear in mind that the intrinsic evaluation results from the manually annotated corpora may not reflect the system performance on real data, since the test corpus often has the same origin as the training corpus (Caporaso *et al.*, 2008). We should thus assess the representativeness of the test corpus for the real world tasks during the phase of corpus construction.

Since there is the cost-dependent limitation in the corpus construction, we may promote our systems by comparing with others. Many workshops have been held for evaluating text mining solutions against shared tasks in the biomedical domain, including the KDD Challenge Cup (Yeh *et al.*, 2002), Biocreative I (Hirschman *et al.*, 2005) and II (Morgan *et al.*, 2008; Krallinger *et al.*, 2008), TREC Genomics (Hersh and Voorhees, 2009), and BioNLP Shared Task on Event Extraction (Kim *et al.*, 2009). U-Compare provides an automated means for comparing text mining systems (Kano *et al.*, 2009).

7.5   Success stories

Text mining empowered by natural language processing provides techniques that contribute to knowledge discovery, as a part of interactive information retrieval applications or integrated into analytic bioinformatics solutions. In this section we present selected examples of success stories that have enabled knowledge discovery based on the literature analysis techniques.

7.5.1   Interactive literature analysis

Many interactive Web solutions have been widely exploited for knowledge discovery tasks such as biomedical curation work. They help users improve their access to information that is difficult to retrieve otherwise. The retrieval engines identify such information from retrieved documents in a pre-processing way to ensure that they can on the fly produce the results that meet user's demands reflected in user queries. Since released, they have been widely used by biomedical researchers and bioinformaticians. Please note that only a small number of the successful systems employ deep NLP techniques like parsing. Other solutions have usually been optimized to meet specific needs in the domain by using reliable but rather unsophisticated methods, for example co-occurrence analysis. Table 7.3 shows the Web addresses of the services.

Table 7.3   Example text mining services and their addresses.

Service	Web address
EBIMed	www.ebi.ac.uk/Rebholz-srv/ebimed/index.jsp
GoPubMed	www.gopubmed.com/web/gopubmed/
iHOP	www.ihop-net.org/UniPub/iHOP/
Textpresso	www.textpresso.org/

Textpresso is an information retrieval system that provides access to the full texts of scientific articles, and tags terms and ontology concepts on phrases and sentences (Mueller *et al.*, 2004). The ontology concepts not only denote entity types (e.g., gene, cell) but also relationships (e.g., regulation, localization) and properties (e.g., biological process). Terms and ontology concepts are integrated into a lexicon and recognized by the system using dictionary-based methods and regular expressions associated with the ontology concepts. This ontological annotation enables users to semantically restrict their queries with the same ontology concepts. Relations such as protein–protein interactions are identified by using co-occurrence analysis. The system allows users to integrate their own lexical resources into the system and can therefore be adapted for individual curation needs.

Two research teams have processed the complete set of MEDLINE abstracts to generate networks of related genes (Jenssen *et al.*, 2001; Hoffmann and Valencia, 2005). Both systems collect co-occurrences of any two gene names in the same sentence and create gene interaction networks based on the co-occurrences. In the case of *iHOP* (Hoffmann and Valencia, 2005), the genes are in addition linked to the corresponding entries of Entrez Gene and UniProt. The interface of iHOP enables navigation from the gene of an initial query to its interaction partners and delivers evidence sentences for the interactions. The complete network of iHOP contains more than 80 000 genes. Altogether, the canned results gathered from the comprehensive analysis of MEDLINE enable efficient knowledge discovery for any bioinformatician working in the analysis of genetic interactions. For example, the content of iHOP can be screened for seeking interactions that have not yet been reported elsewhere. A small gene network from an expression analysis can be compared against the most frequently mentioned interactions of iHOP to identify regulatory mechanisms in the network.

Another Web application, *GoPubMed*, provides a hierarchical classification of retrieved documents (Doms and Schroeder, 2005). Given a user query, it retrieves matched documents from MEDLINE, classifies them based on GO concepts and MeSH terms assigned to them in an offline manner, and visualizes the classification according to the hierarchical structures of Gene Ontology and MeSH. The service also provides a statistical overview of the whole search results in terms of, for example, publication year, journal, and top terms, and generates a textual summary for each document according to the given query. Navigation of the documents is directed by the user's selection of semantic categories, which fit to his interest.

EBIMed is a text mining system that summarizes co-occurrences of terms from retrieved documents (Rebholz-Schuhmann *et al.*, 2007b). Given a user query, it retrieves relevant MEDLINE abstracts together with all mentions of proteins, Gene Ontology concepts, drugs, and species in the abstracts from a pre-processed index. Co-occurrences of the terms are gathered on the fly from sentences of the MEDLINE abstracts and displayed in the form of a table. Altogether, the tool shows an overview of the domain, which is represented by the query, with regard to term associations, and helps us identify important term relations, either explicitly or implicitly stated, within the domain. EBIMed has been extensively exploited for knowledge discovery, for example to generate the STITCH database (Kuhn *et al.*, 2008).

7.5.2 Integration into bioinformatics solutions

The integration of text mining solutions into bioinformatics applications has contributed to a number of knowledge discovery tasks. One of the most successful applications is the identification of protein–protein interactions from the literature for constructing a protein network (Blaschke *et al.*, 1999; Jenssen *et al.*,

2001; Hoffmann and Valencia, 2005; Krallinger *et al.*, 2008). Further work is ongoing to prove the relevance of the generated data (see BioCreative[3]) and to construct a pathway from interactions (Oda *et al.*, 2008).

Another important application is the functional annotation of genes and proteins with ontology concepts found in the context of the gene and protein mentions in text. We distinguish solutions that identify explicitly stated associations between the genes and the ontology concepts from text, which is a text mining task (Couto *et al.*, 2006; Gaudan *et al.*, 2008), from solutions that annotate 'unknown' genes with concepts that are already annotated to similar 'known' genes, which is a data mining task (Groth *et al.*, 2008).

Text mining is being successfully applied to specific knowledge discovery tasks, meeting the needs of lab scientists at hand. For instance, large scale literature analyses have been performed for a number of protein families, including transcription factors, ubiquitin-protein ligases, protein kinases, and G protein-coupled receptors (Rodriguez-Penagos *et al.*, 2007; Lee *et al.*, 2008; Krallinger *et al.*, 2009; Winnenburg *et al.*, 2009). Recent works deal with the functional annotation of individual mutations of genes and the changes of protein residues from the literature (Rebholz-Schuhmann *et al.*, 2004; Nagel *et al.*, 2009). Some of them focus on a particular protein family, for example lipase and amylase enzymes to predict the stability and the disease relevance of mutations in proteins (Yeniterzi and Sezerman, 2009).

7.5.3   Discovery of knowledge from the literature

We here introduce several text mining applications that have led to the discovery of new knowledge in biomedicine.

The simple but significant task of relevant term recommendation has been successfully applied to many domains. Spasic and colleagues (Spasic *et al.*, 2008) used 495 seed terms that describe methods for the analysis of metabolic processes and identified 2612 new relevant terms. Waagmeester and colleagues (Waagmeester *et al.*, 2009) used a similar approach to extract 37 terms relevant to the carotenoid pathway from a total of 89 086 terms. Thirteen terms among the relevant terms were eventually added to the official resource of the carotenoid pathway by curators.

The association of the literature with the results from expression array experiments has also led to many scientific discoveries. Hettne and colleagues (Hettne *et al.*, 2007) showed that NF-kB is located at the center of a network whose nodes are all involved in the complex regional pain syndrome. They concluded that NF-kB must have an important role in the genesis of the syndrome. Jelier and colleagues (Jelier *et al.*, 2007) used text-derived concept profiles to annotate the results from microarray experiments. They observed from their experiments that lysosomal proteins are over-expressed in prostate cancer in contrast to the concept profiles of other cancers and concluded that lysosomes may play a significant role

---

[3] www.biocreative.org/

in the development of prostate cancer. More general approaches have generated similarity profiles for genes and proteins to transfer their functions to similar genes and proteins (Sanfilippo *et al.*, 2007) or identify novel nuclear proteins which have not been attributed to this category before (Schuemie *et al.*, 2007).

Text mining has been also applied for chemical and drug analysis. Weeber and colleagues (Weeber *et al.*, 2003) focused on thalidomide, which had become obsolete due to its embryo-pathogenic effects, and identified that it could have a therapeutic effect in acute pancreatitis, chronic hepatitis C, *Helicobacter pylori*-induced gastritis, or myasthenia gravis. Frijters and colleagues (Frijters *et al.*, 2007) annotated different kinds of carcinogens with concept profiles to identify the biological processes that have been modified by the compounds. Unfortunately, they could only identify those toxicity mechanisms that were already known. Other researchers read structural information from the scientific literature that helps to characterize chemical compounds with regards to their therapeutic or toxicological potentials (Banville, 2006).

The pioneers in the field of knowledge discovery from the scientific literature simply used the text features, not even terminological resources or concept profiles, to identify unpublished yet potential relations between concepts. Swanson (1986) could automatically extract from the literature, and verify with experiments, that fish oil has a therapeutic effect on Raynaud's disease. With the same approach, he also showed that a depletion of magnesium could be the cause of migraine (Swanson, 1988) and that estrogen could be involved in the onset of Alzheimer's disease (Smalheiser and Swanson, 1996).

7.6   Conclusion

Knowledge discovery is a complex task which requires us to take into consideration the background knowledge of the recipient of novel information. Knowledge discovery is the primary goal of automatic literature analysis. We can distinguish interactive use of text processing tools that enable efficient access to textual information from fully automatic text processing that delivers normalized information. The two approaches are complementary to each other and a large number of public solutions make use of both.

In recent years, the text analysis community has made significant progress in providing reliable and accurate solutions for the biomedical informatics domain, where a number of successful solutions are being widely used. Still, this does not exempt bioinformatics researchers from learning how to utilize and benefit from existing solutions. In other words, good background knowledge about the underlying technology supports selecting the right solutions for the tasks under scrutiny.

The automatic analysis of the full body of scientific literature requires access to all contents, including full texts of scientific papers. Great efforts are leading to better access to the scientific literature, supported by public funds. As a result, we expect that in the near future the scientific literature will be fully integrated into the bioinformatics data resources.

The full integration of the scientific literature along with the ready availability of automatic text processing solutions will enable researchers to efficiently test novel hypotheses against the scientific literature and also speed up the turnover of new knowledge.

7.7   References

Abney, S.P. (1996) Part-of-speech tagging and partial parsing, in *Corpus-Based Methods in Language and Speech* (eds K. Church, S. Young, and G. Bloothooft), Kluwer Academic Publishers, Dordrecht, pp. 118–36.

Ahlers, C.B., Fiszman, M., Demner-Fushman, D., *et al.* (2007) Extracting semantic predications from Medline citations for pharmacogenomics. *Pac Symp Biocomput*, 209–20.

Alex, B., Grover, C., Haddow, B., *et al.* (2008) Assisted curation: does text mining really help? *Pac Symp Biocomput*, 556–67.

Ananiadou, S. and Nenadic, G. (2006) Automatic terminology management in biomedicine, in *Text Mining for Biology and Biomedicine* (eds B. Stapley and S. Ananiadou), Artech House Publishers Inc., Norwood, MA, pp. 67–98.

Baeza-Yates, R. and Ribeiro-Neto, B. (1999) *Modern Information Retrieval*, Addison-Wesley, New York.

Banville, D.L. (2006) Mining chemical structural information from the drug literature. *Drug Discov. Today*, **11**(1), 36–42.

Blaschke, C., Andrade, M.A., Ouzounis, C., and Valencia, A. (1999) Automatic extraction of biological information from scientific text: protein-protein interactions. *Proc Int Conf Intell Syst Mol Biol*, **1999**, 60–7.

Blaschke, C., Oliveros, J.C., and Valencia, A. (2001) Mining functional information associated to expression arrays. *Funct. Integr. Genomics*, **4**, 256–68.

Bradley, R.K., Pachter, L., and Holmes, I. (2008) Specific alignment of structured RNA: stochastic grammars and sequence annealing. *Bioinformatics*, **24**(23), 2677–83.

Brady, S. and Shatkay, H. (2008) Epiloc: a (working) text-based system for predicting protein subcellular location. *Pac Symp Biocomput*, 604–15.

Briscoe, E., Carroll, J., and Watson, R. (2006) The second release of the RASP system, in Proceedings of the COLING/ACL 2006 Interactive Presentation Sessions, Sydney, Australia, 17–18 July 2006. Association for Computational Linguistics, Stroudsburg, PA, pp. 77–80.

Cakmak, A. and Ozsoyoglu, G. (2007) Annotating genes using textual patterns. *Pac Symp Biocomput*, **2007**, 221–32.

Caporaso, J.G., Deshpande, N., Fink, J.L., *et al.* (2008) Intrinsic evaluation of text mining tools may not predict performance on realistic tasks. *Pac Symp Biocomput*, **2008**, 640–51.

Chen, E.S., Hripcsak, G., Xu, H., *et al.* (2008) Automated acquisition of disease–drug knowledge from biomedical and clinical documents: an initial study. *J Am Med Inform Assoc*, **15**(1), 87–98.

Cheng, D., Knox, C., Young, N., *et al.* (2008) PolySearch: a web-based text mining system for extracting relationships between human diseases, genes, mutations, drugs and metabolites. *Nucleic Acids Res.*, **36**(Web server), W399–405.

Chowdhary, R., Zhang, J., and Liu, J.S. (2009) Bayesian inference of protein–protein interactions from biological literature. *Bioinformatics*, **25**(12), 1536–42.

Chun, H.W., Tsuruoka, Y., Kim, J.D., *et al.* (2006) Automatic recognition of topic-classified relations between prostate cancer and genes using MEDLINE abstracts. *BMC Bioinformatics*, **7**, S4.

Cimiano, P., Reyle, U., and Sarić, J. (2005) Ontology-driven discourse analysis for information extraction. *Data Knowl Eng*, **55**, 59–83.

Clark, S. and Curran, J.R. (2007) Formalism-independent parser evaluation with CCG and DepBank, in Proceedings of the 45th Annual Meeting of the Association for Computational Linguistics, Prague, Czech Republic, 23–30 June 2007. Association for Computational Linguistics, Stroudsburg, PA, pp. 248–55.

Clegg, A.B. and Shepherd, A.J. (2007) Benchmarking natural-language parsers for biological applications using dependency graphs. *BMC Bioinformatics*, **8**, 24.

Cohen, A.M. and Hersh, W.R. (2005) A survey of current work in biomedical text mining. *Brief. Bioinformatics*, **6**(1), 57–71.

Cohen, K.B. and Hunter, L. (2008) Getting started in text mining. *PLoS Comput. Biol.*, **4**(1), e20.

Cohen, K.B., Fox, L., Ogren, P.V., and Hunter, L. (2005) Empirical data on corpus design and usage in biomedical natural language processing. *AMIA Annu Symp Proc*, **2005**, 156–60.

Corney, D. P. A., Buxton, B.F., Langdon, W.B., and Jones, D.T. (2004) BioRAT: extracting biological information from full-length papers. *Bioinformatics*, **20**(17), 3206–13.

Couto, F.M., Silva, M.J., Lee, V., *et al.* (2006) GOAnnotator: linking protein GO annotations to evidence text. *J Biomed Discov Collab*, **1**(1), 19.

Daraselia, N., Yuryev, A., Egorov, S., *et al.* (2004) Extracting human protein interactions from MEDLINE using a full-sentence parser. *Bioinformatics*, **20**(5), 604–11.

Divoli, A., Hearst, M.A., and Wooldridge, M.A. (2008) Evidence for showing gene/protein name suggestions in bioscience literature search interfaces. *Pac Symp Biocomput*, 568–79.

Doms, A. and Schroeder, M. (2005) GoPubMed: exploring PubMed with the Gene Ontology. *Nucleic Acids Res.*, **33**(Web server), W783–6.

Donnelly, M., Bittner, T., and Rosse, C. (2006) A formal theory for spatial representation and reasoning in biomedical ontologies. *Artif Intell Med*, **36**, 1–27.

Dudley, J. and Butte, A.J. (2008) Enabling integrative genomics analysis of high-impact human diseases through text mining. *Pac Symp Biocomput*, 580–91.

Ellis, D. (2005) Ellis's model of information-seeking behavior, in *Theories of Information Behavior* (eds K.E. Fisher, S. Erdelez, and L. E. F. McKechnie), Information Today, Inc., Medford, NJ, pp. 138–42.

Erhardt, R. A. A., Schneider, R., and Blaschke, C. (2006) Status of text-mining techniques applied to biomedical text. *Drug Discov. Today*, **11**(7-8), 315–25.

Fayyad, U.M., Piatetsky-Shapiro, G., Smyth, P., and Uthurusamy, R. (1996) *Advances in Knowledge Discovery and Data Mining*, AAAI Press, Menlo Park, CA.

Frijters, R., Verhoeven, S., Alkema, W., *et al.* (2007) Literature-based compound profiling: application to toxicogenomics. *Pharmacogenomics*, **8**, 1521–34.

Frijters, R., Heupers, B., van Beek, P., *et al.* (2008) CoPub: a literature-based keyword enrichment tool for microarray data analysis. *Nucleic Acids Res.*, **36**(Web server), W406–10.

Fyshe, A., Liu, Y., Szafron, D., *et al.* (2008) Improving subcellular localization prediction using text classification and the gene ontology. *Bioinformatics*, **24**(21), 2512–17.

Gaudan, S., Yepes, A., Lee, V., and Rebholz-Schuhmann, D. (2008) Combining evidence, specificity, and proximity towards the normalization of gene ontology terms in text. *EURASIP J Bioinform Syst Biol*, **2008**, 342746.

Gene Ontology Consortium (2000) Gene Ontology: tool for the unification of biology. *Nature Genet.*, **25**, 25–9.

Gerstein, M. and Krebs, W. (1998) A database of macromolecular motions. *Nucleic Acids Res.*, **26**(18), 4280–90.

Groth, P., Weiss, B., Pohlenz, H.D., and Leser, U. (2008) Mining phenotypes for gene function prediction. *BMC Bioinformatics*, **9**, 136.

Hahn, U., Tomanek, K., Buyko, E., *et al.* (2009) How feasible and robust is the automatic extraction of gene regulation events? A cross-method evaluation under lab and real-life conditions, in Proceedings of the Workshop on BioNLP, Boulder, Colorado, 4–5 June 2009. Association for Computational Linguistics, Stroudsburg, PA, pp. 37–45.

Hakenberg, J., Plake, C., Leaman, R., *et al.* (2008) Inter-species normalization of gene mentions with GNAT. *Bioinformatics*, **24**(16), i126–32.

Hao, Y., Zhu, X., Huang, M., and Li, M. (2005) Discovering patterns to extract protein-protein interactions from the literature: Part II. *Bioinformatics*, 21(15), 3294–300.

Hearst, M.A., Divoli, A., Guturu, H. *et al.* (2007) BioText search engine: beyond abstract search. *Bioinformatics*, **23**(16), 2196–7.

Heinze, D.T., Morsch, M.L., and Holbrook, J. (2001) Mining free-text medical records. *Proc AMIA Symp.*, **2001**, 254–8.

Hersh, W. and Voorhees, E. (2009) TREC genomics special issue overview. *Inf Retr Boston*, **12**(1), 1–15.

Hettne, K.M., de Mos, M., de Bruijn, A.G., *et al.* (2007) Applied information retrieval and multidisciplinary research: new mechanistic hypotheses in complex regional pain syndrome. *J Biomed Discov Collab*, **2**, 2.

Hirschman, L., Yeh, A., Blaschke, C., and Valencia, A. (2005) Overview of BioCreAtIvE: critical assessment of information extraction for biology. *BMC Bioinformatics*, **6**(1), S1.

Hirst, G. (2004) Ontology and the lexicon, in *Handbook on Ontologies* (eds S. Staab and R. Studer), Springer, Karlsruhe, pp. 209–30.

Hoffmann, R. and Valencia, A. (2005) Implementing the iHOP concept for navigation of biomedical literature. *Bioinformatics*, **21**(Suppl 2), ii252–8.

Howe, D. and Rhee, S.Y. (2008) The future of biocuration. *Nature*, **455**(4), 47–50.

Huang, M., Zhu, X., Hao, Y., *et al.* (2004) Discovering patterns to extract protein-protein interactions from full texts. *Bioinformatics*, **20**(18), 3604–12.

Hunter, L., Lu, Z., Firby, J., *et al.* (2008) OpenDMAP: an open source, ontology-driven concept analysis engine, with applications to capturing knowledge regarding protein transport, protein interactions and cell-type-specific gene expression. *BMC Bioinformatics*, **9**, 78.

Jaeger, S., Gaudan, S., Leser, U., and Rebholz-Schuhmann, D. (2008) Integrating protein-protein interactions and text mining for protein function prediction. *BMC Bioinformatics*, **9**(8), S2.

Jelier, R., Jenster, G., Dorssers, L.C., *et al.* (2007) Text-derived concept profiles support assessment of DNA microarray data for acute myeloid leukemia and for androgen receptor stimulation. *BMC Bioinformatics*, **8**, 14.

Jensen, L.J., Saric, J., and Bork, P. (2006) Literature mining for the biologist: from information retrieval to biological discovery. *Nat. Rev. Genet.*, **7**, 119–29.

Jenssen, T.K., Laegreid, A., Komorowski, J., and Hovig, E. (2001) A literature network of human genes for high-throughput analysis of gene expression. *Nature Genet.*, **28**(1), 21–8.

Jimeno-Yepes, A., Jimenez-Ruiz, E., Lee, V., *et al.* (2008) Assessment of disease named entity recognition on a corpus of annotated sentences. *BMC Bioinformatics*, **9**(3), S3.

Jimeno-Yepes, A., Berlanga-Llavori, R., and Rebholz-Schuhmann, D. (2009) Ontology refinement for improved information retrieval. *Inf Process Manag*, in press.

Johnson, H.L., Baumgartner, W.A., Krallinger, M., *et al.* (2007) Corpus refactoring: a feasibility study. *J Biomed Discov Collab*, **2**, 4.

Kano, Y., Baumgartner Jr, W.A., McCrohon, L., *et al.* (2009) U-Compare: share and compare text mining tools with UIMA. *Bioinformatics*, **25**(15), 1997–8.

Karamanis, N., Seal, R., Lewin, I., *et al.* (2008) Natural language processing in aid of FlyBase curators. *BMC Bioinformatics*, **9**, 193.

Kim, J.D., Ohta, T., Pyysalo, S., *et al.* (2009) Overview of BioNLP'09 shared task on event extraction, in Proceedings of the Workshop on BioNLP, Boulder, Colorado; Companion Volume: Shared Task on Event Extraction, 5 June 2009. Association for Computational Linguistics, Stroudsburg, PA, pp. 1–9.

Kim, J.J. (2006) Bidirectional incremental approach to efficient information extraction: applications to biomedicine. PhD dissertation. Department of Computer Science, KAIST, South Korea.

Kim, J.J. and Rebholz-Schuhmann, D. (2008) Categorization of services for seeking information in biomedical literature: a typology for improvement of practice. *Brief. Bioinformatics*, **9**(6), 452–65.

Kim, J.J., Zhang, Z., Park, J.C., and Ng, S.K. (2006) BioContrasts: extracting and exploiting protein-protein contrastive relations from biomedical literature. *Bioinformatics*, **22**(5), 597–605.

Kim, J.J., Pezik, P., and Rebholz-Schuhmann, D. (2008) MedEvi: retrieving textual evidence of relations between biomedical concepts from Medline. *Bioinformatics*, **24**(11), 1410–12.

Kirsch, H., Gaudan, S., and Rebholz-Schuhmann, D. (2006) Distributed modules for text annotation and IE applied to the biomedical domain. *Int J Med Inform*, **75**(6), 496–500.

Klein, D. and Manning, C.D. (2003) Accurate unlexicalized parsing, in Proceedings of the 41st Annual Meeting of the Association for Computational Linguistics, Sapporo

Convention Center, Sapporo, Japan, 7–12 July 2003. Association for Computational Linguistics, Stroudsburg, PA, pp. 423–30.

Klinger, R., Kolárik, C., Fluck, J., et al. (2008) Detection of IUPAC and IUPAC-like chemical names. Bioinformatics, 24(13), i268–76.

Knudsen, B. and Hein, J. (2003) Pfold: RNA secondary structure prediction using stochastic context-free grammars. Nucleic Acids Res., 31(13), 3423–8.

Korbel, J.O., Doerks, T., Jensen, L.J., et al. (2005) Systematic association of genes to phenotypes by genome and literature mining. PLoS Biology, 3(5), e134.

Krallinger, M., Leitner, F., Rodriguez-Penagos, C., and Valencia, A. (2008) Overview of the protein-protein interaction annotation extraction task of BioCreative II. Genome Biol., 9(Suppl 2), S4.

Krallinger, M., Izarzugaza, J. M. G., Rodriguez-Penagos, C., and Valencia, A. (2009) Extraction of human kinase mutations from literature, databases and genotyping studies. BMC Bioinformatics, 10(Suppl 8), S1.

Küffner, R., Fundel, K., and Zimmer, R. (2005) Expert knowledge without the expert: integrated analysis of gene expression and literature to derive active functional contexts. Bioinformatics, 21(Suppl 2), ii259–67.

Kuhn, M., von Mering, C., Campillos, M., et al. (2008) STITCH: interaction networks of chemicals and proteins. Nucleic Acids Res., 36, D684–8.

Lage, K., Karlberg, E.O., Størling, Z.M. et al. (2007) A human phenome-interactome network of protein complexes implicated in genetic disorders. Nat. Biotechnol., 25, 309–16.

Lee, H., Yi, G.S., and Park, J.C. (2008) E3Miner: a text mining tool for ubiquitin-protein ligases. Nucleic Acids Res., 36(Suppl 2), W416–22.

Leitner, F. and Valencia, A. (2008) A text-mining perspective on the requirements for electronically annotated abstracts. FEBS Lett., 582(8), 1178–81.

Leroy, G. and Chen, H. (2005) Genescene: an ontology-enhanced integration of linguistic and co-occurrence based relations in biomedical texts. J Am Soc Inf Sci Technol, 56(5), 457–68.

Light, M., Qiu, X.Y., and Srinivasan, P. (2004) The language of bioscience: facts, speculations, and statements in between, in Proceedings of HLT-NAACL 2004 Workshop: BioLINK 2004, Linking Biological Literature, Ontologies and Databases, Boston, MA, 6 May 2004. Association for Computational Linguistics, Stroudsburg, PA, pp. 17–24.

Lisacek, F., Chichester, C., Kaplan, A., and Sandor, A. (2005) Discovering paradigm shift patterns in biomedical abstracts: application to neurodegenerative diseases, in Proceedings of the 1st International Symposium on Semantic Mining in Biomedicine SMBM 2005, Hinxton, Cambridgeshire, UK, 10–13 April 2005. EBI, Hinxton, pp. 41–50.

Lussier, Y., Borlawsky, T., Rappaport, D., et al. (2006) PhenoGO: assigning phenotypic context to gene ontology annotations with natural language processing. Pac Symp Biocomput, 2006, 64–75.

McNaught, J. and Black, W.J. (2006) Information extraction, in Text Mining for Biology and Biomedicine (eds B. Stapley and S. Ananiadou) Artech House Publishers Inc., Norwood, MA, pp. 143–78.

Mitkov, R. (2002) Anaphora Resolution, Longman, London.

Miyao, Y., Sagae, K., Saetre, R., et al. (2009) Evaluating contributions of natural language parsers to protein-protein interaction extraction. Bioinformatics, 25(3), 394–400.

Morgan, A.A., Lu, Z., Wang, X., *et al.* (2008) Overview of BioCreative II gene normalization. *Genome Biol.*, **9**(2), S3.

Mottaz, A., Yip, Y.L., Ruch, P., and Veuthey, A.L. (2008) Mapping proteins to disease terminologies: from UniProt to MeSH. *BMC Bioinformatics*, **9**(5), S3.

Mueller, H.M., Kenny, E.E., and Sternberg, P.W. (2004) Textpresso: an ontology-based information retrieval and extraction system for biological literature. *PLoS Biology*, **2**(11), e309.

Nagel, K., Jimeno-Yepes, A., and Rebholz-Schuhmann, D. (2009) Annotation of protein residues based on a literature analysis: cross-validation against UniProtKb. *BMC Bioinformatics*, **10**(8), S4.

Narayanaswamy, M., Ravikumar, K.E., and Vijay-Shanker, K. (2005) Beyond the clause: extraction of phosphorylation information from Medline abstracts. *Bioinformatics*, **21**(1), i319–27.

Netzel, R., Perez-Iratxeta, C., Bork, P., and Andrade, M.A. (2003) The way we write. *EMBO Rep.*, **4**(5), 446–51.

Neveol, A., Shooshan, S.E., Humphrey, S.M., *et al.* (2007) Multiple approaches to fine-grained indexing of the biomedical literature. *Pac Symp Biocomput*, **2007**, 292–303.

Ng, S.K. (2006) Integrating text mining with data mining, in *Text Mining for Biology and Biomedicine* (eds B. Stapley and S. Ananiadou), Artech House Publishers Inc., Norwood, MA, pp. 247–66.

Oda, K., Kim, J.D., Ohta, T., *et al.* (2008) New challenges for text mining: mapping between text and manually curated pathways. *BMC Bioinformatics*, **9**(3), S5.

Ozgür, A., Vu, T., Erkan, G., and Radev, D.R. (2008) Identifying gene-disease associations using centrality on a literature mined gene-interaction network. *Bioinformatics*, **24**(13), i277–85.

Park, J.C. and Kim, J.J. (2006) Named entity recognition, in *Text Mining for Biology and Biomedicine* (eds B. Stapley and S. Ananiadou), Artech House Publishers Inc., Norwood, MA, pp. 121–42.

Pavlopoulos, G.A., Pafilis, E., Kuhn, M., *et al.* (2009) OnTheFly: a tool for automated document-based text annotation, data linking and network generation. *Bioinformatics*, **25**(7), 977–8.

Pezik, P., Yepes, A.J., and Rebholz-Schuhmann, D. (2009) Information retrieval in biomedicine, in *Natural Language Processing for Knowledge Integration. Using Biomedical Terminological Resources for Information Retrieval* (ed. V. Prince), IGI Global Publishing, pp. 58–77.

Pyysalo, S., Airola, A., Heimonen, J., *et al.* (2008) Comparative analysis of five protein-protein interaction corpora. *BMC Bioinformatics*, **9**(3), S6.

Rajagopala, S.V., Goll, J., Gowda, N.D., *et al.* (2008) MPI-LIT: a literature-curated dataset of microbial binary protein-protein interactions. *Bioinformatics*, **24**(22), 2622–7.

Rebholz-Schuhmann, D., Marcel, S., Albert, S., *et al.* (2004) Automatic extraction of mutations from Medline and cross-validation with OMIM. *Nucleic Acids Res.*, **32**(1), 135–42.

Rebholz-Schuhmann, D., Kirsch, H., and Nenadic, G. (2006) IeXML: towards a framework for interoperability of text processing modules to improve annotation of semantic types in biomedical text. SIG BioLink, ISMB 2006, Fortaleza, Brazil.

Rebholz-Schuhmann, D., Cameron, G., Clark, D., *et al.* (2007a) SYMBiomatics: synergies in medical informatics and bioinformatics – exploring current scientific literature for emerging topics. *BMC Bioinformatics*, **8**(Suppl 1), S18.

Rebholz-Schuhmann, D., Kirsch, H., Arregui, M., *et al.* (2007b) EBIMed – text crunching to gather facts for proteins from Medline. *Bioinformatics*, **23**(2), e237–44.

Rebholz-Schuhmann, D., Kavaliauskas, S., and Pezik, P.. (2010) PaperMaker: validation of biomedical scientific publications. *Bioinformatics*, **26**(7), 982–4.

Rhodes, J., Boyer, S., Kreulen, J., *et al.* (2007) Mining patents using molecular similarity search. *Pac Symp Biocomput*, **12**, 304–15.

Rindflesch, T., Libbus, B., Hristovski, D. *et al.* (2003) Semantic relations asserting the etiology of genetic diseases. *AMIA Annu Symp Proc*, **2003**, 554–8.

Roberts, P.M. (2006) Mining literature for systems biology. *Brief. Bioinformatics*, **7**(4), 399–406.

Rodriguez-Penagos, C., Salgado, H., Martinez-Flores, I., and Collado-Vides, J. (2007) Automatic reconstruction of a bacterial regulatory network using Natural Language Processing. *BMC Bioinformatics*, **8**, 293.

Rzhetsky, A., Iossifov, I., Loh, J.M., and White, K.P. (2006) Microparadigms: chains of collective reasoning in publications about molecular interactions. *Proc. Natl. Acad. Sci. U.S.A.*, **103**(13), 4940–5.

Rzhetsky, A., Seringhaus, M., and Gerstein, M. (2008) Seeking a new biology through text mining. *Cell*, **134**(1), 9–13.

Sagae, K., Miyao, Y., and Tsujii, J. (2007) HPSG parsing with shallow dependency constraints, in Proceedings of the 45th Annual Meeting of the Association for Computational Linguistics, Prague, Czech Republic, 23–30 June 2007. Association for Computational Linguistics, Stroudsburg, PA, pp. 624–31.

Sanchez-Graillet, O. and Poesio, M. (2007) Negation of protein-protein interactions: analysis and extraction. *Bioinformatics*, **23**, i424–32.

Sanfilippo, A., Posse, C., Gopalan, B., *et al.* (2007) Combining hierarchical and associative gene ontology relations with textual evidence in estimating gene and gene product similarity. *IEEE Trans Nanobioscience*, **6**, 51–9.

Saric, J., Jensen, L.J., and Rojas, I. (2005) Large-scale extraction of gene regulation for model organisms in an ontological context. *In Silico Biol. (Gedrukt)*, **5**(1), 21–32.

Sasaki, Y., Montemagni, S., Pezik, P. *et al.* (2008) BioLexicon: a lexical resource for the biology domain, in Proceedings of the Third International Symposium on Semantic Mining in Biomedicine (SMBM 2008), Turku, Finland, 1–3 September 2008 (eds T. Salakoski, D. Rebholz-Schuhmann, S. Pyysalo), Turku Centre for Computer Science (TUCS), pp. 109–16.

Scherf, M., Epple, A., and Werner, T. (2005) The next generation of literature analysis: integration of genomic analysis into text mining. *Brief. Bioinformatics*, **6**(3), 287–97.

Schuemie, M., Chichester, C., Lisacek, F., *et al.* (2007) Assignment of protein function and discovery of novel nucleolar proteins based on automatic analysis of MEDLINE. *Proteomics*, **7**, 921–31.

Seki, K. and Mostafa, J. (2007) Discovering implicit associations between genes and hereditary diseases. *Pac Symp Biocomput*, **2007**, 316–27.

Shatkay, H., Pan, F., Rzhetsky, A., and Wilbur, W. (2008) Multi-dimensional classification of biomedical text: toward automated, practical provision of high-utility text to diverse users. *Bioinformatics*, **24**(18), 2086–93.

Smalheiser, N.R. and Swanson, D.R. (1996) Linking estrogen to Alzheimer's disease: an informatics approach. *Neurology*, **47**, 809–10.

Smith, B., Ashburner, M., Rosse, C., *et al.* (2007) The OBO Foundry: coordinated evolution of ontologies to support biomedical data integration. *Nat. Biotechnol.*, **25**(11), 1251–5.

Smith, T.F. and Waterman, M.S. (1981) Identification of common molecular subsequences. *J. Mol. Biol.*, **147**, 195–7.

Spasic, I., Ananiadou, S., McNaught, J., and Kumar, A. (2005) Text mining and ontologies in biomedicine: making sense of raw text. *Brief. Bioinformatics*, **6**(3), 239–51.

Spasic, I., Schober, D., Sansone, S.A., *et al.* (2008) Facilitating the development of controlled vocabularies for metabolomics technologies with text mining. *BMC Bioinformatics*, **9**, S5.

Swanson, D.R. (1986) Fish oil, Raynaud's syndrome, and undiscovered public knowledge. *Perspect. Biol. Med.*, **30**, 7–18.

Swanson, D.R. (1988) Migraine and magnesium: eleven neglected connections. *Perspect. Biol. Med.*, **31**, 526–57.

Theodosiou, T., Darzentas, N., Angelis, L., and Ouzounis, C.A. (2008) PuReD-MCL: a graph-based PubMed document clustering methodology. *Bioinformatics*, **24**(17), 1935–41.

Thomas, P.D., Mi, H., and Lewis, S. (2007) Ontology annotation: mapping genomic regions to biological function. *Curr Opin Chem Biol*, **11**, 4–11.

Trieschnigg, D., Pezik, P., Lee, V., *et al.* (2009) MeSH Up: effective MeSH text classification for improved document retrieval. *Bioinformatics*, **25**(11), 1412–18.

Tsujii, J. and Ananiadou, S. (2005) Thesaurus or logical ontology, which one do we need for text mining? *Language Res Eval*, **39**, 77–90.

Tsuruoka, Y., Tsujii, J., and Ananiadou, S. (2008) FACTA: a text search engine for finding associated biomedical concepts. *Bioinformatics*, **24**(21), 2559–60.

Tuncbag, N., Kar, G., Keskin, O., *et al.* (2009) A survey of available tools and web servers for analysis of protein-protein interactions and interfaces. *Brief. Bioinformatics*, **10**(3), 217–32.

Waagmeester, A., Pezik, P., Coort, S., *et al.* (2009) Pathway enrichment based on text mining and its validation on carotenoid and vitamin A metabolism. *OMICS*, **13**(5), 367–79.

Weeber, M., Vos, R., Klein, H., *et al.* (2003) Generating hypotheses by discovering implicit associations in the literature: a case report of a search for new potential therapeutic uses for thalidomide. *J Am Med Inform Assoc*, **10**, 252–9.

Wermter, J., Tomanek, K., and Hahn, U. (2009) High-performance gene name normalization with GeNo. *Bioinformatics*, **25**(6), 815–21.

Wilbur, W.J., Hazard, G.F., Divita, G., *et al.* (1999) Analysis of biomedical text for chemical names: a comparison of three methods. *Proc AMIA Symp*, **1999**, 176–80.

Wilbur, W.J., Rzhetsky, A., and Shatkay, H. (2006) New directions in biomedical text annotation: definitions, guidelines and corpus construction. *BMC Bioinformatics*, **7**, 356.

Winnenburg, R., Wächter, T., Plake, C., *et al.* (2008) Facts from text: can text mining help to scale-up high-quality manual curation of gene products with ontologies? *Brief. Bioinformatics*, **9**(6), 466–78.

Winnenburg, R., Plake, C., and Schroeder, M. (2009) Improved mutation tagging with gene identifiers applied to membrane protein stability prediction. *BMC Bioinformatics*, **10**(Suppl 8), S3.

Yeh, A., Hirschman, L., and Morgan, A. (2002) Background and overview for KDD Cup 2002 task 1: information extraction from biomedical articles. *SIGKDD Explor*, **4**(2), 87–9.

Yeniterzi, S. and Sezerman, U. (2009) EnzyMiner: Automatic identification of protein level mutations and their impact on target enzymes from PubMed abstracts. *BMC Bioinformatics*, **10**(Suppl 8), S2.

Yu, H. and Kaufman, D. (2007) A cognitive evaluation of four online search engines for answering definitional questions posed by physicians. *Pac Symp Biocomput*, **2007**, 328–39.

Yu, S., Vooren, S.V., Tranchevent, L.C., *et al.* (2008) Comparison of vocabularies, representations and ranking algorithms for gene prioritization by text mining. *Bioinformatics*, **24**(16), i119–25.

Yuan, X., Hu, Z.Z., Wu, H.T., *et al.* (2006) An online literature mining tool for protein phosphorylation. *Bioinformatics* **22**(13), 1668–9.

Zhou, W., Smalheiser, N.R., and Yu, C. (2006) A tutorial on information retrieval: basic terms and concepts. *J Biomed Discov Collab*, **1**, 2.

Zweigenbaum, P., Demner-Fushman, D., Yu, H., *et al.* (2007) Frontiers of biomedical textmining: current progress. *Brief. Bioinformatics*, **8**(5), 358–75.

PART II

APPLICATIONS

Section 3

Gene and Protein Information

8

Fundamentals of gene ontology functional annotation

Varsha K. Khodiyar, Emily C. Dimmer, Rachael
P. Huntley, and Ruth C. Lovering

8.1 Introduction

Over the past decade, technological advances such as whole genome sequencing, microarray technology and high-throughput proteomics techniques have revolutionized our approach to life science research. This progress in technology has resulted in the generation of large amounts of data at a relatively low cost. Consequently, although the study of specific pathways or individual molecules remains a major approach to understanding the intricate molecular and cellular details associated with biological processes and disease, the bottleneck in biological sciences has shifted from data generation to data analysis.

Creating ways to organize, archive, and interpret this flood of data has become a major research activity in its own right. Computer scientists are essential in making raw biological data comprehensible and accessible to both humans and computers, and one method has been to provide the raw data in a standardized format, made available through a database. However, there are important, but often-overlooked steps, between the initial design of a database and the point at which the end user can access the data: the addition, standardization, and maintenance

Knowledge-Based Bioinformatics: From Analysis to Interpretation   Edited by Gil Alterovitz and Marco Ramoni
© 2010 John Wiley & Sons, Ltd

of the data. This is the process of biocuration, which involves the review and standardization of the raw data as well as adding valuable additional information.

Biocurators have been described as the 'museum cataloguers of the Internet age' (Bourne and McEntyre, 2006). However, in addition to displaying and preserving raw biological data in a user-friendly manner, biocurators are also required to interpret complex scientific literature and extract the relevant data in an efficient and consistent manner. Thus, there are two types of curation essential to the life sciences: 'data submission curation' and 'value-added curation' (Thornton, 2009).

8.1.1   Data submission curation

Researchers submitting data to a database will often use a variety of input methods, for example by email, as a spreadsheet, a text document, or via a database's own online submission form. Data submission curators ensure the submission is formatted in a consistent manner and will often request further information from the submitters as required. Once the quality assurance checks are complete, the data is displayed and may be credited to the original submitters within the entry. The sequence databases GenBank (www.ncbi.nlm.nih.gov/Genbank/index.html; Benson et al., 2008) and EMBL-Bank (www.ebi.ac.uk/embl/; Sterk et al., 2007) are two examples of databases that primarily consist of raw biological data that is submitted directly, and 'owned' by researchers.

8.1.2   Value-added curation

Value-added curation can be defined as organizing or interpreting biological data in order to add an extra layer of meaning. For example, when a scientist submits a nucleotide sequence to EMBL-Bank, they are encouraged to include additional information on the entry, including gene or protein names, sequence features, taxonomic and citation information. UniProt Knowledgebase (UniProtKB) biocurators will further curate the corresponding translation, annotating additional information including tissue specificity and protein domain information, as well as information from the literature describing the known function, associated disease, post-translational modifications, cofactors, and regulators retrieved from published scientific literature or by cross-referencing specialist external resources. There are numerous value-added biocuration groups around the world such as UniProtKB (UniProt Consortium, 2008), National Center for Biotechnology Information (NCBI; Sayers et al., 2009), Kyoto Encyclopedia of Genes and Genomes (KEGG; Kanehisa et al., 2008), Reactome (Matthews et al., 2009), Human Unidentified Gene-Encoded Large Proteins (HUGE; Kikuno et al., 2004), and Ensembl (Hubbard et al., 2009).

UniProtKB curators also annotate using biological ontologies, including UniProtKB keywords (www.expasy.ch/cgi-bin/keywlist.pl) and Gene Ontology (GO) terms (www.geneontology.org). The Gene Ontology Consortium (GOC) provides a hierarchical controlled vocabulary of terms to describe the accumulated functional knowledge of a gene product, with respect to its molecular functions, the biological processes it is involved in and its sub-cellular location. Many groups, in addition to UniProtKB, use GO to provide value-added curation of gene products[1], for species as varied as *Escherichia coli* and *Arabidopsis thaliana* to *Mus musculus* and *Homo sapiens* (Reference Genome Group of the Gene Ontology Consortium, 2009). This chapter reviews the GO resource, the annotation process, its usage, and its limitations.

8.2   Gene Ontology (GO)

High-throughput methodologies can provide a wide range of information, including the detailed characterization of specific developmental or disease states, the molecular composition of entire tissues, cells or organelles, or overviews of protein interaction networks. These methodologies are now providing researchers with an increasingly detailed overview of complex molecular interactions within a variety of cell or tissue types. In addition, these high-throughput methodologies can be used for the initial characterization of newly sequenced genes. An important aspect to this approach is the integration of results from high-throughput investigations with data accumulated through the intensive study of single genes or pathways; ensuring that data from different experimental approaches can be used to inform other research projects.

For the past 10 years the GOC has been developing GO terms to describe the functional attributes of proteins from all species in a consistent and computer-readable manner. GO enables functional information, derived from a wide range of sources, to be included in individual gene and protein records in biological sequence databases and within high-throughput analysis software. In this way, the GOC enables sequences to be classified and grouped together according to their functional properties, and consequently improves data integration, bridging the gap between data collation and data analysis (Gene Ontology Consortium, 2001; Ashburner *et al.*, 2000; Dimmer *et al.*, 2008; Lomax, 2005).

8.2.1   Gene Ontology and the annotation of the human proteome

In 1998, the GOC was founded through collaborations between three model organism databases: FlyBase, Saccharomyces Genome Database (SGD), and Mouse Genome Informatics (MGI) (Ashburner *et al.*, 2000). Over the last 10 years, the GOC has expanded to include a wide range of curation groups

---

[1] GO annotations can be used to describe all functional gene products including RNAs. However, since the majority of functional, and therefore annotatable, gene products are proteins, the remainder of this chapter refers to GO annotations in general being applied to proteins. Users should also note that GO annotations are displayed in both gene databases such as NCBI Entrez Gene (www.ncbi.nlm.nih.gov/sites/entrez?db=gene), and in protein databases such as UniProtKB (www.uniprot.org/).

annotating GO to many different species (Table 8.1). External groups have also begun to supplement their data sets with GO terms, including AstraZeneca and Celera Genomics.

As shown in Table 8.1, the majority of GOC members are model organism databases. However the annotation of GO terms to human proteins is being achieved in a more distributed manner, with a variety of groups having contributed GO annotations over the years. These groups have previously included Proteome Inc., the Human Genome Database (GDB; Letovsky et al., 1998) and the HUGO Gene Nomenclature Committee (HGNC; Bruford et al., 2008). Currently, a large proportion of human annotations (both manual and electronic) are being contributed by the Gene Ontology Annotation group (GOA-UniProtKB) at the European Bioinformatics Institute (EBI) (Barrell et al., 2009) and the British Heart Foundation-funded Cardiovascular Initiative at University College London (Lovering et al., 2008a) with additional annotations supplied by LIFEdb (Mehrle et al., 2006), Reactome (Matthews et al., 2009), IntAct (Kerrien et al., 2007), and the Human Proteome Atlas (Berglund et al., 2008).

8.2.2   Gene Ontology Consortium data sets

GO terms are provided as three separate vocabularies (ontologies) that provide a descriptive framework for the normal 'molecular functions' of a protein, the 'biological processes' a protein is involved in and the 'sub-cellular locations' ('cellular components') in which the protein is located. For example, annotations for the cytokine interleukin 6 (*IL6*) include the *Molecular Function* term: 'interleukin-6 receptor binding' (GO:0005138), the *Biological Process* term: 'defense response to virus' (GO:0051607), and the *Cellular Component* term: 'extracellular space' (GO:0005576); whereas the annotations for the telomeric binding factor *TERF1* include the *Molecular Function* term: 'double-stranded telomeric DNA binding' (GO:0003691), the *Biological Process* term: 'negative regulation of telomerase activity' (GO:0051974), and the *Cellular Component* term: 'chromosome, telomeric region' (GO:0000781). Depending on the amount of published data available, gene and/or protein identifiers can be annotated with multiple GO terms from any, or all, of the three gene ontologies (Figure 8.1).

The terms in GO are structured as directed acyclic graphs, where each term can have multiple relationships to broader 'parent' and more specific 'child' terms (Figure 8.2). This hierarchical structure produces a representation of biology that allows a greater amount of flexibility in data analysis than would be afforded by a format based on a simple list of terms. Users can exploit this structure to see either a broad overview of the general functional attributes presented by a set of data, or focus in on specific sections of the ontology to investigate in greater detail (discussed in Section 8.6.3).

8.2.3   GO annotation methods

Annotations can be produced either by a curator reading published scientific papers and manually creating each association or by a combination

Table 8.1 Curation groups annotating to GO.

Curation Group	Specific interest	URL
Full GOC Members		
Berkeley Bioinformatics and Ontology Project (BBOP)	*Drosophila* informatics; development of GO database and software; Sequence Ontology development; National Center for Biomedical Ontology biomedical informatics research	www.berkeleybop.org/
British Heart Foundation – University College London (BHF-UCL)	Human cardiovascular system GO annotation	www.cardiovasculargene ontology.com/
dictyBase	*Dictyostelium discoideum* model organism database	http://dictybase.org/
EcoliWiki	*Escherichia coli* model organism database	http://ecoliwiki.net/
FlyBase	*Drosophila melanogaster* model organism database	http://flybase.bio.indiana.edu/
GeneDB	Model organism database for fission yeast *Schizosaccharomyces pombe* and several protozoan parasites, including *Plasmodium falciparum, Leishmania major* and *Trypanosoma brucei*	www.genedb.org/
Gene Ontology Annotation @ EBI (GOA)	GO annotation for the multispecies UniProt Knowledgebase with a manual annotation focus on the human proteome	www.ebi.ac.uk/GOA/
Gramene	Model organism database for grains including rice (*Oryza*)	www.gramene.org/

(continued overleaf)

Table 8.1 (continued).

Curation Group	Specific interest	URL
Mouse Genome Database (MGD) and Gene Expression Database (GXD)	*Mus musculus* model organism database	www.informatics.jax.org/
Rat Genome Database (RGD)	*Rattus norvegicus* model organism database	http://rgd.mcw.edu/
Reactome	Biological processes and pathways	http://reactome.org/
Saccharomyces Genome Database (SGD)	*Saccharomyces cerevisiae* model organism database; maintenance and public access of GO database and Web interfaces	www.yeastgenome.org/
The Arabidopsis Information Resource (TAIR)	*Arabidopsis thaliana* model organism database	www.arabidopsis.org/
Institute for Genome Sciences (IGS)	Data and tools for genomic research in a variety of model systems.	www.igs.umaryland.edu/
The J. Craig Venter Institute (JCVI)	Databases on several bacterial species	www.jcvi.org/
WormBase	*Caenorhabditis elegans* model organism database	www.WormBase.org/
Zebrafish Information Network (ZFIN)	*Danio rerio* model organism database	http://zfin.org/
GO Associates		
AgBase	Agricultural plant and animals	www.agbase.msstate.edu/
CGD: The Candida Genome Database	*Candida albicans* model organism database	www.candidagenome.org/
Muscle TRAIT	Human skeletal muscle	http://muscle.cribi.unipd.it/
Plant-Associated Microbe Gene Ontology (PAMGO) consortium	Plant-associated microbes	http://pamgo.vbi.vt.edu/

Full GOC Members are expected to show a significant and ongoing commitment to the utilization and further development of the Gene Ontology. GO Associates are groups who make a notable contribution to the GO project. This information was correct at the time of writing (March 2010). For the latest information see http://www.geneontology.org/GO.consortiumlist.shtml.

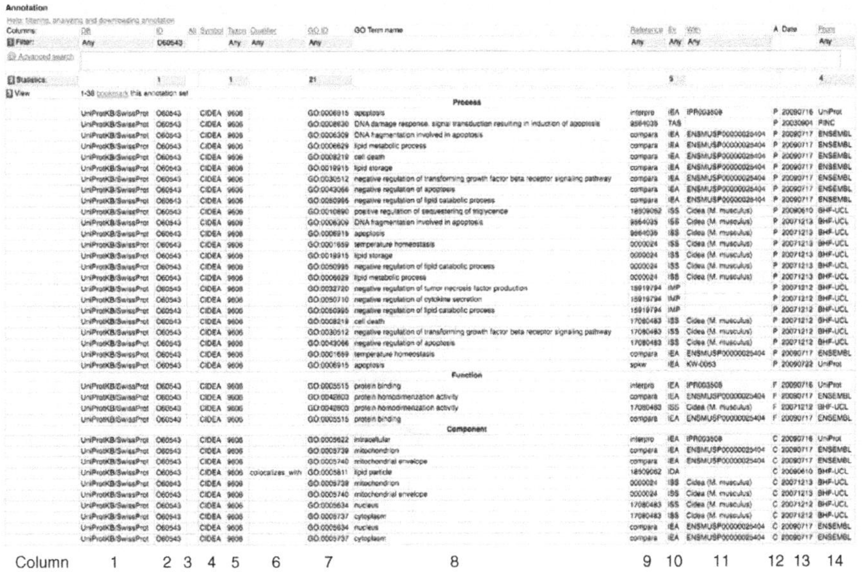

Figure 8.1   Annotations associated with the protein CIDEA. This is a screenshot from the QuickGO gene ontology browser (www.ebi.ac.uk/QuickGO/) of the GO annotations associated with the human protein CIDEA. This protein has multiple annotations associated to it in all three ontologies. Column key: 1, Type of protein ID; 2, Protein ID; 3, Splice variant ID; 4, Gene symbol; 5, Taxon ID; 6, Annotation qualifier; 7, GO term ID; 8, GO term name; 9, Evidence source reference; 10, Evidence code; 11, Homologous sequence for transferred annotations; 12, Ontology code; 13, Annotation date; 14, Acronym of the group that made the annotation.

of computational techniques to produce electronic annotation sets (Camon *et al.*, 2003). Both methods have their own advantages and disadvantages, and are carefully monitored and revised to ensure that conservative, high-quality annotations are created.

The GOC has established standardized annotation procedures, and full details can be obtained from the annotation guide on the GOC website (www .geneontology.org/GO.annotation.shtml). Manual GO annotations provide comprehensive, accurate, and information-rich summaries of the functional knowledge for proteins. These detailed annotations are created by trained curators, through the evaluation of experimental data available in published scientific papers (Barrell *et al.*, 2009; Hill *et al.*, 2008; www.ebi.ac .uk/GOA/annotationexample.html). For example, Nofer *et al.* demonstrated that a mutation in the *ABCA1* gene was associated with Tangier disease and that fibroblasts derived from this patient had reduced APOA1-induced intracellular signaling (Nofer *et al.*, 2006). Tangier disease is characterized by

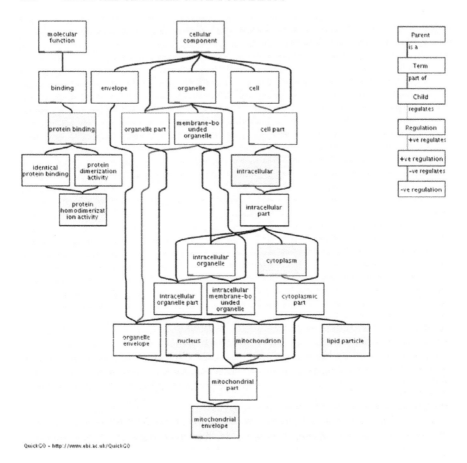

Figure 8.2   Gene Ontology is structured as a directed acyclic graph. This is a screenshot from QuickGO showing the Molecular Function *and* Cellular Component *terms associated with the human CIDEA protein as a chart. This graph view illustrates the directed acyclic nature of the GO. For example, the* Molecular Function *term 'protein homodimerization activity' (GO:0042803) is a type of 'identical protein binding' (GO:0042802). Online the different colors of the linking lines help to visualise the specific nature of the relationship between parent and child terms. For example 'organelle envelope' (GO:0031967) is part of a 'membrane-bounded organelle' (GO:0043227). Whereas the* Cellular Component *term 'nucleus' (GO:0005634) is a type of 'intracellular membrane-bounded organelle' (GO:0043231). Likewise the* Cellular Component *term 'mitochondrion' (GO:0005739) also is a type of 'intracellular membrane-bounded organelle' (GO:0043231). Thus the GO is able to accurately represent biological information in a qualitative manner.*

severe high-density lipoprotein deficiency, leading to cholesterol deposition in tissue macrophages and premature atherosclerosis. The experimental data presented in this paper led to the annotation of the ABCA1 protein with three *Biological Process* GO terms 'G-protein coupled receptor protein signaling pathway' (GO:0007186), 'cholesterol efflux' (GO:0033344), and 'Cdc42 protein signal transduction' (GO:0032488), and three *Molecular Function* GO terms 'small GTPase binding' (GO:0031267), 'apolipoprotein A-I receptor activity' (GO:0034188) and 'apolipoprotein A-I binding' (GO:0034191). For well-studied proteins like TGFB1, the manual annotation process could take many days (www.ebi.ac.uk/QuickGO/GProtein?ac=P01137); in contrast, the annotation of a more recently described protein, like CIDEA, may only take a few hours (Figure 8.1; www.ebi.ac.uk/QuickGO/GProtein?ac=O60543).

8.2.3.1 Evidence codes

All GO annotations refer to the source of evidence that supports the creation of the annotation (such as a NCBI PubMed identifier or a reference to a computational method), as well as an 'evidence code', which indicates the category of evidence that was identified in the associated reference. Each GO annotation also includes the date the annotation was made and an acronym for the group submitting the annotation data (column 14, Figure 8.1).

At the time of writing, there are 18 evidence codes (Table 8.2), which can be split into three broad categories: evidence codes which indicate the annotation is based on experimental results (such as an enzyme assay); from non-experimental statements provided by an author or inferred by a curator (for instance inferring a nuclear localization for an *in vitro*-characterized transcription factor or transferring annotations based on sequence similarity); and finally using evidence from computational predictions. Users can utilize evidence codes to filter annotation sets so that their data analysis includes only experimentally evidenced annotations. However, the usefulness of filtering using evidence codes will, of course, be dependent on the genome of interest. For well-annotated genomes, such as yeast, evidence code-based filtering might be appropriate, whereas for other genomes such as pig, bovine or even human, users often need to rely on the complete annotation set to ensure that their sequences of interest are provided with sufficient annotation data.

8.2.3.2 Annotation qualifiers

Occasionally, manual annotations include 'qualifiers,' to provide an additional layer of information about the relationship between a protein and its associated GO term. Three qualifiers are currently available:

(1) 'colocalizes_with' (which indicates a transient or peripheral association of a protein with an organelle or protein complex (column 6, Figure 8.1));

(2) 'contributes_to' (which implies that a protein facilitates, but does not directly carry out, the function of the protein complex of which it is a subunit);

(3) 'NOT' (used when there is conflicting published data, or when a protein is found to not have a particular activity, location or process involvement, which is in contrast to previous assumptions).

Although qualifiers are rarely used, these statements have an important impact in changing the meaning of the associated annotation. Most importantly, the 'NOT' qualifier leads to an opposite interpretation of an annotation, and users of large data sets are advised to ensure 'NOT' annotations are appropriately applied (or excluded) by the large-scale functional analysis tool of their choice.

Table 8.2   Gene Ontology evidence codes.

Evidence code	Description
Based on published experimental evidence	
EXP	inferred from EXPeriment
IDA	Inferred from Direct Assay
IEP	Inferred from Expression Pattern
IGI	Inferred from Genetic Interaction
IMP	Inferred from Mutant Phenotype
IPI	Inferred from Physical Interaction
Based on computational analysis evidence	
IEA[a]	Inferred from Electronic Annotation
IGC	Inferred from Genomic Context
ISA	Inferred from Sequence Alignment
ISO	Inferred from Sequence Orthology
ISM	Inferred from Sequence Model
ISS	Inferred from Sequence or structural Similarity
RCA	inferred from Reviewed Computational Analysis
Not supported by experimental evidence within the reference used	
IC	Inferred by Curator
TAS[b]	Traceable Author Statement
NAS[c]	Non-traceable Author Statement
ND	No biological Data available
NR[d]	Not Recorded

[a]This code denotes an electronic annotation based on an automated computational orthology statement or an automated computational prediction of function. It denotes an automated electronic annotation and cannot be applied to manual annotations.
[b]The reference provides information about the source of the evidence supporting the statement used for the annotation.
[c]The reference does not provide information about the source of the evidence supporting the statement used for the annotation.
[d]This code is now obsolete.

8.2.4   Different approaches to manual annotation

A variety of approaches have been taken to provide good quality manual annotations to the human proteome. After the initial fast-track approach based on reading abstracts, a more in-depth approach has developed in which the GO curator reads the method and results sections of each paper in detail. The curator will confirm the species of the investigated gene or protein (often in referenced papers) and identify the experimental data supporting the statements made in the abstract, introduction or discussion.

For proteins with a large number of associated papers, the curator will select a recent review and use that as a basis for identifying papers with experimental data for annotation. However, due to the large volume of literature associated with mammalian proteins, curators often need to focus on providing the broadest range of annotations from the most recent papers. Interpretation of experimental data and the subsequent selection of appropriate GO terms to associate with a protein record requires expertise. In addition, the majority of curators are reading papers from a wide range of biological fields, covering the detailed information of biochemical reactions and signal transduction to more complex developmental processes and disease phenotypes. It is therefore challenging to consistently evaluate results from different biological methods and associate all of the appropriate GO terms to each protein record. Annotation consistency is an important focus of the GOC, and a number of methods have been applied, such as the Gene Ontology Normal Usage Tracking System (GONUTS, http://gowiki.tamu.edu/wiki/index.php/Main_Page), which has enabled GO curation projects to compare GO terms associated with one protein across multiple species. For example, during the annotation of proteins involved in reverse cholesterol transport, over 50 annotations were identified as missing from 14 protein records through the use of GONUTS (Table 8.3).

8.2.5   Ontology development

At the time of writing, over 30 000 GO terms exist across the three *Molecular Function, Biological Process* and *Cellular Component* ontologies (March 2010). Although the three ontologies are in a constant state of expansion and revision, GO term identifiers remain stable and revisions can be tracked. And despite this wealth of existing terms, requesting new terms is an integral aspect of both manual and electronic GO curation. For example, during the annotation of human apolipoproteins, a series of GO terms needed to be requested to enable a more detailed description of the extracellular location of these proteins (e.g., 'high-density lipoprotein particle' (GO:0034364) was requested, rather than use the pre-existing, more general term 'plasma lipoprotein particle' (GO:0034358)), their function as receptor ligands (e.g., 'very-low-density lipoprotein receptor binding' (GO:0070326) was created, rather than apply the term 'receptor binding' (GO:0005102)), and their roles within the cardiovascular system (e.g., 'chylomicron remnant clearance' (GO:0034382) was requested, rather than apply 'macromolecular complex disassembly' (GO:0032984)).

Table 8.3 GONUTS comparison of six proteins involved in reverse cholesterol transport.

Term	HUMAN: ABCA1	HUMAN: LIPC	HUMAN: LCAT	HUMAN: CETP	HUMAN: SCSCARB1	HUMAN: ABCG1
triglyceride catabolic process		IDA				
intracellular cholesterol transport	IMP					
positive regulation of cholesterol transport						IMP
cholesterol efflux	IDA				ISS	IDA
phospholipid efflux	IDA					IMP
response to lipid						IDA
very-low-density lipoprotein particle remodeling		IDA[a]	IDA[a]	IDA		
intermediate-density lipoprotein particle remodeling		TAS				
low-density lipoprotein particle r...		IMP[a]		IDA		ISS[a]
high-density lipoprotein particle remodeling		IMP	IDA[a]	IMP	ISS[a]	ISS[a]
high-density lipoprotein particle assembly	IMP[a]					
chylomicron remnant clearance		TAS[a]				
high-density lipoprotein particle clearance					IDA	
cholesterol esterification			IDA			
phosphatidylcholine catabolic process		TAS				

cholesterol homeostasis	IDA	IMP[a]	IDA	IMP	ISS[a]	IDA
reverse cholesterol transport	IMP	IC[a]	IDA	IC	IEP	ISS
positive regulation of cholesterol biosynth...						ISS
phosphatidylcholine metabolic process				IDA		
acylglycerol homeostasis				IDA		
phospholipid homeostasis	IMP			IDA		IMP
triglyceride homeostasis		IMP[a]		IDA[a]	ISS[a]	
					ISS[a]	
cholesterol import				IMP		

The type of evidence supporting the associated GO term is indicated by GO evidence codes (IC, IDA, IEP, IMP, ISS, TAS).
[a] Indicates GO terms added to protein records following the annotation of the process *reverse cholesterol transport*.

New GO terms are requested through the GO editorial team using the online project software SourceForge (http://sourceforge.net/). SourceForge enables all GO users to request changes or additions to GO, and to contribute to discussions. However, there is no automated system available to associate the new GO terms to relevant proteins. Therefore, after a new GO term has been created, the curator or bioinformatician needs to ensure it is consistently associated with the relevant protein records.

8.3   Comparative genomics and electronic protein annotation

Manual annotation sourced directly from published experimental data is often only available for proteins from a limited number of model organisms, and often investigations on different sets of orthologs apply non-overlapping functional assays. Additionally, the exponential rise in the numbers of genomes being sequenced means it is not possible for manual annotation methods alone to provide sufficient annotation coverage. This situation is exacerbated by the expense of funding manual curators and means that there is often an inconsistent set of experimentally evidenced annotations applied to genes from different model organism species. Therefore, methods that project experimentally evidenced annotations to orthologous or homologous genes have been developed. Transferring annotations between proteins can be performed either manually or electronically. However, electronic methods of annotation are essential in enabling the annotation of non-model organism species, where little direct experimental evidence is available, and where it is unlikely that a curation effort would be funded. In all cases it is important that the user is able to identify those annotations which have been transferred in this manner, that they understand the methodology used to create an annotation set, and finally are able to trace back to the original annotation which was applied so that a full understanding of the origins of such annotation predictions can be obtained.

8.3.1   Manual methods of transferring functional annotation

Manual transfers of GO annotations are often carried out by curators to ensure that complete and consistent annotation sets are provided for genes from closely related species. In such circumstances, curators are responsible for reviewing the evidence to ensure that a conservative and accurate transfer of function is carried out. Such annotations are often created by a curator after all the published experimental data has been captured, using the manual evidence code 'Inferred by Sequence or structural Similarity' (ISS) for the projected annotations, to highlight where such manually verified transfers of function have been carried out.

As well as simple projections by sequence similarity, annotation groups are increasingly involved in larger, manual annotation transfers using comparative genomics. Such transfer methods often infer functions of an ancestral gene, based on the annotations that exist for modern descendants, and propagate such ancestral annotations to other descendant genes by inheritance relationships. This latter method allows a more systematic and conservative transfer of functional annotations. This approach has been taken by the Reference Genomes project, which has applied a phylogenetic tree-based approach to identify 'ortholog clusters' to aid in providing comprehensive GO annotation for 12 key model organism genomes in the GOC (Reference Genome Group of the Gene Ontology Consortium, 2009).

8.3.2   Electronic methods of transferring functional annotation

Manually projecting annotations does help produce a consistent data set for curators needing to provide a comprehensive set of annotations for specific model organism species. However, such an approach is inadequate when attempting to provide sufficient annotation coverage for the exponentially increasing number of sequences now available from economically important, non-model organism species. Electronic methods, which transfer annotations across species, provide essential annotations for non-model organism species. Such methods also aid in the generation of a consistent annotation set for manually curated model organism species.

8.3.2.1   Electronic annotation transfer based on homology

Computational annotation techniques often use homology to project annotations and are used by software engineers to cope with the ever-increasing annotation backlog. The Brassica ASTRA team has done exactly this, incorporating the TAIR GO annotation from *Arabidopsis* sequences that had highly similar BLASTx matches to *Brassica* sequences. Using this method, 43% of *Brassica* sequences were supplied with electronic GO annotation (Love *et al.*, 2005).

A number of software teams have also released tools to provide this type of service, which often uses BLAST algorithms to find homologs to GO-annotated sequences. Matching sequences are scored (often using the BLAST E-value), and GO terms for the matched sequences are retrieved and applied to the input sequence. Tools allow submission of either single or multiple cDNA or protein sequences (GoFigure (Khan *et al.*, 2003), GOtcha (Martin *et al.*, 2004), GOblet (Groth *et al.*, 2004), and OntoBlast (Zehetner, 2003)), or accept stretches of genomic sequence (Blast2GO (Conesa *et al.*, 2005)). When annotations are transferred automatically between homologs the evidence code 'Inferred from Electronic Annotation' (IEA) is used.

8.3.2.2   Electronic annotation transfer based on phylogeny

It has been recognized that large-scale functional transfer from homologous sequences can cause high levels of error (Artamonova *et al.*, 2007), so electronic annotation methods which transfer annotations based on phylogenetic relationships are increasingly favored. The Ensembl group supply GO annotations based on their Compara phylogenetic method (Flicek *et al.*, 2008) by transferring experimentally evidenced annotations between over 40 different species, resulting in the creation of 455,000 annotations (GOA UniProt release 81.0). Tools such as SIFTER (Engelhardt *et al.*, 2005) and GOAnno (Chalmel *et al.*, 2005) also offer users the ability to transfer annotations between protein sequences based on phylogenetic relationships.

Whether the transfer of annotations is manual or electronic, care does need to be taken as, of course, differences do exist between orthologs of very closely related species. Transferred annotations must be kept up to date and the accuracy of the GO term should be continually checked to ensure unsuitable terms are not transferred between species; for example, biological process terms which describe lactation should not be transferred from mammalian proteins to avian orthologs.

8.3.3   Electronic annotation methods

Almost 99% of the total number of GO annotations available from the GOC are derived from automated computational prediction pipelines. These electronic annotations are very important for the growing number of proteins from non-model organism species, which often have had no experimental characterization, but are also invaluable for well-studied genomes, providing consistent high level annotations across all species. One of the widely used methods of generating automated electronic annotations is to firstly manually link (or map) GO terms to corresponding concepts in the controlled vocabularies used by the external databases. The resulting files provide translation tables or 'mappings' between these two vocabularies, which are then used to generate GO annotations for those gene/protein identifiers which have previously been curated with the external controlled vocabulary. UniProtKB keywords (www.uniprot.org/keywords/), Enzyme Commission (EC) numbers (NC-IUBMB and Webb, 1992), and Inter-Pro domains (Hunter *et al.*, 2009) are three external, controlled vocabularies used for mapping to GO, which generate large numbers of annotations for many species. The second widely used electronic annotation method is the automatic transfer of manual annotations to orthologs in closely related species (see Section 8.3.2.2).

Each GO annotation from an automated electronic method such as mapping or Compara (Section 8.3.2.2) is given the evidence code 'Inferred from Electronic Annotation' (IEA) (Table 8.2). Each IEA annotation is referenced with a 'GO_REF' identifier from the GO Reference Collection (www.geneontology.org/cgi-bin/references.cgi). The GO reference collection is a publicly available set of abstracts that describe specific methods or sources of data that have resulted in the generation of annotations where a specific literature reference is unavailable.

With a conservative usage, electronic annotation predictions can be highly accurate; the UniProtKB keyword, EC number, and InterPro to GO mappings

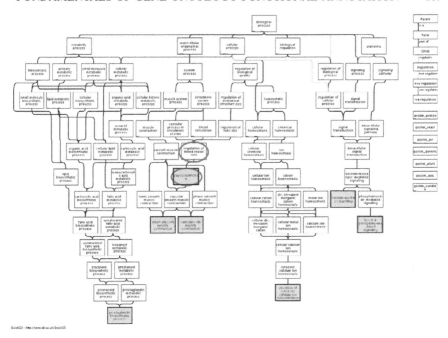

Figure 8.3   Manual annotations tend to be more granular than electronic annotations. In this screenshot, the graph displays the non-regulation Biological Process *terms for the human protein EDN2. The circled terms are electronic annotations and the boxed terms are manual annotations. The term 'vasoconstriction' (GO:0042310) has been associated both manually and electronically. The manually annotated terms 'artery smooth muscle contraction' (GO:0014824) and 'vein smooth muscle contraction' (GO:0014826) are descendants of the two electronically associated terms, illustrating the point that manual annotations tend to be more granular than electronic annotations. The other manual annotations are unrelated to the electronic annotations, thus in this case the manual annotations are also giving a broader coverage of the* Biological Process *ontology.*

have been found to predict an appropriate GO term 91–100% of the time (Camon *et al.*, 2005). However, the GO terms predicted by computational methods are, in general, less specific (or less granular) than those chosen manually (Figure 8.3), and are highly reliant on the quality and breadth of the manual annotation work carried out in external databases.

8.4   Community annotation

Biocurators are highly trained in the interpretation of the literature to create annotations using a controlled vocabulary. However, biocurators cannot expect to understand each gene and protein as thoroughly as a lab-based scientist who has made a career out of researching a particular protein or understanding a particular investigative technique. Thus getting the input of lab-based scientists continues to be an ongoing concern for the majority of curation groups. As such, there are a number of recent outreach methods that curation groups have begun to utilize with varying degrees of success, to bring about community involvement in the annotation process.

8.4.1   Feedback forms

Most curation groups have some kind of feedback form on their websites. Although this method is not widely used by researchers to comment on the annotations associated with their proteins of interest, the *Arabidopsis* curation group, TAIR, have successfully implemented a process whereby authors submit functional information directly to them at the time of publication in the Plant Physiology journal via an online submission form (Berardini, 2009; Ort and Grennan, 2008). TAIR found that they received a 20% author response rate with no author prompting, and this rose to 75% when curators prompted the authors. The curation workload was not necessarily decreased as the authors tended to put in all the information they knew about the protein, and not just the evidence described in the submitted paper, requiring curators to spend time finding evidence for the additional annotations. However, this system has the benefit of providing curators with an author contact for specific proteins in case further information is needed for annotation. This system also has the added benefit of promoting the annotation process to authors, which hopefully means the authors may be more aware of the detail they need to include in their publications, in order to assist the curation process.

8.4.2   Wiki pages

The GOC has created an editable Wiki system to enable research scientists, and other interested parties, to review the GO annotation of their favorite protein and suggest additional information or changes, which would improve their annotation.

The protein-specific wiki pages provide links to general information and existing GO annotation for a given human protein and its orthologs in other species. Once registered, scientists can use the edit page option to add comments about the protein annotation and to suggest additional references or missing terms. These protein-specific pages are accessible via the GOC Wiki Community Annotation Pages (http://wiki.geneontology.org/index.php/Main_Page) (Lovering *et al.*, 2008b, Lovering *et al.*, 2009). However, at the time of writing, no lab-based researchers have utilized these pages.

8.4.3   Community annotation workshops

There are two types of community annotation workshops that GOC members undertake, ontology development workshops and annotation jamborees. Ontology development workshops are held when there is a need to develop a particular area of the ontology on a large scale, and involve GO editors and expert researchers from the relevant field. These workshops often result in a publication to publicize the ontology changes, and this is a good incentive for researchers to participate.

The idea behind the annotation jamboree is to populate the data set with annotations provided by researchers actively studying the proteins of interest. The idea is that, as they are familiar with the research in their area, they will be able to identify the key experimental papers easily. The National Institute of Allergy and Infectious Diseases-funded Pathema group held a two-day annotation-training jamboree in 2007 with the idea that trained scientists would continue to provide annotation updates thereafter. Contributors are given recognition on the protein page; however, Pathema found that after two years they had been sent only four updates (Brinkac, 2009). Thus, biocurators continue to search for new ways to encourage the research community to contribute to curation activities.

8.5   Limitations

As with any scientific data, it is important to appreciate the limitations associated with GO data. Curation groups in the GOC acknowledge the limitations discussed here, and where possible efforts are being made to address these issues.

8.5.1   GO cannot capture all relevant biological aspects

GO allows the curator to record the *Biological Process, Molecular Function* and *Cellular Component* aspects associated with a protein. However, GO does not allow the capture of every single relevant biological aspect for each annotation.

For example, a study by Uronen-Hansson *et al.* showed that the human Toll-like receptor 4 (TLR4) is present on the cell surface of monocytes, but immature dendritic cells only express the protein intracellularly (Uronen-Hansson *et al.*, 2004). The curator annotating this paper was able to make the cellular component annotations: 'integral to plasma membrane' (GO:0005887) and 'perinuclear region of cytoplasm' (GO:0048471), but was unable to state which cell type each component annotation refers to. This distinction would be important for a researcher using immature dendritic cells to study TLR4. GO allows the annotation of both cellular components, but does not allow the curator to capture the cell line context for each annotation.

Another aspect that is currently not captured is the target of the GO annotation. For example, the protein SH3D19 is involved in the ectodomain shedding of several proteins, including HBEGF, TGFA, AREG, and EREG (Tanaka *et al.*, 2004). From the publication by Tanaka *et al.*, the curator was able to annotate the process term 'positive regulation of membrane protein ectodomain proteolysis' (GO:0051044). However, GO does not allow the curator to also capture the protein targets of this process annotation.

Inclusion of such data would provide highly valuable cross-ontology annotations, and the GOC is currently investigating the possibility of allowing the addition of such information into GO annotation.

8.5.2 The ontology is always evolving

As GO develops, curators and editors are able to create new, more specific terms, which describe the biology in greater detail. GO editors approve around 200 new terms each month. Therefore, papers annotated prior to the development of new terms could contain data that may have been captured more accurately if the newer terms had existed at the time of the initial curation. With the volume of literature available, it is unlikely that curators will go back and re-annotate papers that have already been curated. However, these more specific terms will gradually be included in the GO data sets through the continued annotation of the scientific literature.

8.5.3 The volume of literature

The huge volume of literature associated with well-characterized proteins in particular organisms (especially seen for mammalian genes and proteins) can mean that often it is impossible to annotate every single paper associated with each protein. Curators attempt to cover as many different papers as possible; however it is always possible that a publication showing evidence to support a novel annotation has been missed. Thus GO users should bear in mind that whilst the annotations manually associated with each protein are correct, they do not necessarily represent all that is known about the protein.

8.5.4 Missing published data

The annotations a curator is able to make are dependent on the information contained in the literature. Authors occasionally miss out information such as species origin of the proteins and cDNAs they describe, and without this information, the paper cannot be annotated. For example, Meissner *et al.* described the calmodulin binding function of the ryanodine receptors (Meissner *et al.*, 2009). However, the publication does not state the species of the calmodulin they use in their experiments, in all likelihood because this is not important to the outcome of the experiments. In this case, the curator contacted the authors, who responded very promptly, thus enabling the paper to be annotated accurately. Due to time limitations and the huge volume of literature associated with many proteins, a curator will often skip papers that do not provide sufficient data to enable full annotation, and instead focus on papers that have comprehensive protein descriptions. Unfortunately, descriptions which clarify the identity and source of a specific protein are often missing from papers published in high-impact journals, such as *Nature* and *Science*, due to their strict word limits. Consequently, many of the key publications describing the properties of a protein remain un-annotated.

8.5.5 Manual curation is expensive

Manual curation is an expensive undertaking, but unfortunately funding for manual curation activities is not easily secured. Therefore, annotation targets are often prioritized to ensure that the curation project covers the specific aims of the funding body (Lovering *et al.*, 2008a). This means that even well-characterized proteins may not have made it into an annotation priority list, and so there can be a large difference between the annotation sets available for two well-studied proteins.

Thus, research community participation, through the community annotation approaches outlined in Section 8.4, by authors including as much detail as possible in their publications and by direct contact with curators, is an important component of accurate and complete curation.

8.6   Accessing GO annotations

Given the number of GO annotations (at the time of writing – March 2010 – there are over 58 million annotations to over 223 000 taxonomic groups in the GOA-UniProtKB database), it is vital that users are able to easily extract useful information from this data. Each GOC member provides file(s) of annotation data called Gene Association Files (GAFs), which are tab-delimited files containing a complete set of annotations released by an annotation group. These files are available from the GOC website (www.geneontology.org/GO.current. annotations.shtml), the GOC ftp site (www.geneontology.org/GO.downloads.ftp. cvs.shtml), and often additionally from individual GOC members' websites. However, in order to extract useful information from these files, a certain degree of computational knowledge is required. Many users simply do not possess or have access to this type of expertise; it is for this reason that a number of resources have been developed to assist users with their data analysis.

Disparate groups around the world have created resources for analyzing GO; they are all subtly different and they all have their own strengths and weaknesses. Just choosing a tool can be a daunting task for most people, and this is reflected in some of the enquiries that are received by the GO helpdesk. Unfortunately, the GOC is unable to test each and every resource available to determine whether it provides good quality results and is reliable. The user must take it upon themselves to investigate the relevant tools to decide which would work best for them and in doing this it is important to be aware of certain key questions which should be asked when evaluating a tool, including:

- *Is the tool actively maintained?*   This will not only impact on whether the user would receive any support if they encounter problems, but can also affect the results that could be obtained. Tools using out-of-date information can give vastly different results to a tool which is updated regularly with the latest GO terms and annotations, both of which are constantly being created, deleted or refined.

- *Does the tool exploit the hierarchical structure of the GO?*   This will allow users to create an overview of the functional attributes of their list of proteins by 'mapping-up' annotations to less-specific (parent) GO terms.

- *What statistical model does the tool use and are there other options to choose from?* Different statistical models can produce widely differing results.

Other questions to be considered are outlined in Table 8.4 (reproduced from Dimmer *et al.* (2008)). GO tools can be divided into two broad categories: those for searching and browsing GO and/or its associated annotations, and those for categorizing a list of genes or proteins from microarray or proteomic analyses.

8.6.1   Tools for browsing the GO

GO browsers are designed for searching and browsing terms in the three ontologies. Some browsers additionally provide access to GO annotations. A number of different browsers are briefly described below.

8.6.1.1   The Ontology Lookup Service

The Ontology Lookup Service (OLS; www.ebi.ac.uk/ontology-lookup/; Cote *et al.*, 2006) is a Web interface for searching Open Biomedical Ontology (OBO) format ontologies, including GO. It does not include GO annotations, but individual GO terms can be searched or the entire GO hierarchy can be browsed. When a term is selected, the information attached to that term is displayed, including definition and synonyms. A graph is displayed showing either the path of the term up to the root node, including the relationships between these terms, or there is an option to view a graph of the selected term's child terms. The OLS also provides an extensive range of Web services for automatically querying ontology structure, through both REST and SOAP interfaces.

8.6.1.2   OBO-Edit

As well as being an ontology editing tool, the GOC-developed OBO-Edit (http://oboedit.org) can also be used as an OBO browser. OBO-Edit is a graph-based tool into which a user can load an OBO ontology of their choice, including GO, and then view or edit it. As the user loads the ontologies, the ontology files used can be either the most up-to-date or an older, archived version. OBO-Edit is the tool used by the GOC editorial team to develop and maintain GO (Day-Richter *et al.*, 2007).

8.6.1.3   Specialized browsers

There are also many browsers which combine viewing of protein annotation with browsing of GO. Some of these only supply a subset of GO annotations, for instance to a particular species, such as the MGI GO browser (www.informatics. jax.org/searches/GO_form.shtml; Bult *et al.*, 2008), which supplies annotations to mouse proteins, or the TAIR browser (www.arabidopsis.org/servlets/Search? action=new_search&type=keyword; Swarbreck *et al.*, 2008), which provides annotations to *Arabidopsis* proteins.

Table 8.4   Questions to consider when choosing a GO analysis tool.

Key questions	Reasons
Does the tool enable the hierarchical structure of GO to be exploited?	GO analysis tools should be designed to improve the identification of functional groups within a data set by allowing the user to manually consolidate genes associated with highly specific (child) GO terms to those with the higher (parent) GO terms in order to formulate and test specific biological hypotheses.
What is the release date of the data used by the tool?	Each month an average of 240 GO terms is added to the Gene Ontology and 1500 GO annotations are added to the human GO data set. Tools that infrequently download GO data will restrict analyses. Remember to include the release date(s) of the GO annotation data set and ontology file used by the tool in any resulting publications. The tool should provide this information.
Does the tool correctly treat the GO annotations with the qualifier 'NOT'?	This qualifier reverses the meaning of an annotation, so these annotations should be either removed from the analysis or used to calculate the amount of evidence against certain hypotheses involving the terms annotated with NOT.
Does the analysis tool enable concomitant functional profiling for all three GO categories?	In addition, an increasing number of tools also display other annotation data such as TRANSFAC regulatory motifs, BioCarta, KEGG and Reactome pathways.
Is the type of identifier used in the assay directly accepted by the tool (e.g., probe IDs, RefSeq protein IDs) or will it be necessary to map one identifier type to another?	You may need to convert the identifiers of your gene list into those accepted by the tool. This can be an important source of errors since up to 20% of the identifiers can be routinely lost or incorrectly mapped during identifier translations (Draghici *et al.*, 2006). Even if the type of identifier is directly accepted, is this the native identifier used in the analysis or is an internal identifier mapping being performed? If internal identifier mappings are being carried out, what are the sources of data and their release dates?

(continued overleaf)

Table 8.4   *(continued)*.

Key questions	Reasons
Does the tool test for both enrichment AND depletion of the GO terms?	Some tools only test for over-representation of the differentially expressed genes within the given GO term. However, both significantly enriched as well as depleted GO terms can be biologically meaningful.
Does the tool enable the user to submit their own GO annotation data set or select specific evidence code-supported annotations for the analysis?	The facility to submit a data set enables the users to apply the most recent GO annotation data sets to their analysis. However, there are only a few species where filtering out certain evidence code-supported annotations is appropriate.
What is the statistical model used and are there several alternative models that the user may choose from?	One serious and widely neglected problem in GO profiling is that the same data submitted to different tools can provide widely differing results for the same GO terms. Having the ability to specify the model allows the user to eliminate one variable to verify their analysis.
What choice of correction factors is available?	To compensate for the propagation of gene associations from each GO term to all their parent GO terms, many tools give a choice of correction factors, such as Bonferroni, Holmes, false discovery rate (FDR) and Šidák (Rhee *et al.*, 2008). Bonferroni or Šidák are suitable when less than 50 unrelated GO categories are involved, Holmes is more appropriate for larger numbers of unrelated GO categories, and FDR is a good choice if several GO categories are related, for example contain several GO terms with a common parent.

Based on table 2 from Dimmer, E.C. *et al*: The Gene Ontology – Providing a functional role in proteomic studies. *Proteomics*. Published online 17 Jul 2008. Copyright Wiley-VCH Verlag GmbH & Co. KGaA. Reproduced with permission.

Others are even more specialized; the Comparative Toxicogenomics Database (CTD; http://ctd.mdibl.org/; Mattingly *et al.*, 2006) integrates a GO browser to enable users to search for functional information about proteins that are the targets of chemical exposure, thereby providing a resource for environmental health research.

8.6.1.4   AmiGO and QuickGO

Two browsers that provide a broader set of annotations are: AmiGO (http://amigo. geneontology.org/cgi-bin/amigo/go.cgi; Carbon *et al.*, 2009), the official GOC browser, and QuickGO (www.ebi.ac.uk/QuickGO/; Barrell *et al.*, 2009), developed by the GOA group at the EBI. These two are similar, in that GO can be searched and browsed; GO terms and their relationships can be viewed in context with GO hierarchy, either in chart or graphical views; annotations are provided to a large number of species; and annotations can be mapped-up to more general GO terms using a GO slim facility in each of the tools. They are both Web-based browsers and so are straightforward for novice users to start using; they also use the most current ontology and annotation data and are actively maintained. A comparison of the different displays of the GO term hierarchy and GO annotations between AmiGO and QuickGO can be seen in Figure 8.4.

However, AmiGO and QuickGO differ in a number of aspects. AmiGO incorporates a number of analysis tools including a Term Enrichment feature that finds significant terms, which are shared between a group of proteins, and a BLAST feature which allows a user to enter a sequence and be provided with a list of closely related sequences in the GO database with their associated annotations. AmiGO currently only displays manual GO annotations for the majority of proteins, whereas QuickGO displays both manually and electronically assigned annotations; at the time of writing (March 2010), the GOA database contains more than 57 million electronic annotations. QuickGO additionally features extensive filtering capabilities allowing users to create a customized annotation set. QuickGO also incorporates a facility to find terms which are commonly co-annotated with a chosen GO term; this is useful not only for curators to ensure they have made consistent annotations, but also for users who can discover, for example, which *Molecular Functions* are involved in which *Biological Processes*, or in which *Cellular Component* a particular *Biological Process* occurs (Figure 8.5). For example, the *Biological Process* term 'cholesterol transport' is co-annotated with related *Biological Processes* cholesterol homeostasis and lipoprotein catabolic process; *Molecular Functions* high-density lipoprotein binding and cholesterol transporter activity; and the *Cellular Components* caveola and high-density lipoprotein particle. Finally, QuickGO also provides a REST style query interface for programmatic retrieval of GO term information and annotation data. These Web services are fully integrated so that the filtering options and data sets available are fully synchronized between the browsable and Web service interfaces. Results are visible in tab-separated, OBO or XML formats.

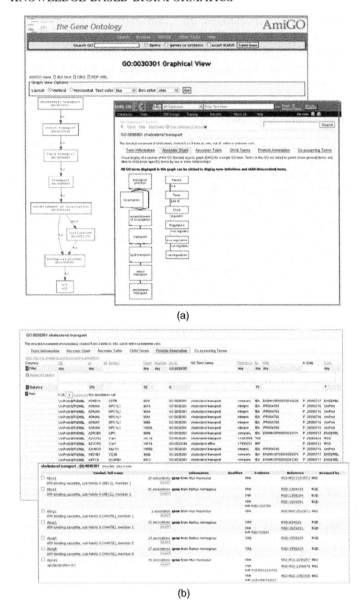

(a)

(b)

Figure 8.4   A comparison of AmiGO and QuickGO views of the GO term 'choles-
terol transport' (GO:0030301). (a) Graphical or ancestry views; GO browsers
generally contain the same ontology information but display this in different ways.
These views show the graphical or ancestry views, which allow users to easily
see the parentage for each term. The AmiGO view shows the root term at the
bottom, whereas the QuickGO shows the root term at the top. (b) Annotation
views; GO browsers often contain different subsets of GO annotations according
to their focus. AmiGO (front insert) does not currently contain electronic (IEA)
annotations, whereas QuickGO does include this data set.

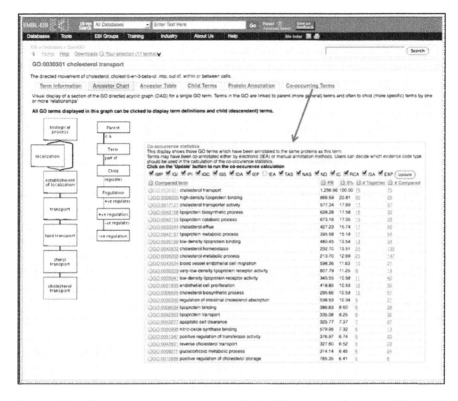

Figure 8.5   Co-occurring terms view in QuickGO. A novel feature of QuickGO is its ability to calculate which GO terms are commonly co-annotated with a selected term. After clicking on the 'Co-occurring Terms' tab in a GO term page, a user can choose which evidence codes used to make annotations should be included in the calculation for co-occurring terms. In this example, the resulting table displays a ranked list of the most common terms which have been anno-tated alongside the GO term 'cholesterol transport' (GO:0030301) using only manual evidence codes (i.e., not IEA evidenced annotations), the most commonly co-annotated terms appearing at the top of the table.

8.6.2   Functional classification

In addition to tools used for viewing GO and its associated annotations, there are many resources available to perform analyses and functionally categorize proteins using GO.

There are tools for characterization of unknown sequences, such as Blast2GO (Conesa and Götz, 2008; www.blast2go.org/) and AmiGO Blast (http://amigo. geneontology.org/cgi-bin/amigo/blast.cgi), which take nucleotide or protein sequences and perform a BLAST search against the GO database for similar sequences that have been annotated using GO, thereby allowing a user to infer

GO attributes for their unknown sequence. In a similar vein, the InterProScan tool (www.ebi.ac.uk/Tools/InterProScan/; Hunter *et al.*, 2009) allows users to input a protein sequence to retrieve protein domains which are present in the sequence and any associated GO terms that the domains may have been mapped to, again allowing inferences about the functionality of an unknown protein sequence.

The most popular type of GO analysis tool is that which groups genes and/or proteins according to their shared annotation to GO terms. There are many of these in the public domain, each with slightly different capabilities, so it can be quite difficult to determine which tool best fits a user's requirement. A comprehensive survey of bioinformatics enrichment tools has been written by Huang da *et al.*, who have classified 68 enrichment tools into three groups; Singular Enrichment Analysis (SEA), Gene Set Enrichment Analysis (GSEA) and Modular Enrichment Analysis (MEA), depending on whether or not the input gene lists have been preordered (SEA vs. GSEA) or whether or not the tool considers GO term relationships in its calculations (MEA) (Huang da *et al.*, 2009a). The tools are classified according to the enrichment algorithms the tool uses, and the features and limitations of each class of tool are also provided. Alternatively, a useful resource to help the user to decide which analysis tool best fits their requirements is SerbGO (http://estbioinfo.stat.ub.es/apli/serbgov131/index.php; Mosquera and Sanchez-Pla, 2008). This Web-based tool compares the functionality of 36 tools and allows users to select the criteria required from the tool, including parameters such as: Web-based or locally downloadable, type of identifier, and frequency of data updates. (See Table 8.4 for questions to consider when selecting a GO tool.)

Many GO analysis tools are similar in that they accept an input of a list of gene or protein identifiers from a specified organism and compare the enrichment of GO terms in this list with a background set. Any terms that are more frequently present in the study set compared with the background set are given a p-value to describe how significant the enrichment is. Some tools will use the rest of the genome as a background set, whereas others allow the user to upload their own background set. Additionally, these tools often provide several options for the statistical method to be used in the analysis. The other difference between tools is the output view of results; most display the results as a table of GO terms with their associated significance, and may list the genes identified in each group; some also provide graph views of the significant GO terms.

Some analysis tools are available directly on the Web, for instance FatiGO (Al-Shahrour *et al.*, 2008), FatiScan (Al-Shahrour *et al.*, 2007), DAVID (Huang da *et al.*, 2009b), Onto-Express (Draghici *et al.*, 2003), while others can only be accessed by downloading the program to your own computer and so require a little more computer knowledge, such as BiNGO (www.psb .ugent.be/cbd/papers/BiNGO/; Maere *et al.*, 2005) and the Ontologizer (Bauer *et al.*, 2008).

A word of caution: during the preparation of this chapter it was noted that many of the GO browsers and GO analysis tools available online have not been actively maintained and, because of this, they are unlikely to have the most recent GO data and GO annotations that are available. This could seriously impact any results obtained from these tools; so, before using a tool, it is worth investigating whether it provides the most up-to-date GO data. We will now give brief descriptions of some of the GO analysis tools available, which we have some experience with and which are regularly updated.

8.6.2.1   FatiGO

FatiGO   (http://babelomics.bioinfo.cipf.es/EntryPoint?loadForm=fatigo;   Al-Shahrour *et al.*, 2008) is a functional enrichment tool, belonging to the SEA class of bioinformatic tools, which compares a list of genes (such as a list of differentially expressed genes or proteins), with a background list of genes (e.g., the rest of the genes on the microarray or the rest of the genome) using a Fisher's exact test with the option to analyze the lists using a variety of knowledge bases including GO, KEGG pathways, and UniProtKB Keywords. Its sister application, FatiScan (http://bioinfo.cipf.es/babelomicswiki/tool:fatiscan), belongs to the GSEA class of bioinformatics tools, and as such it categorizes ordered lists of genes which have been ranked by any experimental or theoretical criteria (e.g., differential expression in disease versus healthy samples) and aims to find groups of genes which share functional properties. FatiScan can find significantly over- or under-represented functional classes whereas many other tools only find over-represented functions.

8.6.2.2   DAVID

DAVID (http://david.abcc.ncifcrf.gov/home.jsp; Huang da *et al.*, 2009b) belongs to the SEA and MEA classes of tools, meaning that it takes a list of unordered genes as input and it takes the relationships of GO terms into account. It uses a modified Fisher's exact test called an EASE score to measure enrichment of GO terms.

8.6.2.3   Onto-Express

Onto-Express (Draghici *et al.*, 2003) is part of a suite of tools (Onto-Tools, http://vortex.cs.wayne.edu/projects.htm) that categorizes lists of differentially expressed genes from microarray experiments. Onto-Express catalogs more than 300 microarrays for use as a reference or background set, or the user can upload their own. The user can also specify the statistical method to be used for the analysis. As is the case for DAVID, Onto-Express also belongs to the SEA/MEA classes of tools.

8.6.2.4   The Ontologizer

The Ontologizer (http://compbio.charite.de/index.php/ontologizer2.html; Bauer *et al.*, 2008) is unique in that the user is able to upload an ontology file and an annotation file; this is incredibly useful as it enables the user to use the most up-to-date ontology and annotations available – the drawback of most other tools is that the user is reliant on the tool providers loading the most recent data files. This flexibility means that the user can determine how the analysis for their set of genes has varied over time by analyzing them with an older version of the ontology and annotation files. The Ontologizer is in the GSEA class of GO tools, allowing the user to provide a ranked list of genes for analysis. The analysis can be performed either 'term-for-term' or 'parent-child' where it takes into account the relationships between GO terms.

8.6.3   GO slims

A useful way of using GO to summarize results from large-scale data sets that has become popular is the GO slim. A GO slim is a subset of high-level GO terms from the ontology that have been specifically chosen to give an overview of general biological features or even just a particular area of biology. Annotations to these terms or their child terms are 'mapped-up' to the slim terms by using the true path rule of the ontology. This rule states that if a protein is annotated to a particular term, it must also be true that the protein could also be annotated to the parent terms. For example, a protein annotated to the GO term 'lamin depolymerization' (GO:0007078) could also, theoretically, be annotated to the parent term 'cell cycle' (GO:0007049). By applying this 'slimming' procedure to a large set of proteins, a researcher could get a broad overview of the functions or processes that the genes or proteins are involved in, without the detail of the more specific GO terms. Many users prefer to make their own GO slim so that they can tailor it to their particular area of research.

Some things to keep in mind when making a GO slim are:

- The terms must be general, high-level terms representing major cellular processes, functions or sub-cellular compartments (depending on what part of the ontology is of interest).

- If constructing a slim to encompass all major biological processes, the terms must provide a good overview of the biology of the organism.

- The slim should contain terms to which a reasonable number of proteins have been annotated; it is pointless to have a GO term in a slim that is only applied by a few proteins.

- Be careful not to choose a term which is so high-level that it represents many biologically different processes. It may be better to choose two terms, or more granular terms.

- Ideally, when your set of proteins have been mapped-up to your GO slim, all proteins should be represented somewhere in the slim.

For example, a recent study utilized a GO slim to characterize the plant, *Thellungiella halophila*. Taji *et al.* exposed the halophyte to various environmental stresses, such as high salinity, freezing, and abscisic acid treatment, and then classified them using the plant GO slim available from the GOC website (www.geneontology.org/GO.slims.shtml). Compared to *Arabidopsis, Thellungiella* had a similar characterization of proteins except that the latter species has 1.5 times as many proteins involved in transport. This result, together with conclusions from other studies, showed that *Thellungiella* has more efficient sodium and potassium homeostasis and prompted the authors to suggest that this species has a unique ion transportation system (Taji *et al.*, 2008).

8.6.4 GO displays in other databases

GO annotations are displayed in a number of different databases, in combination with manually and/or electronically curated data from other curation efforts. For example, the Entrez Gene database (www.ncbi.nlm.nih.gov/sites/entrez?db =gene) displays GO data for each gene alongside protein interaction data from the Human Protein Reference Database (HPRD; www.hprd.org/), BioGRID (www.thebiogrid.org/index.php), the Biomolecular Interaction Network Database (BIND; www.bind.ca/), pathway data from Reactome (www.reactome.org/), and the Kyoto Encyclopedia of Genes and Genomes (KEGG; www.genome.jp/kegg/).

Not all databases display GO in the same manner. For example, Entrez Gene displays all unique GO terms regardless of evidence code, thus for a GO term that has both manual and electronic evidence, Entrez Gene might show the *IEA* code, rather than the *IDA* evidence. This is an accurate display; however, if there is some manually annotated evidence for a particular GO term, the user will not necessarily find it from looking at Entrez Gene. In addition, due to stringent production schedules, many databases are only able to update their GO cross-references on an infrequent basis and this can result in a disparity of GO annotation displays from different resources. As with all information resources, it is best to use data obtained from the primary sources. In the case of GO annotations and terms, the primary source is the GOC website (www.geneontology.org/index.shtml).

8.7 Conclusions

- There are two types of biocuration: data submission and value-added. GO is an example of a value-added curation effort.

- GO is a controlled vocabulary structured as a directed acyclic graph.

- Many biocuration groups use GO, and many of these are focused on a single genus or species.

- GO currently utilizes 18 evidence codes to describe experimental evidence, non-experimental evidence, and computational evidence.

- New terms are created, existing terms revised, and the ontology refined as an ongoing process to improve GO.

- Manual curation generates granular annotations, whereas electronic annotations tend to be less specific.

- Annotations are transferred from well-annotated proteomes to less well-studied proteomes by homologous and phylogenetic methods.

- Community participation is an important aspect of accurate biocuration, but on the whole biocurators have as yet been unable to reliably involve lab-based scientists in curation activities.

- GO does not currently capture all relevant biological aspects, but may soon include data from other biomedical ontologies, so as to provide greater detail for its users.

- GO annotations can be accessed from several GO browsers, and are also displayed in a number of gene or protein databases.

- There are numerous third-party tools that have been developed to assist in the GO analysis of large data sets. However, users must check whether these use the latest ontology and annotation files.

In the authors' experience, researchers are often unaware of the manual effort that goes on behind the scenes of their favorite database, imagining the majority of the work to be electronic. Thus the benefits of biocuration need to be expounded to the life science research community. However, the availability of GO and other functional annotation efforts has had a profound effect on the way in which life science research is carried out today. There are now numerous websites available where a researcher can input a novel sequence and quickly establish a fairly accurate idea of its function, thanks to the ongoing curation of multiple species, by many different biocuration groups. Thus the products of biocuration activities (annotations displayed in user-friendly databases) are now an essential part of the lab-based researcher's tool set.

8.8   References

Al-Shahrour, F., Arbiza, L., Dopazo, H., *et al.* (2007) From genes to functional classes in the study of biological systems. *BMC Bioinformatics*, **8**, 114.

Al-Shahrour, F., Carbonell, J., Minguez, P., *et al.* (2008) Babelomics: advanced functional profiling of transcriptomics, proteomics and genomics experiments. *Nucleic Acids Res*, **36**, W341–6.

Artamonova, Ii., Frishman, G., and Frishman, D. (2007) Applying negative rule mining to improve genome annotation. *BMC Bioinformatics*, **8**, 261.

Ashburner, M., Ball, C.A., Blake, J.A., *et al.* (2000) Gene ontology: tool for the unification of biology. The Gene Ontology Consortium. *Nat Genet*, **25**, 25–9.

Barrell, D., Dimmer, E., Huntley, *et al.* (2009) The GOA database in 2009 – an integrated Gene Ontology Annotation resource. *Nucleic Acids Res*, **37**, D396–403.

Bauer, S., Grossmann, S., Vingron, M., and Robinson, P.N. (2008) Ontologizer 2.0 – a multifunctional tool for GO term enrichment analysis and data exploration. *Bioinformatics*, **24**, 1650–1.

Benson, D.A., Karsch-Mizrachi, I., Lipman, D.J., *et al.* (2008) GenBank. *Nucleic Acids Res*, **36**, D25–30.

Berardini, T. (2009) A report from the Journal-MOD Collaboration Front: TAIR and Plant Physiology. 3rd International Biocuration Conference, Berlin, Germany, 16–19 April 2009.

Berglund, L., Bjorling, E., Oksvold, P., *et al.* (2008) A genecentric Human Protein Atlas for expression profiles based on antibodies. *Mol Cell Proteomics*, **7**, 2019–27.

Bourne, P.E. and McEntyre, J. (2006) Biocurators: contributors to the world of science. *PLoS Comput Biol*, **2**, e142.

Brinkac, L., Madupu, R., Caler, E. *et al.* (2009) Expert assertions through Community Annotation Jamborees. 3rd International Biocuration Conference, Berlin, Germany, 16–19 April 2009.

Bruford, E.A., Lush, M.J., Wright, M.W., *et al.* (2008) The HGNC Database in 2008: a resource for the human genome. *Nucleic Acids Res*, **36**, D445–8.

Bult, C.J., Eppig, J.T., Kadin, J.A., *et al.* (2008) The Mouse Genome Database (MGD): mouse biology and model systems. *Nucleic Acids Res*, **36**, D724–8.

Camon, E., Barrell, D., Brooksbank, C., *et al.* (2003) The Gene Ontology Annotation (GOA) Project-Application of GO in SWISS-PROT, TrEMBL and InterPro. *Comp Funct Genomics*, **4**, 71–4.

Camon, E.B., Barrell, D.G., Dimmer, E.C., *et al.* (2005) An evaluation of GO annotation retrieval for BioCreAtIvE and GOA. *BMC Bioinformatics*, **6**(Suppl 1), S17.

Carbon, S., Ireland, A., Mungall, C.J., *et al.* (2009) AmiGO: online access to ontology and annotation data. *Bioinformatics*, **25**, 288–9.

Chalmel, F., Lardenois, A., Thompson, J.D., *et al.* (2005) GOAnno: GO annotation based on multiple alignment. *Bioinformatics*, **21**, 2095–6.

Conesa, A. and Götz, S. (2008) Blast2GO: A comprehensive suite for functional analysis in plant genomics. *Int J Plant Genomics*, **2008**, 619832.

Conesa, A., Gotz, S., Garcia-Gomez, J.M., *et al.* (2005) Blast2GO: a universal tool for annotation, visualization and analysis in functional genomics research. *Bioinformatics*, **21**, 3674–6.

Cote, R.G., Jones, P., Apweiler, R., and Hermjakob, H. (2006) The Ontology Lookup Service, a lightweight cross-platform tool for controlled vocabulary queries. *BMC Bioinformatics*, **7**, 97.

Day-Richter, J., Harris, M.A., Haendel, M., and Lewis, S. (2007) OBO-Edit – an ontology editor for biologists. *Bioinformatics*, **23**, 2198–200.

Dimmer, E.C., Huntley, R.P., Barrell, D.G., *et al.* (2008) The Gene Ontology - providing a functional role in proteomic studies. *Proteomics*. doi: 10.1002/pmic.200800002

Draghici, S., Khatri, P., Bhavsar, P., *et al.* (2003) Onto-Tools, the toolkit of the modern biologist: Onto-Express, Onto-Compare, Onto-Design and Onto-Translate. *Nucleic Acids Res*, **31**, 3775–81.

Draghici, S., Sellamuthu, S., and Khatri, P. (2006) Babel's tower revisited: a universal resource for cross-referencing across annotation databases. *Bioinformatics*, **22**, 2934–9.

Engelhardt, B.E., Jordan, M.I., Muratore, K.E., and Brenner, S.E. (2005) Protein molecular function prediction by Bayesian phylogenomics. *PLoS Comput Biol*, **1**, e45.

Flicek, P., Aken, B.L., Beal, K., *et al.* (2008) Ensembl 2008. *Nucleic Acids Res*, **36**, D707–14.

Gene Ontology Consortium (2001) Creating the gene ontology resource: design and implementation. *Genome Res*, **11**(8), 1425–33.

Groth, D., Lehrach, H., and Hennig, S. (2004) GOblet: a platform for Gene Ontology annotation of anonymous sequence data. *Nucleic Acids Res*, **32**, W313–7.

Hill, D.P., Smith, B., McAndrews-Hill, M.S., and Blake, J.A. (2008) Gene Ontology annotations: what they mean and where they come from. *BMC Bioinformatics*, **9** (Suppl 5), S2.

Huang da, W., Sherman, B.T., and Lempicki, R.A. (2009a) Bioinformatics enrichment tools: paths toward the comprehensive functional analysis of large gene lists. *Nucleic Acids Res*, **37**, 1–13.

Huang da, W., Sherman, B.T., and Lempicki, R.A. (2009b) Systematic and integrative analysis of large gene lists using DAVID bioinformatics resources. *Nat Protoc*, **4**, 44–57.

Hubbard, T.J., Aken, B.L., Ayling, S., *et al.* (2009) Ensembl 2009. *Nucleic Acids Res*, **37**, D690–7.

Hunter, S., Apweiler, R., Attwood, T.K., *et al.* (2009) InterPro: the integrative protein signature database. *Nucleic Acids Res*, **37**, D211–5.

Kanehisa, M., Araki, M., Goto, S., *et al.* (2008) KEGG for linking genomes to life and the environment. *Nucleic Acids Res*, **36**, D480–4.

Kerrien, S., Alam-Faruque, Y., Aranda, B., *et al.* (2007) IntAct – open source resource for molecular interaction data. *Nucleic Acids Res*, **35**, D561–5.

Khan, S., Situ, G., Decker, K., and Schmidt, C.J. (2003) GoFigure: automated Gene Ontology annotation. *Bioinformatics*, **19**, 2484–5.

Kikuno, R., Nagase, T., Nakayama, M., *et al.* (2004) HUGE: a database for human KIAA proteins, a 2004 update integrating HUGEppi and ROUGE. *Nucleic Acids Res*, **32**, D502–4.

Letovsky, S.I., Cottingham, R.W., Porter, C.J., and Li, P.W. (1998) GDB: the Human Genome Database. *Nucleic Acids Res*, **26**, 94–9.

Lomax, J. (2005) Get ready to GO! A biologist's guide to the Gene Ontology. *Brief. Bioinform*, **6**, 298–304.

Love, C.G., Robinson, A.J., Lim, G.A., *et al.* (2005) Brassica ASTRA: an integrated database for Brassica genomic research. *Nucleic Acids Res*, **33**, D656–9.

Lovering, R.C., Dimmer, E., Khodiyar, V.K., *et al.* (2008a) Cardiovascular GO annotation initiative year 1 report: why cardiovascular GO? *Proteomics*, **8**, 1950–3.

Lovering, R.C., Camon, E.B., Blake, J.A., and Diehl, A.D. (2008b) Access to immunology through the Gene Ontology. *Immunology*, **125**, 154–60.

Lovering, R.C., Dimmer, E.C., and Talmud, P.J. (2009) Improvements to cardiovascular gene ontology. *Atherosclerosis*, **205**, 9–14.

Maere, S., Heymans, K., and Kuiper, M. (2005) BiNGO: a Cytoscape plugin to assess overrepresentation of gene ontology categories in biological networks. *Bioinformatics*, **21**, 3448–9.

Martin, D.M., Berriman, M., and Barton, G.J. (2004) GOtcha: a new method for prediction of protein function assessed by the annotation of seven genomes. *BMC Bioinformatics*, **5**, 178.

Matthews, L., Gopinath, G., Gillespie, M., *et al.* (2009) Reactome knowledgebase of human biological pathways and processes. *Nucleic Acids Res*, **37**, D619–22.

Mattingly, C.J., Rosenstein, M.C., Davis, A.P., *et al.* (2006) The comparative toxicogenomics database: a cross-species resource for building chemical-gene interaction networks. *Toxicol Sci*, **92**, 587–95.

Mehrle, A., Rosenfelder, H., Schupp, I., *et al.* (2006) The LIFEdb database in 2006. *Nucleic Acids Res*, **34**, D415–8.

Meissner, G., Pasek, D.A., Yamaguchi, N., *et al.* (2009) Thermodynamics of calmodulin binding to cardiac and skeletal muscle ryanodine receptor ion channels. *Proteins*, **74**, 207–11.

Mosquera, J.L. and Sanchez-Pla, A. (2008) SerbGO: searching for the best GO tool. *Nucleic Acids Res*, **36**, W368–71.

NC-IUBMB and Webb, E.C. (eds) (1992) *Enzyme Nomenclature 1992: Recommendations of the Nomenclature Committee of the International Union of Biochemistry and Molecular Biology on the Nomenclature and Classification of Enzymes*, Academic Press.

Nofer, J.R., Remaley, A.T., Feuerborn, R., *et al.* (2006) Apolipoprotein A-I activates Cdc42 signaling through the ABCA1 transporter. *J Lipid Res*, **47**, 794–803.

Ort, D.R. and Grennan, A.K. (2008) Plant Physiology and TAIR partnership. *Plant Physiol*, **146**, 1022–3.

Reference Genome Group of the Gene Ontology Consortium (2009) The Gene Ontology's Reference Genome Project: a unified framework for functional annotation across species. *PLoS Comput Biol*, **5**, e1000431.

Rhee, S.Y., Wood, V., Dolinski, K., and Draghici, S. (2008) Use and misuse of the gene ontology annotations. *Nat Rev Genet*, **9**, 509–15.

Sayers, E.W., Barrett, T., Benson, D.A., *et al.* (2009) Database resources of the National Center for Biotechnology Information. *Nucleic Acids Res*, **37**, D5–15.

Sterk, P., Kulikova, T., Kersey, P., and Apweiler, R. (2007) The EMBL Nucleotide Sequence and Genome Reviews Databases. *Methods Mol Biol*, **406**, 1–21.

Swarbreck, D., Wilks, C., Lamesch, P., *et al.* (2008) The Arabidopsis Information Resource (TAIR): gene structure and function annotation. *Nucleic Acids Res*, **36**, D1009–14.

Taji, T., Sakurai, T., Mochida, K., *et al.* (2008) Large-scale collection and annotation of full-length enriched cDNAs from a model halophyte, Thellungiella halophila. *BMC Plant Biol*, **8**, 115.

Tanaka, M., Nanba, D., Mori, S., *et al.* (2004) ADAM binding protein Eve-1 is required for ectodomain shedding of epidermal growth factor receptor ligands. *J Biol Chem*, **279**, 41950–9.

Thornton, J. (2009) Data curation in biology – past, present and future. 3rd International Biocuration Conference, Berlin, Germany, 16–19 April 2009.

UniProt Consortium (2008) The universal protein resource (UniProt). *Nucleic Acids Res*, **36**(Database issue), D190–5.

Uronen-Hansson, H., Allen, J., Osman, M., *et al.* (2004) Toll-like receptor 2 (TLR2) and TLR4 are present inside human dendritic cells, associated with microtubules and the Golgi apparatus but are not detectable on the cell surface: integrity of microtubules is required for interleukin-12 production in response to internalized bacteria. *Immunology*, **111**, 173–8.

Zehetner, G. (2003) OntoBlast function: from sequence similarities directly to potential functional annotations by ontology terms. *Nucleic Acids Res*, **31**, 3799–803.

9

Methods for improving genome annotation

Jonathan Mudge and Jennifer Harrow

9.1 The basis of gene annotation

9.1.1 Introduction to gene annotation

The ultimate value of a genome sequence depends entirely on the quality of the accompanying annotation. The term 'genome annotation' refers to the identification and description of any element on the genome to which a biological functionality can be attached. Most importantly this annotation is required to describe the gene content of a particular organism's genome, and in this chapter we will focus mainly on the identification of functional transcribed gene structures. However, it should be noted that a total understanding of a genome sequence would also require the identification of a wide range of other functional elements, for example gene promoters and splicing signals.

How do you annotate the gene content of a genome? There are many annotation or 'gene finding' processes that can be used, as summarized in Figure 9.1, and the choice of strategy is very much dependent on the data available. A scientist provided with no other resources than a single genome sequence would face a difficult task in describing its gene content. In this scenario, the scientist would have no option but to construct gene structures *ab initio*, using knowledge or assumptions regarding the gene architecture of the organism to design a gene finding pipeline. While such single-genome *ab initio* approaches have their limitations when used in isolation, they have proved highly useful and indeed

Knowledge-Based Bioinformatics: From Analysis to Interpretation   Edited by Gil Alterovitz and Marco Ramoni

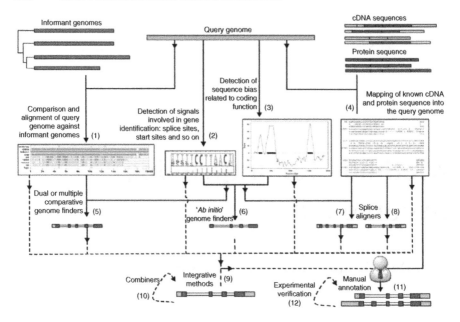

Figure 9.1   A variety of strategies for gene finding. Information on the location of genes can be obtained from four main sources: (1) a conservation-based analysis using other genomes; (2) the identification of sequence sites and motifs that indicate the presence of a gene structure; (3) the intrinsic statistical properties of coding sequences; (4) known transcript and protein sequences. Frequently these sources are used in combination. Whilst basic ab initio gene finders consider only information contained within the target genome sequence (6), more sophisticated programs have been designed to use information from other genomes (5) and/or transcriptional evidence (7). The use of cDNAs, ESTs and proteins is preferable where available, and annotation pipelines such as Ensembl often begin with the mapping of such sequences to the genome, with the refined construction of exon/intron boundaries provided by the use of 'splice alignment algorithms' (8). Incomplete gene models constructed in this way can be improved by combining data from ab initio *or comparative gene finders (9), and programs are available which combine the outputs of many gene finders into a single coherent gene set (10). Ultimately, manual genome annotation is the most accurate and comprehensive gene building process available (11), and gene models which lack strong support can be subjected to experimental confirmation.*

necessary in combination with other methods. Access to transcriptional data from the organism of interest greatly enhances the gene annotation process, since these sequences direct us to regions of the genome undergoing transcription and allow us to accurately identify the exonic structures of the underlying genes. Such 'evidence-based' gene construction methodologies either depend entirely on the speed and consistency of computer algorithms (automated annotation), or else integrate a degree of manual assessment, for example, to improve the description of nonstandard gene elements (manual annotation). Finally, if the scientist could compare their organism's genome to others which had already been annotated, the task would be easier still, since evolutionary conservation provides a powerful tool for the identification of gene structures (comparative annotation, or 'phylogenomics'). Currently, the most successful gene building processes combine all these different approaches to produce a reliable gene set. As we shall discuss, flexibility is critical to such processes in order that the gene set can be updated as appropriate when new information becomes available.

9.1.2   Progression in *ab initio* gene prediction

A number of computer algorithms are available to perform single-genome *ab initio* gene prediction. The most popular of these is GENSCAN (Burge and Karlin, 1997), due to its high exon specificity (81%) and sensitivity (78%) in comparison to other related algorithms (Burset and Guigo, 1996). Each gene prediction algorithm shares the same essential functionality: it searches for sequence within the genome that implies the presence of gene structures, typically branch site motifs and splice sites, open reading frames (ORFs), codon usage statistics, and transcription signals (see Figure 9.1); an organism-specific training set is usually used to calibrate the underlying heuristics. In the context of the human genome project, single-genome *ab initio* gene prediction algorithms were a valuable aid to genome annotation in the early stages of data production (Hattori *et al.*, 2000; Dunham *et al.*, 1999). Today, however, their usefulness is limited compared with the superior transcription-based approaches described below; many *ab initio* algorithms have in fact been further developed to combine transcriptional or comparative data. Such models include NSCAN, which supplements *ab initio* functionality with information derived from multispecies genome alignments (van Baren *et al.*, 2007), and AUGUSTUS, which can be adapted to use a variety of external information including alignments, ESTs, and protein sequences (Stanke *et al.*, 2006).

9.1.3   Annotation based on transcribed evidence

Evidence for the transcription of certain genomic regions takes the form of expressed sequence tags (ESTs), cDNAs, and protein sequences submitted to the public databases such as EMBL/GenBank/DDBJ and Uniprot (Benson *et al.*, 2009; Kulikova *et al.*, 2007; Sugawara *et al.*, 2008; UniProt Consortium, 2009). Typical 'evidence-based' gene building methodologies involve two stages: the alignment of the total transcriptomics data available for that species against the genome sequence, followed by the generation of gene models based on these alignments either by automated or manual approaches (see Figure 9.1). In the case of the human genome, extensive transcriptomics data is available, making evidence-based gene prediction a powerful tool. Several large-scale projects are currently funded to describe the human gene set using transcriptional evidence as a basis, and as a result there are a number of different gene 'builds' available to researchers.

The Ensembl genebuild process, generated in collaboration between the Wellcome Trust Sanger Institute (WTSI) and the European Bioinformatics Institute (EBI), begins with the alignment of all available protein data for that species against the genome sequence (the 'targeted stage'), which is forced into a predicted gene structure using the Genewise algorithm; this is therefore automated annotation (Hubbard *et al.*, 2009). At this stage, proteins from closely related species are also aligned and used to build structures in regions not covered by same-species evidence; the pipeline is thus a combination of transcriptional and comparative annotation processes. Following this, cDNA evidence is aligned to provide further refinement, before transcript models are merged to eliminate redundancy. The resulting Ensembl gene – including a list of its supporting evidence – can be viewed in the Ensembl gene web browser (www.ensembl.org). Typically, a global genebuild takes four months to complete. This is not the end of the process; the Ensembl build is updated regularly in order to accommodate changes to the genome sequence, increased coverage of the transcriptome, and underlying improvements to the annotation pipeline. Note that the methodology described is only suitable for good-quality, high-coverage genomes; the annotation of low-coverage genomes will be discussed in Section 9.2.1.

At this stage, the Ensembl genebuilds provided for human and mouse are combined with a gene set accessible via the Vertebrate Genome Annotation (Vega) database (Wilming *et al.*, 2008; http://vega.sanger.ac.uk), a central repository for the manual annotation of vertebrate sequences. This annotation, generated by the HAVANA team at the WTSI, is described as manual since each gene model is manually reviewed. In short, this involves the confirmation and recording of the genomic coordinates of the start and stop codons of each gene, as well as every splice site. Transcriptional evidence remains central to this process, and since objects can be constructed based on ESTs alone, the Vega gene set includes a large number of objects representing alternative spliceforms. Furthermore, annotators also classify transcripts as retained introns where the RNA has not finished the splicing process, and as artifacts where a cDNA appears to represent poor quality sequence. Finally, the annotator is also free to consider information of any kind available in constructing a model. This may include the results of *ab initio* programs as well as automated gene-building methodologies such as Ensembl. Further support often comes from published literature relevant to a particular gene, and the identification of genomic motifs in the form of CpG islands, and polyadenylation (polyA) signals and sites.

In comparison to the Ensembl genebuilds, the manual annotation gene sets for the human and mouse genomes in Vega are not yet complete. In order to provide the best current representation for these genomes, HAVANA genes from the Vega database are now merged with the Ensembl genebuild. In human, the resulting genebuild is referred to as the GENCODE data set. GENCODE represents an effort to identify and map all protein coding genes within the boundaries of the ENCODE (Encyclopedia of DNA Elements) project, which itself seeks to understand all functional elements in the human genome. The initial pilot phase of the ENCODE project used 1% of the genome as a testing ground to compare different strategies of element discovery, and these regions were subjected to HAVANA annotation (Harrow *et al.*, 2006). The predicted gene models were then subjected to experimental validation: RACE data supported the 5′ extension of 30 loci and the addition of novel splice variants to 50 loci, whereas RT-PCR confirmed 47 out of 161 transcripts which had not been classed as 'known' based on correspondence to curated UniProt proteins. Finally, RT-PCR was also used to check *ab initio* models that were not supported by HAVANA annotation, that is, lacked transcriptional support; only 1.2% of splice junctions were confirmed, adding confidence in the completeness of the GENCODE gene set.

The NCBI (National Center for Biotechnology Information) is also combining manual and automated annotation in the construction of its RefSeq collection, although the methodology used is different to that used by Ensembl/HAVANA (Pruitt *et al.*, 2005). Ultimately, RefSeq aims to complete a non-redundant data set of all naturally occurring nucleotide and protein molecules for the organisms of major scientific interest. Rather than using the genomic sequence as a starting point, however, the core RefSeq process focuses on extracting transcripts from GenBank and linking them to the correct protein translations, which are then in turn linked to the chromosome (and to a wide variety of other genome resources). This alignment stage focuses on the use of full-length cDNAs; unlike for HAVANA annotation, ESTs are not commonly used. In the case of human and mouse, the RefSeq collection contains both models that were constructed computationally before being subjected to manual curation (prefixed as NMs), and models that remain as computational predictions (XMs), constructed using the Gnomon gene prediction tool (www.ncbi.nlm.nih.gov/genome/guide/gnomon.shtml). Again, external data such as publications or expert opinions may be consulted during manual curation. Finally, the RefSeq data set is also central to the fully automated human genebuild constructed by the University of California, Santa Cruz (UCSC), since RefSeq models are represented as UCSC genes where successfully aligned to the genome (Hsu *et al.*, 2006). UCSC genes are also constructed based on GenBank RNAs where at least one further piece of supporting evidence is found, typically CCDS (consensus coding sequence) models or human or mouse ESTs. This genebuild, as well as the alignments of other data sources such as transcriptional evidence, can be visualized using the UCSC genome browser (http://genome.ucsc.edu).

9.1.4   A comparison of annotation processes

The ideal annotation pipeline would have two key characteristics: speed and accuracy. Automated annotation processes such as Ensembl are significantly faster than the manual approaches used by HAVANA and RefSeq; the less sophisticated *ab initio* programs are faster still. On the other hand, a fully manual approach is the most accurate in describing a particular locus since an annotator has the flexibility to integrate and filter information available on that locus from a wide variety of sources. However, the major drawback to manual annotation is that the process is slow and labor intensive, and thus expensive. Today, the number of genome sequences that require annotation is increasing rapidly; in 2006 there were over 800 eukaryotic genome projects underway, many of which have limited funding and a small research community (Liolios *et al.*, 2006). Automatic annotation is absolutely necessary in such cases, and this process will typically take a comparative approach whereby gene models are constructed based on the genomic alignment of existing models from related species; this is discussed further in Section 9.2.1.

How close, then, can automated annotation processes get to the accuracy standards of the manual efforts? The accuracy of a variety of automated gene prediction methodologies was judged in 2005 as part of the ENCODE genome annotation project (EGASP), via a comparison with HAVANA manual annotation (Guigo *et al.*, 2006; Guigo and Reese, 2005). This study included single-genome *ab initio* algorithms, programs using transcriptional evidence or multiple genomes, as well as more complex annotation pipelines such as Ensembl. Overall, automated annotation was seen to be quite effective at identifying genes, with the best methods predicting over 70% of the loci that had been identified manually. For all processes considered, however, the actual description of the underlying transcript structures was less accurate than for manual annotation, achieving 40–50% accuracy. In particular, it became clear that automated annotation is weaker in the description of alternative splicing, a process which affects the majority of mammalian genes (see also Section 9.2.3). Overall, methodologies utilizing transcriptional data were seen to be superior in general to those based on only genome comparisons, which were in turn more successful than the single-genome *ab initio* algorithms.

In the future, single-genome based *ab initio* gene prediction is likely to be the first choice option only for genomes which are both (1) significantly diverged from other well-studied genome sequences (thus ruling out a comparative approach), and (2) lacking in accompanying transcription data. Such genomes, however, do exist; amongst prokaryotes in particular. Furthermore, GENSCAN and GENEID played a vital role in the recent initial annotation of the fish *Tetraodon nigroviridis* (Jaillon *et al.*, 2004). Finally, a large proportion of genome projects currently underway will generate draft sequence of low coverage (as low as 2×; see Section 9.2.1). Prior to a full Ensembl genebuild, the assembly of a new low-coverage genome is released with a GENSCAN prediction set in the PreEnsembl site (Hubbard *et al.*, 2009). In contrast, reference genomes such as human and mouse that have high-quality finished sequence will be subjected to the most sophisticated gene build annotation processes available, with the focus placed on achieving a 'reference' gene set.

9.1.5   The CCDS project

Having dissected the differences between these reference annotation strategies, it should be noted that the Ensembl/HAVANA, RefSeq, and UCSC groups are in fact collaborating to produce a unified coding gene annotation set available via the major genome browsers. This initiative, the consensus CDS (CCDS) project, aims to provide a single set of gene translations for the human and mouse genomes for which agreement has been reached regarding the placement of the ATG and termination codons (Pruitt *et al.*, 2009). It began with the alignment of all human and mouse RefSeq models to the respective genome, followed by an ongoing comparison of these models with the equivalent Ensembl/HAVANA genes. Where agreement is found, the coding transcript is promoted to the CCDS database. Where there is a discrepancy, a discussion is entered into, and a consensus reached as to the best way to represent the locus.

The nature of the disagreements typically identified provides further insights into the different nature of these annotation processes. In many cases the underlying issue relates to the transcriptional evidence; for example, the lack of a full-length transcript may leave a gene structure truncated. Many such loci remain in both the human and mouse genomes. Further problems can be encountered when annotating the ATG codon of the CDS (coding sequence); again, this often results from an interpretation of incomplete transcriptional evidence. For example, where the gene model has been constructed based on a single cDNA, it is often possible to extend the object further at the 5' end using ESTs. In turn, this can lead to the introduction of a potential ATG codon upstream of that which was previously suspected to mark the 5' end of the CDS. However, this ATG codon is not selected simply on the basis that it demarcates the longest CDS, since it is known that the most 5' ATG is not always that actually used in translation. An upstream ATG may gain favor if it has a strong Kozak consensus sequence: a conformation of base pairs flanking the codon known to promote its usage (Kozak, 1984). Evolutionary conservation, however, provides an argument for change with more weight: if an upstream ATG in human, for example, is also seen to be supported in other species, typically mouse, then this is taken as strong evidence for the functionality of this codon.

9.1.6   Pseudogene annotation

Pseudogenes are sequence elements formed from protein coding genes, with translational potential having been lost due to a combination of truncations, frameshifts, and nonsense mutations. Their correct identification and classification is essential in fully describing the genome of a species, since they are present in high abundance; 22 600 are currently predicted to exist in the human genome, suggesting that they outnumber protein coding genes (www.pseudogene.org). In particular, a properly annotated pseudogene collection is critical in recapitulating the evolutionary history of an organism, and also a valuable resource in population genetics. Pseudogenes can be formed by three mechanisms, as summarized in Figure 9.2: retrotransposition, duplication, and inactivation. In the first instance, an mRNA sequence is transcribed from a particular gene, before being fully or partially retrotransposed back into the genome sequence; the inserted sequence is referred to as a 'processed pseudogene.' In the second instance a gene becomes

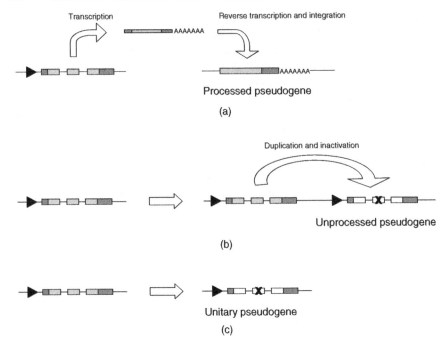

Figure 9.2   Three mechanisms of pseudogene formation. (a) Processed pseudo-genes are formed where an mRNA undergoes reverse transcription and integrates into the genomic sequence; the site of integration is essentially random. Processed pseudogenes lack promoter elements (black triangles) and may be truncated at one or both ends (UTRs are shown in dark grey). They can be recognized due to their lack of intronic sequences and common integration of the polyA tail. (b) Unprocessed pseudogenes are formed by the full or partial duplication of a progenitor gene, typically in tandem. If the new gene is intact on arrival it may either acquire functionality as an additional copy, or else undergo pseudogenization by mutation (shown as X). (c) Unitary pseudogenes result from the deactivation of an existing gene by mutation; the locus is subsequently fixed in a population due to genetic drift or natural selection.

duplicated, either partially or entirely, and the duplicated sequence either cannot function or loses the ability to function due to mutation over the course of time; this is an 'unprocessed pseudogene.' Finally, an existing non-duplicated gene may undergo an inactivating mutation, forming a 'unitary' pseudogene.

The annotation of pseudogenes is not straightforward. Automated prediction is most successfully performed when separated from the gene building efforts described in Sections 9.1.2 and 9.1.3; Ensembl and the automated Ref-Seq pipeline, for example, do not target pseudogenes (although they do highlight possible pseudogenization events where gene objects contain frameshifts). Furthermore, the majority of pseudogene pipelines are only designed to target one category of event, typically retrotransposition or duplication. For example, Suyama *et al.* recently used an integrated homology and orthology analysis independent of current gene annotation to identify 9484 and 9017 duplicated genes in human and mouse respectively (Suyama *et al.*, 2006). This data set was further separated into active and inactive genes, thus identifying 1811 duplicated pseudogenes in human and 1581 in mouse. These results, representing the largest collection of non-processed pseudogenes available, can be accessed via the pseudogene.org website; the major resource for both eukaryotic and prokaryotic pseudogenes.

Manual annotation is also suited to pseudogene identification, since each gene alignment is analyzed on a case-by-case basis; this also allows pseudogenes to be categorized according to their method of formation. What are the relative merits of automated and manual pseudogene identification methodologies? This question was investigated by the GENCODE consortium. Their initial goal was to obtain an accurate annotation of pseudogenes in the ENCODE pilot regions; 44 sections of the human reference assembly representing 1% of the total genome sequence (Zheng *et al.*, 2007). The predictions of four automatic pseudogene discovery programs were compared alongside the HAVANA manual annotation: the GIS-PET method from the Genome Institute of Singapore, which infers pseudogenes where paired-end diTags map to multiple genomic locations; PseudoPipe from Yale University, which is based on the BLAST alignments of existing proteins (Zhang *et al.*, 2006); pseudoFinder from the University of California, Santa Cruz (UCSC), which identifies homologies to reference genes using Human Blastz Self Alignment; and pseudoFinder, also from UCSC, which targets retrotransposition events based on the genomic alignment of total human mRNA content. The resulting data sets were quite dissimilar, with less than 20% of the total pseudogene content being identified by all methods. The critical factor in this disagreement was not the actual design of the pseudogene pipelines; rather it was seen to be the annotation quality of the parent genes/proteins which were used to make the predictions. In particular, the data set was seen to be contaminated by pseudogene predictions that were based on parent CDSs seen, in retrospect, to represent spurious translations (i.e., false positives). Clearly, then, the availability of a high-quality gene set to use as a starting point is a crucial resource for pseudogene prediction.

Automated annotation, on the other hand, seems to be advantageous in identifying small degenerate blocks of CDS alignment representing older retrotransposition events that can be overlooked by manual annotators. However, as well as generating false positives due to dubious input data, automated annotation often struggles in making the distinction between pseudogenes and coding genes. Unprocessed pseudogenes generally contain at least partially intact exonic structures, and if a significant CDS is detected, such a locus can be erroneously classed as a coding gene. Similarly, a unitary pseudogene that has become inactivated in recent time – due to perhaps a single mutation – is commonly described as a coding locus (recent pseudogenization events may in fact reflect polymorphism; this is discussed in Sections 9.2.4 and 9.2.5). It is in such situations that the power of the manual annotation approach is apparent, since the rate of false positive predictions is very low.

In summary, there is currently no single prediction method for successfully identifying all categories of pseudogenes. The strategy being used by the GENCODE consortium to describe the human content is therefore a consensus approach: HAVANA manual annotation based on both transcriptional data and the results of automatic pseudogene identification pipelines (Zheng *et al.*, 2007). The results of such human annotation can be viewed in the UCSC genome browser (Kent *et al.*, 2002).

9.1.7   The annotation of non-coding genes

The structures of around 20 000 protein coding genes are contained within fewer than 2% of the human genome sequence (Lander *et al.*, 2001). In recent years, however, it has become apparent that a sizeable proportion of the remaining 98% of the genome sequence is in fact transcribed, producing non-coding RNAs (ncRNAs) (Bertone *et al.*, 2004; Birney *et al.*, 2007). This category includes the 'historical' tRNA and rRNA classes of molecule, as well as snoRNAs, spliceosomal RNAs and miRNAs. Each of these classes is well represented in the Rfam database, which provides a general resource for ncRNAs and other RNA elements (Gardner *et al.*, 2009). Rfam currently contains 1537 RNA families, including, for example, 497 human nuclear tRNA genes. Information on miRNAs is also collected within the miRBase repository; at present 5071 molecules from 58 different species are represented (Griffiths-Jones *et al.*, 2008).

A distinct class of ncRNA has been the recent focus of excitement in the RNA community: large non-coding RNAs, or lncRNAs. The existence of a 'silent majority' of non-coding transcription had in fact been suspected for a number of years; the RIKEN Mouse Gene Encyclopedia Project, for example, fully sequenced over 60 000 mouse cDNA molecules and classified 12 000 as non-coding (Carninci *et al.*, 2005). Indeed, genome-based microarray studies suggest that transcription is essentially ubiquitous across the human genome (Birney *et al.*, 2007; Cheng *et al.*, 2005; Kapranov *et al.*, 2002; Willingham and Gingeras, 2006); the challenge is now to ascribe functionality to these molecules (Ponting *et al.*, 2009). Recently, Guttman *et al.* used global chromatin maps in order to identify mouse lncRNAs that were likely to be functional; of the 1600 lncRNAs identified, 95% showed evidence of purifying selection (Guttman *et al.*, 2009). Even so, lncRNAs as a whole remain poorly understood at present, with only around a dozen individual molecules having a function described with confidence. A noteworthy example is HOTAIRM1, which is an antisense transcript to *HOXA1* and *HOXA2*, originating from the same CpG island promoter; it appears to function in the modulation of HOX cluster gene expression (Rinn *et al.*, 2007).

The major challenge in identifying and annotating lncRNAs is obtaining certainty that the transcript does not code for a protein. This problem is perhaps better dissected by approaching it from the other side: how can we be sure that an annotated CDS is not a spurious prediction built on an ncRNA? Automated annotation methods for gene finding are typically biased towards the construction of coding genes, generating false positive CDSs at reasonable frequencies. These can arise where translations are attached to pseudogenes which still undergo transcription, as discussed above, and also where a CDS is predicted within a single block of alignment – such as non-splicing cDNA or a 3' UTR – simply because the algorithm has identified a decent sized ORF. Even manual annotation can have difficulty in distinguishing certain coding and non-coding loci.

In practice, the majority of mammalian CDSs are easy to confirm simply because an enormous number of protein coding genes are now 'known', that is, already identified and curated by a project such as RefSeq or Swiss-Prot (although it should be noted that neither database is immune to false positives). If a potential CDS is not known, it may be possible to identify either a curated ortholog or a paralogous gene family member. Alternatively, the CDS may have homology to known functional domains, as assessed by querying a resource such as Pfam (Finn et al., 2008). Finally, for a multiexon gene, the presence of a CDS is often immediately apparent simply by its size and exonic structure; a 500 amino acid translation covering 10 exons of a 10-exon gene is unlikely to represent a false prediction. If the potential CDS lacks all of the above, then there are two possibilities: either the locus encodes a legitimate 'orphan gene' (i.e., a completely novel gene which is lineage specific), or the translation is indeed spurious. This represents a common dilemma in genome annotation, and these two possibilities are difficult to distinguish with complete certainty. In lieu of the generation of proteomics data, one approach that has had some success is to consider the codons that make up the potential CDS. It has been shown that true CDSs can often be distinguished from ORFs by the frequency with which certain DNA triplets occur; in short, the DNA content of an ORF is essentially random, whilst that of a CDS is not. It is this basic approach that has been used to classify lncRNAs as noncoding in global analyses such as the chromatin mapping performed by Guttman et al. (2009). In addition, a recent study comparing a set of over 1000 predicted human orphan genes taken from the Ensembl genebuild (filtered to remove artifacts) against the mouse and dog genomes found that over 99% had codon frequency scores that did not indicate CDS potential (Clamp et al., 2007). This investigation therefore suggests that truly new proteins evolve rarely in the mammalian lineage. Such studies are undermined by the lack of equivalent proteomics data to complement the high coverage of transcriptomes available to researchers. However, efforts are currently underway to improve this situation; recent technical advances in mass-spectrometry, for example, are allowing global analyses of protein content to be performed for the first time (Gstaiger and Aebersold, 2009).

In summary, aside from the small number of cases where experimental support for activity is available, it can be seen that the annotation of lncRNAs is currently a process based on negative evidence; annotation projects treat lncRNAs simply as gene models for which a realistic CDS cannot be found. In the future, however, more information about ncRNAs is likely to become available – perhaps in the form of a greater understanding of secondary structures or functional sites – and it may be that the identification of such genes will become a more proactive process.

9.2   The impact of next generation sequencing on genome annotation

The human genome took ten years to sequence, and the HGP remains one of the most expensive scientific endeavors ever conceived (Lander *et al.*, 2001). In contrast, the 1000 genomes project has deposited more than 80-times as much human genome sequence in public databases in under a single year of operation (www.1000genomes.org; see Section 9.2.5). The key to this massive increase in output is the rapid improvements made in DNA sequencing technology (Schuster, 2008; www.solexa.com; www.454.com; www.appliedbiosystems.com). As well as providing more human genome sequences, such 'next generation' sequencing platforms are also being used to rapidly generate genome assemblies for an ever-increasing number of other species. As a result, scientists are gaining a deeper understanding into both the genetic variation that exists within our species, and also those genome changes which demarcate the branches on the evolutionary tree of life. Furthermore, the same essential technology is also being used in the field of transcriptomics, leading to an exponential rise in the amount of transcriptional evidence available with which to construct gene models. In this section, we will discuss the benefits of this next-generation technology to genome annotation.

9.2.1   The annotation of multispecies genomes

Evolutionary conservation represents a powerful tool for genome annotation. It is now known, for example, that around 80% of human protein coding genes share a 1:1 orthologous relationship with mouse genes, and that a significant proportion of the remainder are accounted for by lineage-specific duplications within gene families (Church *et al.*, 2009). Existing genome annotation can be mapped with success between species that are not distantly related; the annotation of new genomes should therefore become progressively easier as more genomes are described.

This theory was recently put into practice in a comparative study within the *Drosophila* genus. Ten *Drosophila* genome sequences were generated at differing levels of coverage for comparison against the pre-existing *Drosophila melanogaster* and *D. pseudoobscura* genomes (Clark *et al.*, 2007). In this project, pair-wise alignments with *D. melanogaster* and *D. pseudoobscura* were essential in converting the new scaffolds into contigs, as well as facilitating comparative gene annotation via the projection of *D. melanogaster* gene models onto each genome. This process was supplemented with *ab initio* gene prediction methodologies, a requirement due to both the quality of the assemblies and also the evolutionary distance between the species; the relative radiation in the *Drosophila* genus is actually greater than of the mammalian clade. Stark *et al.* subsequently used this data set to uncover a vast array of functional elements in the fly, leading to the modification of 438 gene models within the manually annotated FlyBase gene set (Stark *et al.*, 2007). Lin *et al.* further demonstrated the power of this phylogenomics strategy, using the 12 genomes to develop a comparative genomic metric to distinguish protein coding and non-coding gene regions, allowing the identification 142 new genes (Lin *et al.*, 2008).

The Ensembl analysis pipeline operates along similar principles; as noted in Section 9.1.3, the alignment of protein sequences from related organisms is a key stage of every new gene build. Furthermore, Ensembl genebuilds can take this process further where the new genome is particularly close to a high-quality genome such as human or mouse. The majority of gene models in the genome of the orangutan *Pongo pygmaeus*, for example, were extrapolated from the direct projection of the existing human models, thus circumnavigating the relative lack of transcriptional evidence in this species (Hubbard *et al.*, 2009). Such an approach was possible because the orangutan genome sequence is considered to be good quality; it has '6× coverage,' meaning that on average each base in the genome will have been sequenced six times. Whilst the accuracy of the new sequencing technologies is not lower than that of the old methods *per se*, the high-throughput design of many new genome projects means that the coverages achieved in such cases will be lower. The mammalian genome project is an effort coordinated by the Broad Institute to sequence the genomes of 24 additional mammals to 2× coverage, with the ultimate purpose of aiding the identification of functional human elements by comparative mapping (www.broad.harvard.edu/science/projects/mammals-models/mammalian-genome-project). The annotation of such genomes is problematic since the underlying draft sequences are fragmentary compared with high-coverage genomes, meaning that genes are frequently incomplete or split into pieces on distinct sequence 'scaffolds.' Ensembl have therefore developed a modified pipeline for these genomes. Firstly, a whole genome alignment is performed between the low coverage genome and an annotated reference genome (typically human). The gene structures in the reference genome are then used to arrange the scaffolds into contigs containing intact gene sequences, and CDSs in these gene models are extrapolated and corrected as required based on the projection of the total CDS content from the reference genome. In the future, however, it seems fair to assume that the quality of sequencing protocols will improve, as will the strategies used for sequence assembly. In particular, there is some excitement regarding the potential of nanopore technology, which may offer the possibility of directly sequencing single molecules of DNA (Ashkenasy *et al.*, 2005).

Finally, it should be noted that, while comparative annotation allows us to identify the similarities between two genomes, it is less successful at identifying those differences that ultimately distinguish the species at the phenotype level. Two species genomes, for example, will typically be distinguished by subsets of lineage-specific genes; such loci can prove difficult to identify by phylogenomics (Toll-Riera *et al.*, 2009). Furthermore, the problems discussed in Section 9.1.6 regarding automated pseudogene annotation are exacerbated when projecting coding genes from one genome onto another.

9.2.2 Community annotation

Web-based community annotation projects aim to manually annotate or curate sizeable data sets via the remote involvement of a large number of researchers across the globe. The logic behind such projects is simple: the speed of manual annotation can be dramatically elevated by a concomitant increase in the number of annotators involved. These can be completely open, where anyone with an internet connection can submit information of their choosing on a purely voluntary basis, or restricted to a particular community of researchers. Such projects are particularly valuable for species where comparative annotation is of limited usefulness, that is, those without closely related counterparts in other genome sequencing projects. In the case of the fungus *Aspergillus nidulans*, the latter approach is being used; 32 laboratories have been involved in manually annotating and curating over 2500 protein coding genes thus far, providing detailed information on inferred functionality (Wortman *et al.*, 2009). The RNA WikiProject, in comparison, uses the Wikipedia schema to provide an open-source community resource for ncRNA molecules. Researchers with knowledge of ncRNA are invited to submit functional annotation of specific ncRNA molecules in conjunction with the usual publication process (Daub *et al.*, 2008). More recently, a Gene Wiki portal has been set up at the Wikipedia Web site, aiming to provide functional information for every human gene in the form of user-generated content (Huss *et al.*, 2008). This initiative thus seeks to rival the Entrez Gene hub at the NCBI web server, and the advantage of the Wiki approach has been stated as an increase in the speed and flexibility of reporting.

However, several criticisms can be leveled at open community annotation projects. Firstly, there is some skepticism that trained scientists will freely give up their time and effort in suitable numbers to annotate or curate gene articles, a process for which they will achieve no tangible reward or recognition. Secondly, there are questions regarding the accuracy of the entries. While the malicious hijacking of content does not appear to be a major issue in practice, there are concerns that the quality of the annotation may not always be up to the acceptable standard. Furthermore, there may be issues of consistency, even when entries are produced by competent scientists following written guidelines; one researcher may accept a borderline CDS as valid, whereas another one may reject it. One highly desirable characteristic of a complete genome annotation set is that all loci are built along precisely the same criteria. This is of course a natural characteristic of automated annotation methodologies, and it can be achieved for manual annotation efforts such as HAVANA, RefSeq and UniProt, where the number of annotators is small enough to keep the project training centralized in one geographic location.

For community annotation, a common solution to this problem is to bring a large number of researchers together in one location over a number of days, allowing gene building methods to become standardized and a large number of loci to be constructed. The success of such 'jamborees' was demonstrated by the preliminary annotation of the complete *D. melanogaster* genome, which was manually produced by 14 scientists over a two-week period at the Celera institute (Adams *et al.*, 2000). More recently, genome or mRNA annotation jamborees have been held for *Xenopus*, pig and cow (unpublished data); with the rapid increase in genome sequences requiring annotation, such efforts are likely to remain common over the next few years. For such efforts to succeed, it is vital that the data produced is made freely available in a browser or database. Ensembl, for example, allows members of the zebrafish community to add functional data to the reference sequence via the DAS track provided.

9.2.3 Alternative splicing and new transcriptomics data

Traditionally, the transcriptome was regarded as the total set of RNA molecules produced by a particular species. Today, a tighter definition defines it as a specific transcript set for an individual cell at a particular stage of development, including information on the relative abundance of each RNA (Wang *et al.*, 2009). Clearly, the human physiology supports a large number of distinct transcriptomes. The production of these different transcript sets ultimately occurs via gene regulation, which has both a qualitative and quantitative effect on the RNA molecules transcribed. Alternative splicing is one of the major mechanisms by which the action of gene regulation is manifested. The majority of human genes are subjected to alternative splicing, leading to an increase in both the size of the proteome and overall metabolic complexity (Matlin *et al.*, 2005). At present, there is a general lack of experimental data linking specific alternative protein isoforms to alternative functionality. However, it has recently been shown that global splicing patterns in the human genome can frequently be correlated with tissue-specific expression profiles (Wang *et al.*, 2008a). It is therefore clear that a full understanding of human metabolism will depend on a complete description of alternative splicing within our genome.

While certain genomic motifs influencing alternative splicing have been described (Wang and Burge, 2008), our current understanding as to how alternative transcripts are generated at the sequence level remains limited. At present, it is not possible to predict the occurrence, let alone the structure, of alternative transcripts based on the genome sequence alone. As such, while *ab initio* annotation methods can locate individual splice sites with reasonable efficiency, they cannot recapitulate actual alternative splicing patterns. Instead, annotation must depend entirely on transcriptional evidence. Traditionally, as discussed in Section 9.1, this entails constructing gene models based on the alignment of ESTs, mRNAs, and cDNAs (i.e., low-throughput libraries) against the genome sequence, either manually or computationally.

At present, only a small number of other genomes possess a level of EST and mRNA-based transcriptional coverage comparable to that seen for human, largely because generating RNA libraries of appreciable size is expensive and labor-intensive. However, recent years have seen a shift from the production of such low-throughput libraries to the use of new high-throughput methodologies. The first step forward into the new era of 'transcriptional profiling' was taken by the development of the serial analysis and cap analysis of gene expression systems (SAGE and CAGE respectively; Shiraki *et al.*, 2003; Velculescu *et al.*, 1995). These methodologies both generate large sets of 'tags' corresponding to distinct mRNAs isolated within the target cell. While these tags are short, typically around 20 nucleotides long, they are useful for profiling the expression levels of different mRNAs within the sample. Even so, this process is not efficient; the majority of tags within a given set cannot be unambiguously mapped to the genome sequence, and it is thus difficult to distinguish individual splice variants within a particular locus. For these same reasons, the use of CAGE and SAGE tags in genome annotation has thus far met with limited success; in particular, it has proved very difficult to incorporate this information into automated annotation pipelines. CAGE tags have displayed one practical benefit to manual annotation, however; since they are designed to capture information at the 5' end of mRNAs, they often provide support for alternative transcriptional start sites.

Instead, the true transcriptomics revolution looks set to begin with the advent of the RNA-Seq methodology (Morin *et al.*, 2008). This technique, which is still being perfected, utilizes essentially the same 'next generation' sequencing technology that is driving the explosion in genome sequencing described above. In short, one of a number of modern 'deep sequencing' platforms is used to rapidly produce millions of 30–400 bp RNA fragments from a total RNA sample; such fragments can then be aligned to genome sequences in order to facilitate transcriptional profiling. RNA-Seq is significantly faster and cheaper than the CAGE/SAGE methodology, and the generation of reads is not limited to the transcript ends. Furthermore, RNA-Seq libraries provide highly accurate data on levels of transcription; information that, at present, is lacking from all genome annotation projects. Finally, RNA-Seq can be used to provide information on alternative splicing. This technology thus promises to allow us to compare (and annotate) the transcriptional levels of distinct splice variants, and also to identifying novel splice junctions that have not been picked up by the traditional Sanger sequencing pathways (Wang *et al.*, 2009).

The challenge, then, becomes the integration and interpretation of this information into annotation programs. Once again, the major problem is the computational mapping of millions of relatively small RNA fragments (most of which are 70–110 bp) to a genome sequence, and there are several issues to be resolved. Firstly, experience with large genomes indicates that reads are frequently mapped with equivalent fidelity to more than one genomic location. Such reads have typically been discarded as non-informative matches from those global transcriptome analyses published thus far (given that the number of reads in a library typically runs into the millions, this does not necessarily lead to a major loss of coverage). However, this should become less of a problem over time as read lengths inevitably increase; alternatively, a paired-end approach where reads are taken from both ends of an mRNA is proving to be helpful. Secondly, mapping is greatly complicated by the fact that the data sets contain a mixture of splicing and non-splicing sequences. Reads which do not contain intronic sequence are relatively easy to align (assuming they do not align to multiple locations); although in terms of annotation their use is likely to be limited to the fine-mapping of transcriptional start sites and polyadenylation sites (i.e., the start and end points of a transcriptional unit). In contrast, reads which span exon junctions can be difficult to align, since the portion of sequence covering one exon may be small. This problem can be reduced to some extent by mapping the reads to a database of known splice junctions rather than the total genome sequence; while this solution does not support the identification of previously undescribed splice sites, it can successfully identify exon skipping events (which represent one of the most common forms of alternative splicing). However, the identification of novel splice sites within the RNA-Seq data sets is highly desirable, and this has been achieved in mouse with some degree of success by aligning the reads against a set of potential splice junctions extrapolated from within known exonic and intronic sequences (Mortazavi *et al.*, 2008). Alternatively, novel exons have been found by clustering groups of reads alongside the genomic alignment of pre-existing ESTs (Sultan *et al.*, 2008). In spite of the remaining technological issues, a recent RNA-Seq-based survey of human transcription showed that up to 94% of genes undergo alternative splicing, a higher proportion than previously estimated (Wang *et al.*, 2008a). Furthermore, RNA-Seq has been used to survey the existing annotation of the nematode *Caenorhabditis elegans*, identifying both new exons and splice junctions as well as at least 80 putative new genes (Hillier *et al.*, 2009).

In the near future, next-generation genome sequencing will be combined with RNA-Seq in the description of new genomes. While RNA-Seq alone is likely to be of limited value in the *de novo* annotation of new genomes (since mapping the reads successfully depends on pre-existing annotation), a gene set constructed by phylogenomics could be used as the scaffold onto which the reads are aligned; indeed, the alignment of these reads could then be used reciprocally to improve the gene models. In fact, it is now possible to perform in-depth transcriptome surveys of species for which no genome species is available, since RNA-Seq reads can be combined *ab initio* into whole mRNA molecules.

9.2.4 The annotation of human genome variation

To date, annotation of the human genome has been focused almost entirely on the primary 'reference' assembly which was published in 2001 (Lander *et al.*, 2001), in large part simply because until recently this was the only human genome sequence available (with the exception of the competing assembly generated by Celera (Venter *et al.*, 2001)). Whilst numerous changes to the assembly have been made since 2001, these have taken the form of small, localized improvements as opposed to large-scale rebuilds. Historically, therefore, annotators have regarded this reference assembly as an essentially static entity; a mosaic of DNA fragments from several individuals intended to provide a genome sequence representative of our species. Today, it is well established that the genomes of individuals contain both small-scale and large-scale differences when compared against one another. Base pair mutation causes single nucleotide polymorphisms (SNPs), and it is generally estimated that the SNP content of any two human genomes leads to an overall sequence difference of approximately 0.1% (International HapMap Consortium, 2003). SNPs can be silent, that is, without consequence, or lead to phenotypic change, for example where CDSs are affected. However, it is now known that the greater proportion of total genome variation is caused by instability at the chromosome level, leading to sequence duplications, deletions and other rearrangements (Iafrate *et al.*, 2004; Sebat *et al.*, 2004). Duplications and deletions are often classed as copy number variants (CNVs); segments of DNA over 1 kb in size which have been shown to present in differing copy numbers in at least two individuals (Feuk *et al.*, 2006). As with SNPs, CNVs may lead to phenotypic consequences for the individual.

Genomic variation within gene loci often becomes apparent based on the alignment of transcriptional data. EST and cDNA libraries are derived from a wide range of individuals; hence mutations in the reference genome are often apparent as mismatched base pairs. Pertinently, manual annotators have found numerous CDSs in the reference assembly that have been disrupted by one or few mismatches, typically leading to the introduction of a premature STOP codon. If transcriptional evidence exists which does not support this apparent pseudo-genization event – that is, supports an intact CDS – then this locus represents a polymorphic pseudogene. It is possible to extend this process by considering the large set of SNPs genotyped by the HapMap project (International HapMap Consortium, 2003; Frazer *et al.*, 2007); that is, a SNP may be found which supports an intact CDS in another individual. Since the reference assembly is intended to represent the genome of our species and not an individual, it has been decided that polymorphic pseudogenes are to be replaced with intact CDSs where appropriate evidence is found (with specific caveats for certain gene families; see below); this is happening under the jurisdiction of the Genome Reference Consortium (GRC; http://ncbi.nlm.nih.gov/projects/genome/assembly/grc/).

9.2.5   The annotation of polymorphic gene families

Annotation projects are particularly affected by genome variation when tackling certain gene families. Often, a cluster of genes expands in a locus from a single progenitor via tandem duplication, as occurred during the initial evolution of the HOX gene family. The four mammalian HOX clusters can be described as stable, since the individual gene members are largely common to the order, with orthology readily assigned between species (Lemons and McGinnis, 2006). However, other gene clusters are dynamic, displaying polymorphism due to both ongoing base-pair mutation and chromosomal instability. Such rapidly evolving gene families are typically involved in chemosensation, reproduction, or immunity; where there is selective pressure to generate a flexible pool of protein (Church *et al.*, 2009; Horton *et al.*, 2008; Mudge *et al.*, 2008). A major problem in the annotation of unstable gene families is that the underlying genome tiling paths are often incomplete, and may in certain cases be misassembled. This is largely because the duplicative units making up the cluster ('cassettes') tend to be highly similar to one another at the sequence level, making it difficult to select the correct clones for sequencing. This occurs where the formative duplications are recent events, though gene conversion can also act to homogenize the cassettes (Chen et al., 2007). During annotation, the presence of gap regions leads to genes being fragmented, such that partial models have to be constructed. Furthermore, where neighboring cassettes are highly similar at the sequence level, the genes contained within are often seen to be identical. This can cause problems for automated annotation pipelines such as Ensembl, since cDNAs can become aligned so as to span adjacent loci by using a combination of exons from each; the result is the creation of an erroneous gene model. Manual annotation is thus required to correctly represent such loci. Finally, gene clusters frequently contain unprocessed (often polymorphic) pseudogenes that often represent highly recent inactivation events; this causes confusion in automated annotation pipelines, as discussed in Section 9.1.6.

The high sequence similarity between such cassettes also indicates that the cluster is likely to be subject to CNV. The presence of large-scale polymorphism impacts the sequencing process: gene clusters in the reference assembly can consist of sequence combined from multiple haplotypes, giving a composite haplotype that may not exist in nature. This precise problem was observed in the initial analysis of the major histocompatibility complex (MHC). This highly variable locus represents a collection of genes and gene families covering 4 Mb on chromosome 6, predominantly involved in the immune response. The obvious solution was to re-sequence the cluster as a single haplotype; to date, this remains the largest haploid region of the human reference assembly. The problem with this approach is that this haplotype is linked to a single individual, and as such it does not adequately 'summarize' this region over our species as a whole. In other words, the architecture of the MHC does not exist as a 'parental' state against which the structures of other haplotypes provided for comparison could then be seen to exist as 'derived' states. Indeed, the pattern of variation within the locus is so complex as to make deriving a parental assembly impossible. Instead, the region can only be successfully represented as a series of haplotypes; eight distinct MHC haplotypes have in fact already been sequenced, and these were recently subjected to an automated SNP-based variation analysis in combination with HAVANA manual annotation (Horton et al., 2008).

As well as providing a valuable resource to the medical community, the generation and comparison of eight MHC haplotypes will also provide a useful test-model for the analysis of other complex genome regions. This is pertinent, since the GRC intend to provide alternative haplotypes for all regions in the reference assembly subjected to significant structural variation. Notably, the time and labor costs involved in describing some 30 Mb of MHC sequence to the level of detail desired were sizeable. This is significant, since the amount of genetic variation described for our species is set to increase dramatically in the immediate future. The first additional human genomes to be sequenced were those of Craig Venter and James Watson, published in 2007 and 2008 respectively (Levy et al., 2007; Wheeler et al., 2008), and since this time the genome sequences of a Chinese and Korean individual have also been described (Ahn et al., 2009; Wang *et al.*, 2008b). Of greater potential significance is the data from the 1000 genomes project, which will be released to the public in 2009 (www.1000genomes.org). This multinational initiative was set up in 2008 with the goal of rapidly obtaining a complete catalog of human variants that are present in over 1% of the population, this being achieved by sequencing some 1200 genomes from distinct geographic populations using next-generation technology. This enormous data set will contain SNPs (far more than were generated by the HapMap project) and CNVs. Though this information will undoubtedly be hugely beneficial to genome science, the challenge of understanding and displaying this variation is vast. Annotation projects must evolve in order to cope with this exponential increase in genome sequence.

At present, a working strategy to integrate genome annotation with large-scale variation data sets has yet to be finalized, although wide-scale manual annotation is clearly unsuitable for a data set of this size. Instead, integration must depend heavily on the design of effective computational methodologies. We can anticipate that large-scale efforts to annotate genetic variation of any species will be dependent on the availability or production of a single high-quality reference genome annotation set against which changes can be identified. In theory, the subsequent alignment of (for example) 1000 genome sequences is then essentially a problem of computing power, as is the identification of the individual SNPs contained within. Following this, given a set of high-quality gene models as a template for comparison, automated methodologies are likely to prove effective in identifying pseudogenization events and amino acid substitutions where one-to-one orthologies are inferred. However, one problem is that sequencing errors across 1000 genomes are likely to be relatively common, and these will be difficult to distinguish from low-frequency SNPs. Second, complex gene families are once again likely to prove troublesome, and this is particularly significant given the genic variation that exists in such regions. It remains to be seen whether the shotgun-based next-generation sequencing strategies will be able to successfully provide quality assemblies over these regions, given the high degree of sequence similarity that is often found within. Even if quality assemblies are available, it will likely prove difficult to correctly discern the relationship between gene members of different complex haplotypes using computational means; in other words, to match a particular SNP to the relevant gene. Before solutions are found to such problems, initiatives such as the 1000 genomes project are likely to be of compromised value when it comes to the description of complex gene families. Even so, given the advances made in genomics over the last decade and the pace at which sequencing technology is evolving, it seems fair to assume that any setbacks are likely to be temporary. Indeed, if anything, the speed of the genomics revolution is likely to increase over the second decade of the twenty-first century.

9.3   References

Adams, M.D., Celniker, S.E., Holt, R.A., *et al.* (2000) The genome sequence of Drosophila melanogaster. *Science*, **287**(5461), 2185–95.

Ahn, S.M., Kim, T.H., Lee, S., *et al.* (2009) The first Korean genome sequence and analysis: full genome sequencing for a socio-ethnic group. *Genome Res*, **19**(9), 1622–9.

Ashkenasy, N., Sanchez-Quesada, J., Bayley, H., and Ghadiri, M.R. (2005) Recognizing a single base in an individual DNA strand: a step toward DNA sequencing in nanopores. *Angew. Chem. Int. Ed. Engl.*, **44**(9), 1401–4.

van Baren, M.J., Koebbe, B.C., and Brent, M.R. (2007) Using N-SCAN or TWINSCAN to predict gene structures in genomic DNA sequences. *Curr Protoc Bioinformatics*, Chapter 4, Unit 4.8.

Benson, D.A., Karsch-Mizrachi, I., Lipman, D.J., *et al.* (2009) GenBank. *Nucleic Acids Res.*, **37**(Database issue), D26–31.

Bertone, P., Stolc, V., Royce, T.E., *et al.*, (2004) Global identification of human transcribed sequences with genome tiling arrays. *Science*, **306**(5705), 2242–6.

Birney, E., Stamatoyannopoulos, J.A., Dutta, A., *et al.*, (2007) Identification and analysis of functional elements in 1% of the human genome by the ENCODE pilot project. *Nature*, **447**(7146), 799–816.

Burge, C. and Karlin, S. (1997) Prediction of complete gene structures in human genomic DNA. *J. Mol. Biol.*, **268**(1), 78–94.

Burset, M. and Guigo, R. (1996) Evaluation of gene structure prediction programs. *Genomics*, **34**(3), 353–67.

Carninci, P., Kasukawa, T., Katayama, S., *et al.* (2005) The transcriptional landscape of the mammalian genome. *Science*, **309**(5740), 1559–63.

Chen, J.M., Cooper, D.N., Chuzhanova, N., *et al.* (2007) Gene conversion: mechanisms, evolution and human disease. *Nat. Rev. Genet.*, **8**(10), 762–75.

Cheng, J., Kapranov, P., Drenkow, J., *et al.* (2005) Transcriptional maps of 10 human chromosomes at 5-nucleotide resolution. *Science*, **308**(5725), 1149–54.

Church, D.M., Goodstadt, L., Hillier, L.W., *et al.* (2009) Lineage-specific biology revealed by a finished genome assembly of the mouse. *PLoS Biol.*, **7**(5), e1000112.

Clamp, M., Fry, B., Kamal, M., *et al.* (2007) Distinguishing protein-coding and noncoding genes in the human genome. *Proc. Natl. Acad. Sci. U.S.A.*, **104**(49), 19428–33.

Clark, A.G., Eisen, M.B., Smith, D.R., *et al.* (2007) Evolution of genes and genomes on the Drosophila phylogeny. *Nature*, **450**(7167), 203–18.

Daub, J., Gardner, P.P., Tate, J., *et al.* (2008) The RNA WikiProject: community annotation of RNA families. *RNA*, **14**(12), 2462–4.

Dunham, I., Shimizu, N., Roe, B.A., *et al.* (1999) The DNA sequence of human chromosome 22. *Nature*, **402**(6761), 489–95.

Feuk, L., Carson, A.R., and Scherer, S.W. (2006) Structural variation in the human genome. *Nat. Rev. Genet.*, **7**(2), 85–97.

Finn, R.D., Tate, J., Mistry, J., *et al.* (2008) The Pfam protein families database. *Nucleic Acids Res*, **36**(Database issue), D281–8.

Frazer, K.A., Ballinger, D.G., Cox, D.R., *et al.* (2007) A second generation human haplotype map of over 3.1 million SNPs. *Nature*, **449**(7164), 851–61.

Gardner, P.P., Daub, J., Tate, J.G., *et al.* (2009) Rfam: updates to the RNA families database. *Nucleic Acids Res.*, **37**(Database issue), D136–40.

Griffiths-Jones, S., Saini, H.K., van Dongen, S., and Enright, A.J. (2008) miRBase: tools for microRNA genomics. *Nucleic Acids Res.*, **36**(Database issue), D154–8.

Gstaiger, M. and Aebersold, R. (2009) Applying mass spectrometry-based proteomics to genetics, genomics and network biology. *Nat. Rev. Genet.*, **10**(9), 617–27.

Guigo, R. and Reese, M.G. (2005) EGASP: collaboration through competition to find human genes. *Nat. Methods*, **2**(8), 575–7.

Guigo, R., Flicek, P., Abril, J.F., *et al.*, (2006) EGASP: the human ENCODE Genome Annotation Assessment Project. *Genome Biol.*, **7**(Suppl 1), S2.

Guttman, M., Amit, I., Garber, M., *et al.* (2009) Chromatin signature reveals over a thousand highly conserved large non-coding RNAs in mammals. *Nature*, **458**(7235), 223–7.

Harrow, J., Denoeud, F., Frankish, A., *et al.* (2006) GENCODE: producing a reference annotation for ENCODE. *Genome Biol.*, **7**(Suppl 1), S4.

Hattori, M., Fujiyama, A., Taylor, T.D., *et al.* (2000) The DNA sequence of human chromosome 21. *Nature*, **405**(6784), 311–19.

Hillier, L.W., Reinke, V., Green, P., *et al.* (2009) Massively parallel sequencing of the polyadenylated transcriptome of C. elegans. *Genome Res.*, **19**(4), 657–66.

Horton, R., Gibson, R., Coggill, P., *et al.* (2008) Variation analysis and gene annotation of eight MHC haplotypes: the MHC Haplotype Project. *Immunogenetics*, **60**(1), 1–18.

Hsu, F., Kent, W.J., Clawson, H., *et al.* (2006) The UCSC Known Genes. *Bioinformatics*, **22**(9), 1036–46.

Hubbard, T.J., Aken, B.L., Ayling, S., *et al.* (2009) Ensembl 2009. *Nucleic Acids Res.*, **37**(Database issue), D690–7.

Huss, J.W., Orozco, C., Goodale, J., *et al.* (2008) A gene Wiki for community annotation of gene function. *PLoS Biol.*, **6**(7), e175.

Iafrate, A.J., Feuk, L., Rivera, M.N., *et al.* (2004) Detection of large-scale variation in the human genome. *Nat. Genet.*, **36**(9), 949–51.

International HapMap Consortium (2003) The International HapMap Project. *Nature*, **426**(6968), 789–96.

Jaillon, O., Aury, J.M., Brunet, F., *et al.* (2004) Genome duplication in the teleost fish Tetraodon nigroviridis reveals the early vertebrate proto-karyotype. *Nature*, **431**(7011), 946–57.

Kapranov, P., Cawley, S.E., Drenkow. J., *et al.* (2002) Large-scale transcriptional activity in chromosomes 21 and 22. *Science*, **296**(5569), 916–19.

Kent, W.J., Sugnet, C.W., Furey, T.S., *et al.* (2002) The human genome browser at UCSC. *Genome Res.*, **12**(6), 996–1006.

Kozak, M. (1984) Point mutations close to the AUG initiator codon affect the efficiency of translation of rat preproinsulin in vivo. *Nature*, **308**(5956), 241–6.

Kulikova, T., Akhtar, R., Aldebert, P., *et al.* (2007) EMBL nucleotide sequence database in 2006. *Nucleic Acids Res.*, **35**(Database issue), D16–20.

Lander, E.S., Linton, L.M., Birren, B., *et al.* (2001) Initial sequencing and analysis of the human genome. *Nature*, **409**(6822), 860–921.

Lemons, D. and McGinnis, W. (2006) Genomic evolution of Hox gene clusters. *Science*, **313**(5795), 1918–22.

Levy, S., Sutton, G., Ng, P.C., *et al.* (2007) The diploid genome sequence of an individual human. *PLoS Biol.*, **5**(10), e254.

Lin, M.F., Deoras, A.N., Rasmussen, M.D., and Kellis, M. (2008) Performance and scalability of discriminative metrics for comparative gene identification in 12 Drosophila genomes. *PLoS Comput. Biol.*, **4**(4), e1000067.

Liolios, K., Tavernarakis, N., Hugenholtz, P., and Kyrpides, N.C., (2006) The Genomes On Line Database (GOLD) v.2: a monitor of genome projects worldwide. *Nucleic Acids Res.*, **34**(Database issue), D332–4.

Matlin, A.J., Clark, F., and Smith, C.W. (2005) Understanding alternative splicing: towards a cellular code. *Nat. Rev. Mol. Cell. Biol.*, **6**(5), 386–98.

Morin, R., Bainbridge, M., Fejes, A., *et al.* (2008) Profiling the HeLa S3 transcriptome using randomly primed cDNA and massively parallel short-read sequencing. *BioTechniques*, **45**(1), 81–94.

Mortazavi, A., Williams, B.A., McCue, K., *et al.* (2008) Mapping and quantifying mammalian transcriptomes by RNA-Seq. *Nat. Methods*, **5**(7), 621–8.

Mudge, J.M., Armstrong, S.D., McLaren, K., *et al.* (2008) Dynamic instability of the major urinary protein gene family revealed by genomic and phenotypic comparisons between C57 and 129 strain mice. *Genome Biol.*, **9**(5), R91.

Ponting, C.P., Oliver, P.L., and Reik, W. (2009) Evolution and functions of long noncoding RNAs. *Cell*, **136**(4), 629–41.

Pruitt, K.D., Tatusova, T., and Maglott, D.R. (2005) NCBI Reference Sequence (RefSeq): a curated non-redundant sequence database of genomes, transcripts and proteins. *Nucleic Acids Res.*, **33**(Database issue), D501–4.

Pruitt, K.D., Harrow, J., Harte, R.A., *et al.* (2009) The consensus coding sequence (CCDS) project: identifying a common protein-coding gene set for the human and mouse genomes. *Genome Res.*, **19**(7), 1316–23.

Rinn, J.L., Kertesz, M., Wang, J.K., *et al.* (2007) Functional demarcation of active and silent chromatin domains in human HOX loci by noncoding RNAs. *Cell*, **129**(7), 1311–23.

Schuster, S.C. (2008) Next-generation sequencing transforms today's biology. *Nat. Methods*, **5**(1), 16–18.

Sebat, J., Lakshmi, B., Troge, J., *et al.* (2004) Large-scale copy number polymorphism in the human genome. *Science*, **305**(5683), 525–8.

Shiraki, T., Kondo, S., Katayama, S., *et al.* (2003) Cap analysis gene expression for high-throughput analysis of transcriptional starting point and identification of promoter usage. *Proc. Natl. Acad. Sci. U.S.A.*, **100**(26), 15776–81.

Stanke, M., Tzvetkova, A., and Morgenstern, B. (2006) AUGUSTUS at EGASP: using EST, protein and genomic alignments for improved gene prediction in the human genome. *Genome Biol.*, **7**(Suppl 1), S11–18.

Stark, A., Lin, M.F., Kheradpour, P., *et al.* (2007) Discovery of functional elements in 12 Drosophila genomes using evolutionary signatures. *Nature*, **450**(7167), 219–32.

Sugawara, H., Ogasawara, O., Okubo, K., *et al.* (2008) DDBJ with new system and face. *Nucleic Acids Res.*, **36**(Database issue), D22–4.

Sultan, M., Schulz, M.H., Richard, H., *et al.* (2008) A global view of gene activity and alternative splicing by deep sequencing of the human transcriptome. *Science*, **321**(5891), 956–60.

Suyama, M., Harrington, E., Bork, P., and Torrents, D. (2006) Identification and analysis of genes and pseudogenes within duplicated regions in the human and mouse genomes. *PLoS Comput. Biol.*, **2**(6), e76.

Toll-Riera, M., Bosch, N., Bellora, N., *et al.* (2009) Origin of primate orphan genes: a comparative genomics approach. *Mol. Biol. Evol.*, **26**(3), 603–12.

UniProt Consortium (2009) The Universal Protein Resource (UniProt) 2009. *Nucleic Acids Res.*, **37**(Database issue), D169–74.

Velculescu, V.E., Zhang, L., Vogelstein, B., and Kinzler, K.W. (1995) Serial analysis of gene expression. *Science*, **270**(5235), 484–7.

Venter, J.C., Adams, M.D., Myers, E.W., *et al.* (2001) The sequence of the human genome. *Science*, **291**(5507), 1304–51.

Wang, E.T., Sandberg, R., Luo, S., *et al.* (2008a) Alternative isoform regulation in human tissue transcriptomes. *Nature*, **456**(7221), 470–6.

Wang, J., Wang, W., Li, R., *et al.* (2008b) The diploid genome sequence of an Asian individual. *Nature*, **456**(7218), 60–5.

Wang, Z. and Burge, C.B. (2008) Splicing regulation: from a parts list of regulatory elements to an integrated splicing code. *RNA*, **14**(5), 802–13.

Wang, Z., Gerstein, M., and Snyder, M. (2009) RNA-Seq: a revolutionary tool for transcriptomics. *Nat. Rev. Genet.*, **10**(1), 57–63.

Wheeler, D.A., Srinivasan, M., Egholm, M., *et al.* (2008) The complete genome of an individual by massively parallel DNA sequencing. *Nature*, **452**(7189), 872–6.

Willingham, A.T. and Gingeras, T.R. (2006) TUF love for "junk" DNA. *Cell*, **125**(7), 1215–20.

Wilming, L.G., Gilbert, J.G., Howe, K., *et al.* (2008) The vertebrate genome annotation (Vega) database. *Nucleic Acids Res.*, **36**(Database issue), D753–60.

Wortman, J.R., Gilsenan, J.M., Joardar, V., *et al.* (2009) The 2008 update of the Aspergillus nidulans genome annotation: a community effort. *Fungal Genet. Biol.*, **46**(Suppl 1), S2–13.

Zhang, Z., Carriero, N., Zheng, D., *et al.* (2006) PseudoPipe: an automated pseudogene identification pipeline. *Bioinformatics*, **22**(12), 1437–9.

Zheng, D., Frankish, A., Baertsch, R., *et al.* (2007) Pseudogenes in the ENCODE regions: consensus annotation, analysis of transcription, and evolution. *Genome Res.*, **17**(6), 839–51.

10

Sequences from prokaryotic, eukaryotic, and viral genomes available clustered according to phylotype on a Self-Organizing Map

Takashi Abe, Shigehiko Kanaya, and Toshimichi Ikemura

10.1 Introduction

Since the development of next-generation DNA sequencers, complete genome sequences of more than 2000 species have been determined, and metagenomic analyses covering a large number of species in various environments have become common (Amann *et al.*, 1995; Delong and Karl, 2005). It is highly likely that microorganisms in diverse environments contain an abundance of novel genes and, therefore, intense research activities are underway using samples obtained from a wide variety of environments, such as seawater and soil, and human intestines. The number of sequences obtained from metagenomic analyses and registered in the International Nucleotide Sequence Databases (INSD) has soared above 17 million. For most of these genomic sequence fragments, however, it is difficult to estimate the phylogeny of organisms from which individual

Knowledge-Based Bioinformatics: From Analysis to Interpretation   Edited by Gil Alterovitz and Marco Ramoni
© 2010 John Wiley & Sons, Ltd

sequences are derived or to determine the novelty of such sequences. Most of the metagenomic sequences registered in the databases have limited utility without phylogenetic information or functional annotation. This situation has arisen because orthologous sequence sets, which cover a broad phylogenetic range and are required for the creation of reliable phylogenetic trees, are unavailable for sequences of novel genes, making it difficult to estimate the phylogeny of organisms from which subject sequences are derived through conventional sequence homology searches. A method for estimating the phylogeny of organisms and gene functions based on principles different from sequence homology searches is urgently needed.

We previously modified the self-organizing map (SOM) developed by Kohonen's group (Kohonen, 1990; Kohonen *et al.*, 1996; Kohonen, 1997) for genome informatics on the basis of batch-learning SOM (BLSOM), making the learning process and resulting map independent of the order of data input (Kanaya *et al.*, 2001; Abe *et al.*, 2002; Abe *et al.*, 2003). The BLSOM thus developed could recognize phylotype-specific characteristics of oligonucleotide frequencies in a wide rage of genomes, and permitted clustering of genome fragments according to phylotypes with neither the orthologous sequence set nor the troublesome and error-prone processes of sequence alignment (Abe *et al.*, 2002; Abe *et al.*, 2003). Furthermore, the BLSOM was suitable for actualizing high-performance parallel computing with high-performance supercomputers such as 'the Earth Simulator,' and permitted clustering (self-organization) of almost all genomic sequences available in the International DNA Databanks on a single map (Abe *et al.*, 2005; Abe *et al.*, 2006a; Abe *et al.*, 2006b). In practice, by focusing on the frequencies of oligonucleotides (e.g., tri- and tetranucleotides), the BLSOM has allowed highly accurate classification (self-organization) of most genomic sequence fragments on a species basis without providing species-related information during computation. Unlike conventional phylogenetic estimation methods, it requires no orthologous sequence set or sequence alignment and can perform the estimation based only on oligonucleotide frequencies. The present unsupervised and alignment-free clustering method, BLSOM, is thought to be the most suitable one for phylogenetic estimation for sequences from novel unknown organisms (Abe *et al.*, 2005; Hayashi *et al.*, 2005; Uchiyama *et al.*, 2005; Ricke *et al.*, 2005). Other research groups have also proposed alternative methods for performing phylogenetic estimation for environmental sequences obtained from metagenomic analyses (Teeling *et al.*, 2004; Huson *et al.*, 2007; McHardy *et al.*, 2007; McHardy and Rigoutsos, 2007).

In addition to phylogenetic estimation, BLSOM can be widely applied to visualizing its results on a plane and extracting novel sequences from novel species systematically. We actually employed BLSOM for analyses of environmental genomic fragments in joint research with experimental research groups analyzing various environmental and clinical samples (Hayashi *et al.*, 2005; Uchiyama *et al.*, 2005; Ricke *et al.*, 2005). To exemplify various genomics analyses based on the BLSOM, this chapter introduces a strategy for how to efficiently explore the genomic sequences of novel unknown microorganisms by utilizing numerous metagenomic sequences and how to determine the diversity and novelty of genomes in environmental microbial communities.

10.2   Batch-learning SOM (BLSOM) adapted for genome informatics

The neural network algorithms can be supervised or unsupervised. In the unsupervised training, there is no external teacher to oversee the learning process. The learning normally is driven by a similarity measure without specifying target vectors. The self-organizing map modifies the weights so that the most similar vectors are assigned to the same output unit, which is represented by an example vector. SOM is an unsupervised neural network algorithm that implements a characteristic nonlinear projection from the high-dimensional space of input data onto a two-dimensional array of weight vectors (Kohonen, 1990; Kohonen et al., 1996; Kohonen, 1997). In the conventional SOM developed by Kohonen, the map is a two-layered network that can organize a topological map of cluster units from a random starting point. The network combines an input layer with a competitive layer of processing units. During the self-organization process, the cluster unit whose weight vector matches the input pattern most closely (typically based on minimum Euclidean distance) is chosen as the winner. The winning unit and its neighboring units update their weights. After training is complete, pattern relationships and grouping are observed from the competitive layer. This yields the graphical organization of pattern relationships. These maps result from an information compression that retains only the most relevant common features of the set of input signals. This preserves effectively the topology of the high-dimensional data space. It is thought of as a flexible net that is spread into the multidimensional 'data cloud.' Because the net is a two-dimensional array, it can be visualized easily. The weight vectors (w_{ij}) are arranged in the two-dimensional lattice denoted by i $(= 0, 1, \ldots, I - 1)$ and j $(= 0, 1, \ldots, J - 1)$.

The learning process of the SOM was designed to be independent of the order of input of vectors on the basis of batch-learning SOM (BLSOM), as we previously reported (Kanaya et al., 2001; Abe et al., 2002). In the conventional SOM (Kohonen, 1990; Kohonen et al., 1996; Kohonen, 1997), the initial weight vectors w_{ij} are set by random values, but in the present method the vectors are initialized by principal components analysis (PCA: Step 1). For mapping multidimensional space data onto a plane, PCA rotates the vector space with the eigenvectors (the principal components) of the covariance matrix as a new basis. The principal components are orthogonal, and the plane spanned by the two first components, PC1 and PC2, was usually used for linear data projection. Weights in the first dimension (the number of lattice points in the first dimension is denoted by I) were arranged into 250 nodes for 10 kb sequences (Figure 10.1) corresponding to a width of five-times the standard deviation ($5\sigma_1$) of the first principal component; and the second dimension (J) was defined by the nearest integer greater than $(\sigma_2/\sigma_1) \times I$.

Tetra-BLSOM, 10kb window   Tetra-BLSOM, 100kb window

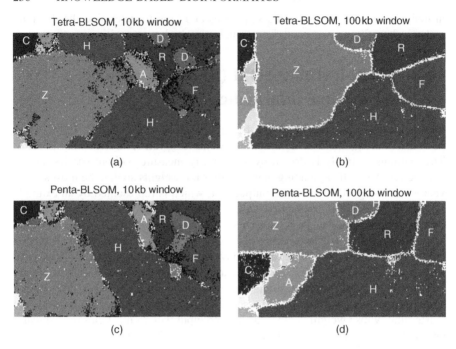

(a)   (b)

Penta-BLSOM, 10kb window   Penta-BLSOM, 100kb window

(c)   (d)

Figure 10.1   BLSOMs for non-overlapping 10kb and overlapping 100kb sequences of 13 eukaryotic genomes. (a) Tetra-BLSOMs. (b) Penta-BLSOMs. Nodes that include sequences from plural species are indicated in black; those that contain no genomic sequences are indicated in white, and those containing sequences from a single species are indicated with different gray levels and with letters as follows: C. elegans (C), Arabidopsis (A), rice (R), Drosophila (D), Fugu (F), zebrafish (Z), and human (H). For color pictures, refer to http://trna.nagahama-i-bio.ac.jp/TakashiAbe_paper_figure/KBB-2009/ Figure_1.pdf.

The weight vector on the ijth lattice (w_{ij}) was represented as follows:

$$w_{ij} = x_{av} + \frac{5\sigma_1}{I}\left[b_1\left(i - \frac{I}{2}\right) + b_2\left(j - \frac{J}{2}\right)\right],\qquad(10.1)$$

where x_{av} is the average vector for oligonucleotide frequencies of all input vectors, and b_1 and b_2 are eigenvectors for the first and second principal components. In Step 2, the Euclidean distances between the input vector x_k and all weight vectors w_{ij} were calculated; then x_k was associated with the weight vector (called $w_{i'j'}$) satisfied in minimal distance. After associating all input vectors with weight vectors, updating was performed according to Step 3.

In Step 3, the ijth weight vector was updated by

$$w_{ij}^{(\text{new})} = w_{ij} + \alpha(r) \left(\frac{\sum\limits_{x_k \in S_{ij}} x_k}{N_{ij}} - w_{ij} \right), \tag{10.2}$$

where components of set S_{ij} are input vectors associated with $w_{i'j'}$ satisfying $i - \beta(r) \leq i' \leq i + \beta(r)$ and $j - \beta(r) \leq j' \leq j + \beta(r)$. Here, $w_{ij}^{(\text{new})}$ is an updated vector. The two parameters $\alpha(r)$ and $\beta(r)$ are learning coefficients for the rth cycle, and N_{ij} is the number of components of S_{ij}. $\alpha(r)$ and $\beta(r)$ are set by

$$\alpha(r) = \max\{0.01, \alpha(1)(1 - r/T)\}, \tag{10.3}$$

$$\beta(r) = \max\{1, \beta(1) - r\}, \tag{10.4}$$

where $\alpha(1)$ and $\beta(1)$ are the initial values for the T-cycle of the learning process. In the present study, we selected 80 for T, 0.6 for $\alpha(1)$, and 60 for $\beta(1)$. The learning process is monitored by the total distance between x_k and the nearest weight vector $w_{i'j'}$, represented as

$$Q(r) = \sum_{k=1}^{N} \left\{ \|x_k - w_{i'j'}\|^2 \right\}, \tag{10.5}$$

where N is the total number of sequences analyzed.

10.3   Genome sequence analyses using BLSOM

10.3.1   BLSOMs for 13 eukaryotic genomes

To initially investigate clustering power of BLSOM for a wide range of eukaryotic sequences, we analyzed tetra- and pentanucleotide frequencies in 300 000 non-overlapping 10 kb sequences and overlapping 100 kb sequences with a 10 kb sliding step from 13 eukaryotic genomes. These genomes included human *Homo sapiens*, puffer fish *Fugu rubripes*, zebrafish *Danio rerio*, rice *Oryza sativa*, *Arabidopsis thaliana*, *Medicago truncatula*, *Drosophila melanogaster*, *Caenorhabditis elegans*, *Dictyostelium discoideum*, *Plasmodium falciparum*, *Entamoeba histolytica*, *Schizosaccharomyces pombe*, and *Saccharomyces cerevisiae*. The BLSOM, which was adapted to genome informatics, was constructed as described previously (Kanaya *et al.*, 2001; Abe *et al.*, 2002; Abe *et al.*, 2003; Abe *et al.*, 2005; Abe *et al.*, 2006a; Abe *et al.*, 2006b). First, oligonucleotide frequencies in the 10 or 100 kb sequences were analyzed by PCA, and the first and second principal components were used to set the initial weight vectors that were arranged as a two-dimensional array. After 80 learning cycles, oligonucleotide frequencies in the sequences could be represented by the final weight vectors in the two-dimensional array, and the resulting BLSOM revealed clear species-specific separations (Figure 10.1). The sequences were clustered (self-organized) primarily into species-specific territories; lattice points that include sequences from a single species are indicated in color, and those that include sequences from more than one species are indicated in black; for color pictures, refer to http://trna.nagahama-i-bio.ac.jp/TakashiAbe_paper_figure/KBB-2009/Figure_1. pdf. Sequences from each species were clustered on both the tetra- and pentanucleotide BLSOMs (Tetra- and Penta-BLSOMs; Figures 10.1(a) and 10.1(b)). For example, 97% and 98% of analyzed human sequences were classified into the human territories ('H' in Figure 10.1) in the 10 kb Tetra- and Penta-BLSOMs, respectively.

In the 10 kb BLSOMs, intra-species separations were evident; for example, human was divided into two major territories in the 10 kb Tetra-BLSOM. In the Penta-BLSOM, however, human sequences were classified into a single continuous territory, indicating that despite wide variations among 10 kb segments of human sequences, the BLSOM recognized common features of pentanucleotide frequencies in human sequences. In the 100 kb BLSOMs, interspecies (but not intra-species) separations were more prominent than in the 10 kb BLSOMs; in the 100 kb Tetra- and Penta-BLSOMs, each species had one major territory. Furthermore, the species territory was surrounded by contiguous white lattices, which contained no genomic sequences. The species borders could be drawn automatically on the basis of the contiguous white lattices, because the vectors of the species-specific lattices that were located even near a territory border were distinct between territories.

10.3.2   Diagnostic oligonucleotides for phylotype-specific clustering

G+C% has long been used as a fundamental parameter for phylogenetic characterization of species and especially of microorganisms. However, G+C% is apparently too simple as a parameter to differentiate a wide variety of species. We previously found the G+C%, which was obtained from the weight vector of each node on BLSOMs for genomic sequence analyses, to be reflected in the horizontal axis (Kohonen, 1997). Supporting this previous finding, G+C% increased from left to right on the Tetra- and Penta-BLSOMs (Figure 10.1); that is, sequences with high G+C% were located on the right side. Importantly, sequences even with the same G+C% were clearly separated on BLSOMs by a complex combination of oligonucleotide frequencies, resulting in accurate phylotype separation. BLSOMs recognized the species-specific combination of oligonucleotide frequencies that is the representative signature of each genome and enabled us to identify the frequency patterns that are characteristic of individual genomes. The frequency of each oligonucleotide in each lattice vector in the 100 kb BLSOMs was calculated and normalized with the level expected from the mononucleotide composition at each lattice point, and the observed/expected ratios are illustrated in red (overrepresented), blue (underrepresented), or white (moderately represented) in Figure 10.2; for color pictures, refer to http://trna.nagahama-i-bio.ac.jp/TakashiAbe_paper_figure/KBB-2009/Figure_2.pdf. This normalization allowed oligonucleotide frequencies in each lattice point to be studied independently of mononucleotide compositions. Transitions between red (over-representation) and blue (under-representation) for various tetra- and pentanucleotides often coincided exactly with species borders. Several diagnostic examples for the species separations are presented in Figure 10.2(a). AATT was overrepresented in rice, *Drosophila*, and *C. elegans*; underrepresented in *Fugu* and zebrafish; and moderately represented in human and *Arabidopsis*. CAGT was overrepresented in all three vertebrates but underrepresented in rice, *Arabidopsis*, *Plasmodium*, and *Dictyostelium*. BLSOMs

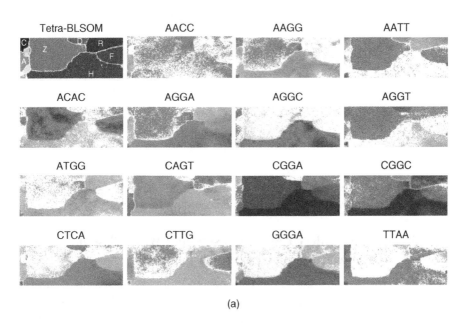

(a)

Figure 10.2   Level of each tetranucleotide (a) and pentanucleotide (b) in 100 kb BLSOMs. Diagnostic examples of species separations are presented. Level of each tetra- and pentanucleotide in each node in the 100 kb Tetra- and Penta-SOMs (Figure 10.1) was calculated and normalized with the level expected from the mononucleotide composition of the node. The observed/expected ratio is indicated with levels of blackness shown at the bottom of the figure. The 100 kb BLSOMs in Figure 10.1(a) and (b) are presented in the first panel with letters indicating species name: C. elegans (C), Arabidopsis (A), rice (R), Drosophila (D), Fugu (F), zebrafish (Z), and human (H). For other species, refer to the legend of http://trna.nagahama-i-bio.ac.jp/TakashiAbe_paper_figure/KBB-2009/Figure_1.pdf, and for color pictures, refer to http://trna.nagahama-i-bio.ac.jp/TakashiAbe_paper_figure/KBB-2009/Figure_2.pdf.

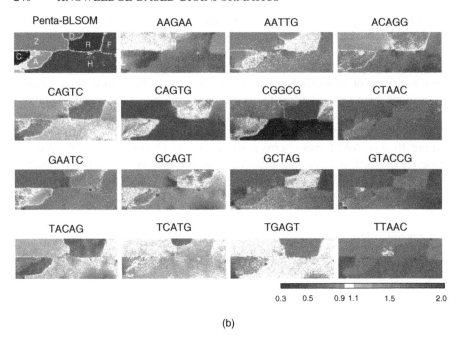

(b)

Figure 10.2   (continued)

utilized a complex combination of many oligonucleotides for sequence separations, which results in classification according to species.

10.3.3   A large-scale BLSOM constructed with all sequences available from species-known genomes

In Figure 10.1, a good separation (self-organization) of eukaryotic sequences according to phylotypes (species, in this case) was observed. We next examined phylogenetic separation of prokaryotic sequences, by applying the BLSOM to the phylogenetic prediction of sequences obtained by metagenome analyses. Large-scale metagenomic studies of uncultivable microorganisms in environmental and clinical samples have recently been conducted to survey genes useful in industrial and medical applications and to assist in developing accurate views of the ecology of uncultivable microorganisms in each environment. Conventional methods of phylogenetic classification of gene/genomic sequences have been based on sequence homology searches and therefore the phylogenetic studies focused inevitably on well-characterized gene sequences, for which orthologous sequences from a wide range of phylotypes are available for constructing a reliable phylogenetic tree. The well-characterized genes, however, often are not industrially attractive. It would be best if microbial diversity and ecology could be assessed during the process of screening for novel genes with industrial and scientific significance. The present unsupervised and alignment-free clustering method, BLSOM, is thought to be the most suitable one for this purpose because there was no need of orthologous sequence sets previously (Kanaya *et al.*, 2001; Abe *et al.*, 2002; Abe *et al.*, 2003; Abe *et al.*, 2005; Abe *et al.*, 2006a; Abe *et al.*, 2006b).

Metagenomic analyses can be applied to not only environmental but also medical samples such as clinical samples, and, therefore, are usable for exploring unknown pathogenic microorganisms that cause novel infectious diseases; mixed genome samples in the medical and pharmaceutical fields may contain DNA from a wide range of eukaryotes, as well as from humans. When we consider phylogenetic classification of genomic sequences derived from species-unknown environmental microorganisms obtained by metagenome studies, it is necessary to construct BLSOMs in advance with all available sequences not only from species-known prokaryotes but also species-known eukaryotes, viruses and organelles compiled in the International DNA Databanks. According to our previous studies of metagenome sequences (Abe *et al.*, 2005), the BLSOM was constructed with oligonucleotide frequencies in 5 kb sequence fragments. In DNA databases, only one strand of a pair of complementary sequences is registered. Our previous analyses revealed that sequence fragments from a single prokaryotic genome are often split into two territories that reflect the transcriptional polarities of the genes present in the fragment (Abe *et al.*, 2003). For phylotype classification of sequences from uncultured microbes, it is not necessary to know the transcriptional polarity of the sequence, and the split into two territories complicates assignment to species. Therefore, we previously introduced a BLSOM in which frequencies of a pair of complementary oligonucleotides (e.g., AACC and GGTT) were summed, and the BLSOMs for the degenerate sets of tetranucleotides were designated DegeTetra-BLSOMs.

Using a high-performance supercomputer, 'the Earth Simulator' (Abe *et al.*, 2006b), we could analyze almost all genomic sequences available from 2813 prokaryotes, 111 eukaryotes, 31 486 viruses, 1728 mitochondria, and 110 chloroplasts. The 2813 prokaryotes were selected because at least 10 kb genomic sequences were registered in the International DNA Sequence Databases. One important target of the phylogenetic classification of metagenome sequences is the sequences derived from species-unknown novel microorganisms. It is necessary to keep good resolution for microorganism sequences on the BLSOM by avoiding excess representation of sequences derived from higher eukaryotes with large genomes. Therefore, in the cases of higher eukaryotes, 5 kb sequences were selected randomly from each large genome up to 200 Mb in Figure 10.3 (a). In this way, the total quantities of prokaryotic and eukaryotic sequences were made almost equal. The separation between eukaryotes and prokaryotes was achieved with a high accuracy of 95%; the separation between organelles and viruses and between nuclear genomes and viruses was also achieved with a high accuracy of approximately 80%; for a color picture, refer to http://trna.nagahama-i-bio.ac.jp/TakashiAbe_paper_figure/KBB-2009/Figure_3.pdf. Clear separation of the species-known prokaryote sequences into 28 major families was also

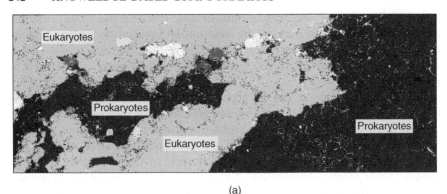

(a)

(b)

Figure 10.3   BLSOM for phylogenetic classification of environmental sequence.
(a) DegeTetra-BLSOM of 5 kb sequences derived from species-known 2813
prokaryotes, 111 eukaryotes, 1728 mitochondria, 110 chloroplasts, and 31 486
viruses. (b) Sargasso sequences longer than 1 kb were mapped on the 5 kb
DegeTetra-BLSOM, after normalization of the sequence length. For color pic-
tures, refer to http://trna.nagahama-i-bio.ac.jp/TakashiAbe_paper_figure/KBB-
2009/Figure_3.pdf.

observed (data not shown). The separation of eukaryotic sequences according to
families was also observed on this 5 kb BLSOM. During BLSOM computation,
no information was given to the computer regarding which species each
sequence fragment belonged to (*unsupervised learning algorithm*).

10.3.4   Phylogenetic estimation for environmental DNA sequences and microbial community comparison using the BLSOM

More than 17 million genomic sequence fragments obtained from various environments through metagenomic analysis have been registered in the International

Nucleotide Sequence Databases. A major portion of them is novel and has a limited utility without phylogenetic and functional annotation. The phylogeny estimation of genomic sequence fragments of novel microorganisms, based on the BLSOM, requires in advance the determination of oligonucleotide frequency of all species-known microorganisms sequenced. Therefore, a large-scale BLSOM map (Figure 10.3 (a)) covering all known sequences, including those of viruses, mitochondria, chloroplasts, and plasmids was constructed. On this BLSOM, numerous sequence fragments derived from an environmental sample could be mapped; that is, the similarity of the oligonucleotide frequency in fragmental sequences from environmental samples with that of sequences from species-known genomes was examined. In Figure 10.3 (b), 210 000 sequences with a fragment size of 1 kb or more, which were collected from the Sargasso Sea near Bermuda (Venter *et al.*, 2004), were mapped; for a color picture, refer to http://trna.nagahama-i-bio.ac.jp/TakashiAbe_paper_figure/KBB-2009/Figure_3 .pdf. The analysis of all sequence fragments obtained from one subject environmental sample can estimate numbers and proportions of species present in the sample. Approximately 70% of sequences from the Sargasso Sea were mapped to the prokaryotic territories, while the rest were mapped to the eukaryotic, viral or organelle territories.

To further identify the detailed phylogenies of the environmental sequences thus mapped to the prokaryotic territories, a BLSOM analyzing 5 kb genomic sequence fragments only from 2389 known prokaryotes was created with degenerate tetranucleotides (Figure 10.4 (a), BLSOM for prokaryotic phylotype groups); for a color picture, refer to http://trna.nagahama-i-bio.ac.jp/TakashiAbe _paper_figure/KBB-2009/Figure_4.pdf. For the 2389 species-known prokaryotes used to create this BLSOM, their separation into 28 phylogenetic groups was examined, revealing that 85% of the sequences separated (self-organized) according to their phylogenetic groups. The reason why 100% separation was not achieved is thought to be mainly because of horizontal gene transfer between the genomes of different microbial species (Abe *et al.*, 2003; Abe *et al.*, 2005). The 140 000 metagenomic sequences from the Sargasso Sea that were mapped previously to the prokaryotic territories in Figure 10.3 (b) were remapped on the BLSOM for the detailed prokaryotic phylotype assignment. They broadly spread across the BLSOM, demonstrating that the sequences belonged to a wide range of phylogenies (Figure 10.4 (b)). Interestingly, there were areas on the map where metagenomic sequences were densely mapped, which may indicate dominant species/genera. In sum, the estimation of prokaryotic phylogenetic groups could provide phylogenetic information for almost half of sequence fragments from the Sargasso Sea. The procedure above can be used to establish the phylogenetic distribution of microbial communities living in the subject environment and thus to understand the diversity of floras (Figure 10.4 (c)); for a color picture, refer to http://trna.nagahama-i-bio.ac.jp/TakashiAbe_paper_figure/KBB-2009/ Figure_4.pdf.

Through successive mapping of the subject sequences on a BLSOM created with the sequences from known genomes of each phylogenetic group, such as

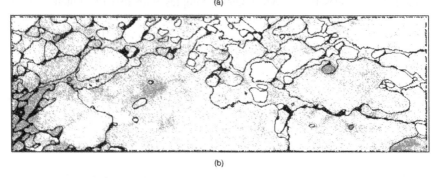

(a)

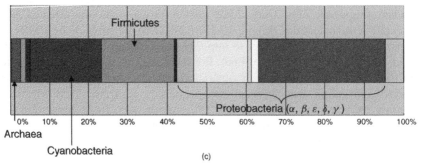

(b)

(c)

Nodes that include sequences from plural species are indicated in black, those that contain no genomic sequences are indicated in white, and those containing sequences from a single species are indicated in color as follows:
Acidobacteria (▪), Actinobacteria (▪), Alphaproteobacteria (▪), Aquificae (▪), Bacteroidetes (▪), Betaproteobacteria (▪), Chlamydiae (▪), Chlorobi (▪), Cenibacterium (▪), Chloroflexi (▪), Crenarchaeota (▪), Cyanobacteria (▪), Deinococcus-Thermus (▪), Deltaproteobacteria (▪), Dictyoglomi (▪), Epsilonproteobacteria (▪), Euryarchaeota (▪), Fibrobacteres (▪), Firmicutes (▪), Fusobacteria (▪), Gammaproteobacteria (▪), Nanoarchaeota (▪), Nitrospirae (▪), Planctomycetes (▪), Spirochaetales (▪), Thermodesulfobacteriales (▪), Thermotogales (▪), Verrucomicrobiae (▪)

Figure 10.4   Phylogenetic classification of sequences from an environmental sample. (a) DegeTetra-BLSOM of 5 kb sequences derived from species-known 2813 prokaryotes. (b) Sargasso sequences that were classified into prokaryotic territories in Figure 10.3 (b) were mapped on the 5 kb DegeTetra-BLSOM constructed with the sequences only from the species-known 2813 prokaryotes. (c) Microbial distribution of Sargasso sequences predicted by BLSOM. For color pictures, refer to http://trna.nagahama-i-bio.ac.jp/TakashiAbe_paper_figure/KBB-2009/Figure_4.pdf.

one family, further detailed phylogenetic estimation at the genus or species level becomes possible. In other words, by tracking phylogeny in a stepwise manner from the domains of organisms (e.g., eukaryotes and prokaryotes), through the phylogenetic groups to the genus or species level, more detailed phylogenetic estimation can be carried out. In addition, this procedure can determine the novelty of obtained environmental sequences at various phylogenetic levels, allowing the efficient detection of sequences with high novelty.

Currently, metagenomic analyses focusing on an abundance of viruses in seawater have been reported (Edwards and Rohwer, 2005). Since virus genomes contain no rDNA, conventional methods of phylogenetic estimation based on rRNA sequences cannot be used. BLSOM analysis for fully sequenced virus genomes showed a separation according to their phylogenies, allowing us to conduct phylogenetic estimation in the viral kingdom without relying on orthologous sequence sets or sequence alignments. The publication of a large-scale BLSOM result obtained by separating (self-organizing) all genomic sequences available, including those of viruses and organelles, will provide a foundation of novel and large-scale genomic information, which is useful for a broad range of life sciences, such as medical and pharmaceutical sciences, and related industrial fields. The mapping of novel sequences on the large-scale BLSOM that was constructed with a high-performance supercomputer can be performed using a PC-level computer; our group has created a PC software program for the BLSOM mapping. It is recommended that readers interested in the PC software program or large-scale BLSOM maps contact the authors.

10.3.5   Reassociation of environmental genomic fragments according to species

When a certain genomic fragment containing a useful gene of scientific or industrial interest is found through metagenomic analyses, it is practically difficult to determine other genes present on the subject genome, which construct a genetic system such as a metabolic pathway along with the respective useful gene. This is because the cloning of genetic fragments derived from a mixed genome sample causes most of the genes to separate from each other, making it seemingly impossible to trace their interrelationships. When dominant species are present in the environmental sample, the entire genome of each dominant species may be reconstructed by shotgun sequencing to accumulate numerous genomic fragmental sequences. However, dominant species are often well-studied culturable species or species closely related to them. If sequence fragments from novel unculturable species other than dominant species can be reassociated according to species *in silico*, a part or the outline of the genetic system (e.g., metabolic pathway) of novel species can be understood.

In order to reassociate sequences obtained through a metagenomic study on a species basis, the creation of a large-scale BLSOM for the mixture of the numerous metagenomic sequences plus all sequences of known species was effective (Abe *et al.*, 2005). For example, Figure 10.5 shows a BLSOM constructed for

Species-known and environmental samples   Biofilm sequences (acid mine: low complexity)

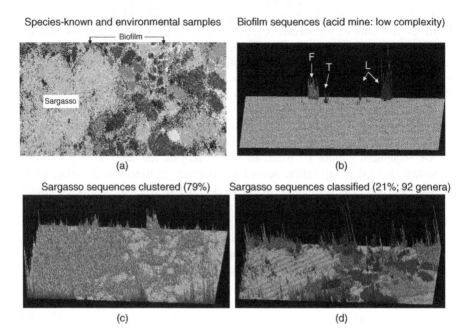

(a)   (b)

Sargasso sequences clustered (79%)   Sargasso sequences classified (21%; 92 genera)

(c)   (d)

Figure 10.5 DegeTetra-BLSOM of species-known, plus environmental sequences. (a) Species-known plus environmental sequences: lattice points that contain sequences from a single phylotype are indicated with different gray levels; those that contain only Sargasso or biofilm sequences are also indicated with different gray levels (▒ or ▪); and those that include environmental and species-known sequences or those from more than one known phylotype are indicated in black. (b) Biofilm sequences: square root of the number of biofilm sequences classified into each lattice point is indicated by the height of the bar distinctively indicated with different gray levels to show the dominant species reported by Tyson et al., (2004): Ferroplasma *(F),* Leptospirillum *(L), and* Thermoplasmatales *(T). (c) Sargasso sequences unclassified: square root of the number of Sargasso sequences classified into each lattice point containing no species-known sequences is indicated by the height of the bars. (d) Sargasso sequences classified: square root of the number of Sargasso sequences that were classified into lattice points containing species-known sequences from a single phylotype is indicated by the height of the bar distinctively indicated with different gray levels to show the phylotype. For color pictures, refer to results of our original paper (Abe et al., 2005) or http://trna.nagahama-i-bio.ac.jp/TakashiAbe_paper_figure/KBB-2009/Figure_5.pdf.*

the mixture of numerous sequence fragments obtained from the Sargasso Sea reported by Venter *et al.* (2004), sequence fragments from biofilms of mine drainage water reported by Tyson *et al.* (2004), and 5 kb sequence fragments obtained from approximately 1500 species-known prokaryotes that have been registered in the DNA databases with sequences longer than 10 kb; for a color picture, refer to http://trna.nagahama-i-bio.ac.jp/TakashiAbe_paper_figure/KBB-2009/Figure_5.pdf. Tyson *et al.* focused on the biofilm of mine drainage water as a sample containing less complex genomes, while Venter *et al.* analyzed samples from the Sargasso Sea as an example containing highly complex genomes. During computation, no species-related information except the tetranucleotide frequency of each sequence was given to the computer. The sequences from the biofilm gathered (self-organized) into some clearly defined small areas, suggesting that the sequences in each small area belonged to the same species; the vertical bar in Figure 10.5 (b) indicates the number of sequences present in one node. However, the sequences from the Sargasso Sea were widely distributed; 79% of them did not overlap with the sequences of known species (Figure 10.5 (c)), suggesting that these sequences belonged to novel genomes; for details, refer to the authors' original paper (Abe *et al.*, 2005). The reassociation of fragmental sequences on a species basis may allow estimation of the outline of the metabolic pathways of each species living in the environment and may help to clarify the collective biological systems built by microbial communities in the subject environment.

10.4   Conclusions and discussion

Large-scale metagenomic analysis covering various environmental samples has been carried out, and the application of bioinformatics to phylogenetic estimation, identification of genetic regions, and estimation of protein functions for numerous environmental sequences will further develop into a new research field. The present BLSOM is an unsupervised algorithm that can separate most sequence fragments based only on the similarity of oligonucleotide frequencies. It can separate sequences on a species basis with no other preliminary information to clarify species-specific characteristics in genomic sequences. Unlike the conventional phylogenetic estimation methods, the BLSOM requires no orthologous sequence set or sequence alignment, and therefore, this method is suitable for phylogenetic estimation for novel gene sequences. It can be used to visualize an environmental microbial community on a plane and to accurately compare it between different environments. Large-scale metagenomic analyses using recently released next-generation sequencers are also underway, and the numerous environmental sequences obtained have been registered and published in the public databases. At present, the sequence length determined by the next-generation sequencers is primarily up to 200 bases. Therefore, the development of approaches to achieving accurate phylogenetic estimation for such shorter sequences is required. We are now trying to improve the accuracy of phylogenetic estimation with the BLSOM for the shorter sequences.

For almost half of genes from novel genomes sequenced, it has become clear that protein functions cannot be estimated through sequence homology search. To complement sequence homology search, the establishment of a protein function estimation method based on different principles is important. We have recently applied BLSOM to protein sequence studies to analyze the frequency of oligopeptides, and found the separation (self-organization) of proteins according to their functions (Abe *et al.*, 2009). This suggests that the BLSOM may be used as a protein function estimation that does not rely on sequence homology search and a novel method to find scientifically or industrially important protein genes that have not been found by sequence homology searches. Large-scale BLSOM analysis covering vast quantities of data from genomic sequences and protein sequences facilitates the efficient extraction of useful information that supports research and development in a broad range of life sciences and industrial fields.

10.5   References

Abe, T., Kanaya, S., Kinouchi, M., *et al.* (2002) A novel bioinformatic strategy for unveiling hidden genome signatures of eukaryotes: self-organizing map of oligonucleotide frequency. *Genome Inform*, **13**, 12–20.

Abe, T., Kanaya, S., Kinouchi, M., *et al.* (2003) Informatics for unveiling hidden genome signatures. *Genome Res.*, **13**, 693–702.

Abe, T., Sugawara, H., Kinouchi, M., *et al.* (2005) Novel phylogenetic studies of genomic sequence fragments derived from uncultured microbe mixtures in environmental and clinical samples. *DNA Res.*, **12**, 281–90.

Abe, T., Sugawara, H., Kanaya, S., *et al.* (2006a) Self-Organizing Map (SOM) unveils and visualizes hidden sequence characteristics of a wide range of eukaryote genomes. *Gene*, **365**, 27–34.

Abe, T., Sugawara, H., Kanaya, S., and Ikemura, T. (2006b) Sequences from almost all prokaryotic, eukaryotic, and viral genomes available could be classified according to genomes on a large-scale Self-Organizing Map constructed with the Earth Simulator. *J Earth Simulator*, **6**, 17–23.

Abe, T., Kanaya, S., Uehara, H., and Ikemura, T. (2009) A novel bioinformatics strategy for function prediction of poorly-characterized protein genes obtained from metagenome analyses. *DNA Res.*, **16**, 287–97.

Amann, R.I., Ludwig, W., and Schleifer, K.H. (1995) Phylogenetic identification and in situ detection of individual microbial cells without cultivation. *Microbiol. Rev.*, **59**, 143–69.

Delong, E.F. and Karl, D.M. (2005) Genomic perspectives in microbial oceanography. *Nature*, **437**, 336–42.

Edwards, R.A. and Rohwer, F. (2005) Viral metagenomics. *Nat. Rev. Microbiol.*, **3**, 504–10.

Hayashi, H., Abe, T., Sakamoto, M., *et al.* (2005) Direct cloning of genes encoding novel xylanases from the human gut. *Can. J Microbiol.*, **51**, 251–9.

Huson, D.H., Auch, A.F., Qi, J., and Schuster, S.C. (2007) MEGAN analysis of metagenomic data. *Genome Res.*, **17**, 377–86.

Kanaya, S., Kinouchi, M., Abe, T., *et al.* (2001) Analysis of codon usage diversity of bacterial genes with a self-organizing map (SOM): characterization of horizontally transferred genes with emphasis on the E. coli O157 genome. *Gene*, **276**, 89–99.

Kohonen, T. (1990) The self-organizing map. *Proc IEEE*, **78**, 1464–80.

Kohonen, T. (1997) *Self-Organizing Maps*, Springer, Berlin.

Kohonen, T., Oja, E., Simula, O., *et al.* (1996) Engineering applications of the self-organizing map. *Proc IEEE*, **84**, 1358–84.

McHardy, A.C. and Rigoutsos, I. (2007) What's in the mix: phylogenetic classification of metagenome sequence samples. *Curr. Opin. Microbiol.*, **10**, 499–503.

McHardy, A.C., Martin, H.G., Tsirigos, A., *et al.* (2007) Accurate phylogenetic classification of variable-length DNA fragments. *Nat. Methods*, **4**, 63–72.

Ricke, P., Kube, M., Nakagawa, S., *et al.* (2005) First genome data from uncultured upland soil cluster alpha methanotrophs provide further evidence for a close phylogenetic relationship to Methylocapsa acidiphila B2 and for high-affinity methanotrophy involving particulate methane monooxygenase. *Appl. Environ. Microbiol.*, **71**, 7472–82.

Teeling, H., Meyerdierks, A., Bauer, M., *et al.* (2004) Application of tetranucleotide frequencies for the assignment of genomic fragments. *Environ. Microbiol.*, **6**, 938–47.

Tyson, G.W., Chapman, J., Hugenholtz, P., *et al.* (2004) Community structure and metabolism through reconstruction of microbial genomes from the environment. *Nature*, **428**, 37–43.

Uchiyama, T., Abe, T., Ikemura, T., and Watanabe, K. (2005) Substrate-induced gene-expression screening of environmental metagenome libraries for isolation of catabolic genes. *Nat. Biotechnol.*, **23**, 88–93.

Venter, J.C., Remington, K., Heidelberg, J.F., *et al.* (2004) Environmental genome shotgun sequencing of the Sargasso Sea. *Science*, **304**, 66–74.

Section 4

Biomolecular Relationships and Meta-Relationships

11

Molecular network analysis and applications

Minlu Zhang, Jingyuan Deng, Chunsheng V. Fang, Xiao Zhang, and Long Jason Lu

11.1 Introduction

In the post-genomic era of systems biology, the cell itself can be viewed as a complex network of interacting proteins, nucleic acids, and other biomolecules (Hartwell *et al.*, 1999; Eisenberg *et al.*, 2000). Graph representation adopted from mathematics and computer science has been widely applied to describe various molecular systems including protein interaction maps, metabolites and reactions, transcriptional regulation maps, signal transduction pathways, and functional association networks (Barabasi and Oltvai, 2004; Girvan and Newman, 2002; Tong *et al.*, 2004; Balazsi *et al.*, 2005). Applications based on network analysis have been proven useful in areas such as predicting protein functions, identifying targets for structural genomics, facilitating drug discovery and design, and expediting novel biomarker identification (Sharan *et al.*, 2007; Chuang *et al.*, 2007; Hopkins, 2008; Huang *et al.*, 2008).

In the first four sections of this chapter, we will review four types of commonly conducted network analyses on large-scale molecular networks and their applications: (1) Topology analysis focuses on the characterization of global network structures using quantitative measures, or network statistics. The applications of topology analysis include important node identification

Knowledge-Based Bioinformatics: From Analysis to Interpretation Edited by Gil Alterovitz and Marco Ramoni
© 2010 John Wiley & Sons, Ltd

and protein function prediction. (2) Motif analysis is the extraction and characterization of frequently occurring small sub-network patterns known as motifs, with the identification of motifs in homogeneous, integrated and dynamic networks as its applications. (3) Modular analysis focuses on the clustering of molecular networks to extract community-like modular structures, while its applications comprise module-assisted protein function prediction and module-based biomarker discovery. (4) Network comparison consists of approaches to find the best matches between two molecular networks, with its applications in the alignment of the same type of networks, the comparison among different types of networks, and the query using a small network against a large-scale network. In the fifth section, we will compare the main functions and features of eight commonly used network tools that can facilitate the visualization and analysis of molecular networks. At the end of each section, challenges as well as future directions of each type of network analysis will be discussed.

11.2   Topology analysis and applications

Topology analysis refers to the characterization of molecular network structure, including the global network topology, using quantitative measures, or network statistics. The applications of topology analysis include the identification of important nodes in large-scale molecular networks according to topology measures such as connectivity and centrality, the target selection based on identified important nodes, as well as protein function prediction based on linkages of the network.

11.2.1   Global structure of molecular networks: scale-free, small-world, disassortative, and modular

It is widely accepted that most large-scale molecular networks, such as PPI networks of major model organisms (Jeong et al., 2001; Yook et al., 2004; Giot et al., 2003; Li et al., 2004; Stelzl et al., 2005; Rual et al., 2005), transcriptional regulatory networks where nodes are transcription factors (TFs) and target genes (TGs) and edges are directed transcriptional regulations (Milo et al., 2002; Shen-Orr et al., 2002), and metabolic networks where nodes are metabolites and edges are reactions (Wagner and Fell, 2001; Jeong et al., 2000), approximate a scale-free topology (Barabasi and Oltvai, 2004). Generally, a network is scale-free if it has a power-law degree distribution, denoted as $P(k) \sim k^{-\gamma}$, where $P(k)$ is the fraction of nodes and k is the node degree, with a constant degree exponent γ often smaller than 3. For example, Figure 11.1 illustrates the degree distribution of the human interactome network (9630 proteins and 36 641 interactions without self-loop) based on the eighth release of the Human Protein Reference Database (HPRD; Mishra et al., 2006), where the degree exponent $\gamma = 1.88$. The power-law distribution of the node degree indicates that the higher connectivity a node has, the fewer number of such nodes exist in the network. One possible hypothesis for the

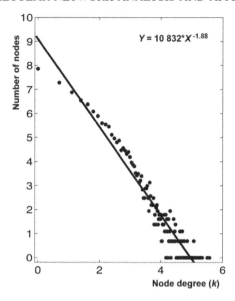

Figure 11.1   The scale-free topology of a human interactome network. A human interactome network consisting of 9630 proteins and 36641 interactions based on the eighth version of the HPRD has a power-law degree distribution. The node degree k and the number of nodes are log-transformed.

origin of the observed scale-free topology is through gene duplication (Pastor-Satorras *et al.*, 2003). Due to the scale-free topology, a small portion of nodes with the highest connectivity, denoted as hubs, are highly influential in the network.

The scale-free topology contributes to the robustness of molecular networks. Due to the sparse connectivity of the majority of nodes in a scale-free network, even removing 80% of randomly selected nodes would not disconnect the rest of the network (Albert *et al.*, 2000). On the other hand, a scale-free network is vulnerable to the selective removal of those highly connected nodes; that is, attacking hubs will shatter the whole system into isolated clusters (Albert *et al.*, 2000).

As a consequence of the scale-free topology, large-scale molecular networks have the small-world property (Cohen and Havlin, 2003). The small-world property of a network, usually depicted by a high average clustering coefficient and short characteristic path length (Table 11.1), means that the majority of node pairs can be connected by a short path consisting of a small number of edges. For example, a typical path length between the most distant metabolites in metabolic networks is four (Wagner and Fell, 2001; Jeong *et al.*, 2000). The small-world property indicates that the majority of network components can quickly respond to any perturbation within the network. Therefore, such a property also contributes to the robustness of molecular networks.

Another global property of large-scale molecular networks is the disassortativity (Barabasi and Oltvai, 2004), which means hubs are more likely to connect

Table 11.1   Summary of network measures/statistics.

Network measures (network statistics)	Description and usage	Calculation
Degree k, in-degree k_{in}, out-degree k_{out}	• k measures the connectivity of a node. In directed networks, k_{in} and k_{out} measure the in-coming and out-going connectivity of a node, respectively. • Hubs are defined as a group of nodes with the highest degree values.	• Undirected networks: $k =$ the number of edges linked to a node. • Directed networks: $k_{in} =$ the number of in-coming edges; $k_{out} =$ the number of out-going edges of a node. • Average degree $<k> = 2L/N$, where L is the total number of edges, N is the total number of nodes, and $<>$ denotes average.
Clustering coefficient C	• C measures how densely connected the neighbors of a node are. • The average clustering coefficient $<C>$ denotes the tendency of nodes in a network to form clusters. It is also an indicator of the modularity in a network.	• For a node v, the clustering coefficient is $$C = \frac{2n}{k(k-1)},$$ where k is the number of its neighbors in the network, and n is the total number of edges.
Shortest path length l_{uv}	• l_{uv} between two nodes is the number of edges of the shortest path between the two nodes.	• $l_{uv} =$ the smallest number of edges between two nodes u and v.
Characteristic path length $<l>$	• $<l>$ is the median of the means of the shortest path lengths, which measures the overall navigability of the network.	• l_{uv} is infinite if nodes u and v are not linked by a path.

Table 11.1 *(continued)*.

Network measures (network statistics)	Description and usage	Calculation
	• Short characteristic path length and high average clustering coefficient indicate that a network has the small-world property.	
Eccentricity ε	• ε of a node is the distance (the longest shortest path) between this node and any node in the network. • Average eccentricity $<\varepsilon>$ of all nodes in a network may indicate whether a network has the small-world property.	• The eccentricity of a node v is $$\varepsilon = \mathrm{Max}\{l_{vi}\},$$ where i can be any node in the network, and l_{vi} is the shortest path length between nodes v and i.
Node betweenness centrality $C_B(v)$ Edge betweenness centrality $C_B(e)$	• $C_B(v)$ or $C_B(e)$ measures the number of shortest paths that go through the node or the edge. • Bottlenecks are defined as the nodes with the highest betweenness centrality values (Yu et al., 2007). • Edge betweenness can be used to cluster a network (Girvan and Newman, 2002).	• $C_B(v)$ for a node v is $$C_B(v) = \sum_{i \neq v \neq j \in V, i \neq j} \frac{\sigma_{ij}(v)}{\sigma_{ij}},$$ where σ_{ij} is the number of shortest paths from i to j, and $\sigma_{ij}(v)$ is the number of shortest paths from i to j that go through the node v. • The edge betweenness $C_B(e)$ is calculated accordingly.

to less-connected nodes, or non-hubs. Maslov and Sneppen found such a property in a yeast PPI network and a yeast transcriptional regulatory network, by systematically comparing these two networks with random networks that have the same degree distributions but fully rewired edges (Maslov and Sneppen, 2002). The disassortativity increases the overall robustness of a network, as well as suppressing crosstalk between different sub-network components, which may correspond to functional modules that perform distinct functions (Hartwell et al., 1999; Maslov and Sneppen, 2002).

Molecular systems consist of functional units known as functional modules which carry out various distinct functions during biological processes (Hartwell et al., 1999; Eisenberg et al., 2000). In molecular networks, functional modules are expected to correspond to topologically densely connected and loosely interconnected sub-networks (Spirin and Mirny, 2003; Girvan and Newman, 2002). The coexistence of the scale-free topology and topologically and functionally self-contained modules results in the hierarchical modularity, that is, hierarchically organized modular structures, which is universally observed in large-scale molecular networks (Ravasz et al., 2002, Yook et al., 2004; Wagner et al., 2007; Ravasz, 2009).

11.2.2   Network statistics/measures

In order to quantitatively characterize networks, a number of network statistics/measures have been proposed, such as degree, clustering coefficient, shortest path length, eccentricity, and betweenness centrality. These network statistics, summarized in Table 11.1, are applied to large-scale molecular networks to characterize various molecular systems and help extract biologically meaningful information.

11.2.3   Applications of topology analysis

Major applications of topology analysis in molecular networks are important node identification and targeting, and protein function prediction. Topologically important nodes in a network, such as hubs and bottlenecks, can be inferred by network statistics. Pair-wise linkage and known functional annotations have provided a handful of approaches to predict protein functions. Another category of protein function prediction methods based on functional modules in the network will be discussed in Section 11.4.

11.2.3.1   Important node identification: hubs and bottlenecks

In highly heterogeneous scale-free molecular networks, nodes are not created equal with respect to connectivity. Nodes with the highest local connectivity and the highest global centrality measured by degree and betweenness centrality are defined as hubs and bottlenecks, respectively (Figure 11.2). Hubs and bottlenecks are the most commonly studied important nodes in molecular networks.

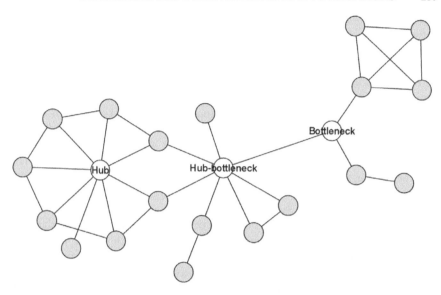

Figure 11.2   Hubs and bottlenecks in a PPI network. In the network, each node is a protein and each edge is a physical interaction between two proteins. A hub represents a node with high local connectivity, and a bottleneck represents a node with high betweenness centrality in a network. The white node in the center of the figure is both a hub and a bottleneck.

Hubs are pivotal to the integrity of molecular systems. Jeong *et al.* reported that, in a large-scale yeast PPI network, proteins that have more interactions are more likely to be essential, which are vital to the survival of the organism (Jeong *et al.*, 2001). Using five large-scale PPI networks of different model organisms, including human, mouse, yeast, worm, and mouse-ear cress, Wuchty confirmed the correlation between protein connectivity and essentiality (Wuchty *et al.*, 2003). In addition, assessed by a method called Evolutionary Excess Retention, hubs in all five PPI networks were observed to be more evolutionarily conserved than non-hubs (Wuchty *et al.*, 2003). Similarly, Said *et al.* reported a correlation between connectivity and essentiality in a large-scale yeast PPI network from DIP (Database of Interacting Proteins; Said *et al.*, 2004; Xenarios *et al.*, 2002). The authors also found that toxicity-modulating proteins were similar in network measures/statistics to essential proteins. Based on these important proteins, protein complexes and signaling pathways related to toxicology were identified (Said *et al.*, 2004).

Yu *et al.* defined bottlenecks as the proteins with the highest betweenness centrality values in an integrated conglomerate yeast PPI network combining available binary interactions from various data sources (Yu *et al.*, 2007). Bottlenecks can be viewed as bridge-like connectors between different modules/clusters in a network, and are generally more likely to be essential in both PPI and regulatory networks. In directed networks such as transcriptional regulatory networks, betweenness is a more significant indicator of essentiality than degree, based on further classification of nodes into hub-bottleneck, hub-non-bottleneck, non-hub-bottleneck, non-hub-non-bottleneck and their essentiality (Yu et al., 2007). In addition, bottlenecks were found to correspond to the dynamic components of the PPI network, based on the evidence that they were more poorly co-expressed with their neighbors than non-bottlenecks (Yu et al., 2007).

Though defined elegantly as a centrality measure, betweenness considers only the number of the shortest paths that go through a node, and ignores any other sub-optimal path. This approximation of true centrality in a network may lead to errors in identifying the global centers in a network. Missiuro et al. recently proposed an information flow score as an alternative centrality measure (Missiuro et al., 2009). This model views a global network with weighted edges (with weights indicating the reliability of interactions) as a circuit, and calculates the information flow score as the current going through each of the nodes. The nodes with the highest information flow scores were considered as centers of the network. Similar to bottlenecks, such center proteins were likely to be essential or phenotype-encoding (Missiuro et al., 2009). The advantage of this centrality measure is that it not only takes all paths in the network into account, but allows the consideration of weighted networks as well, which is important due to the incompleteness and errors in currently identified large-scale networks, especially PPI networks (Yu et al., 2008; von Mering et al., 2002).

11.2.3.2   Targeting identified important nodes

Focusing on cancer proteins in a predicted human interactome with edge weights indicating the reliability of interactions, Jonsson and Bates found that the connectivity of cancer-related proteins is significantly higher than proteins that are not related to cancer (Jonsson and Bates, 2006), consistent with the notion that phenotypically important proteins are likely to be hubs (Said et al., 2004; Jeong et al., 2001). The study also reported, through domain analysis, that cancer-related proteins contained a higher ratio of highly promiscuous structural domains compared with proteins unrelated to human cancer; as well as reporting on the central roles of these cancer-related proteins in the network through a network clustering modular analysis (Jonsson and Bates, 2006).

Huang et al. overlaid human cancer pathways from KEGG (Kanehisa and Goto, 2000) onto an integrated large-scale human PPI network, and built the Human Cancer Pathway Protein-Interaction Network (HCPIN; Huang et al., 2008). In HCPIN, the top central proteins that are hubs as well as bottlenecks include cancer-related essential proteins, such as p53, Grb2, Raf1, EGFR, and so on. In addition, the authors summarized a general framework called BioNet target selection strategy to systematically study the physical, structural and topological properties of pathway-involved proteins, such as proteins in cancer-related pathways or particular complexes (Huang *et al.*, 2008).

11.2.3.3   Protein function prediction based on direct connections

The available large-scale molecular networks, especially PPI networks from high-throughput experiments, have enabled the automatic prediction of protein function in the context of networks. Network-based protein function prediction methods can be classified into two categories: methods that utilize direct connections, and approaches based on extracted clusters/modules (Sharan *et al.*, 2007). The latter category of methods and their applications will be discussed in Section 11.4.5.1.

The simplest and most straightforward method is to assign annotations to an unknown protein based on the functions of its interacting partners. Using a yeast PPI network and the annotations of direct neighbors, Schwikowski *et al.* predicted up to three functions for each of the 364 previously uncharacterized proteins in the network (Schwikowski *et al.*, 2000). One limitation of this method is that no statistical significance is assessed during the process. Another limitation is that the method only considers its level-1 neighbors, that is, direct interacting partners, to predict the annotation for a protein, and ignores the global topology of the whole PPI network. Hishigaki *et al.* improved the above method by computing a χ^2-like score to assess each function assignment, and assessed the predictive accuracy for protein functions to be over 50% (Hishigaki *et al.*, 2001). Chua *et al.* proposed, tested and confirmed that a protein is likely to share functions with not only its level-1 neighbors but also its level-2 neighbors, which are proteins that have the same level-1 interacting partners (Chua *et al.*, 2006). Based on this notion, the authors devised a functional similarity score that assigns weights to proteins according to their distance to the target protein, and scored each function based on its weighted frequency in the level-1 and level-2 neighbors (Chua *et al.*, 2006).

In addition to considering the pair-wise linkages for protein function prediction, another group of methods takes the global topology connections into account. Vazquez *et al.* proposed a method to assign functions to unclassified proteins based on not only the functions of known neighbor proteins but also possible functions of unknown neighbor proteins (Vazquez *et al.*, 2003). Specifically, to assign a function to each unclassified protein, their method aims to globally minimize a scoring function E:

$$E = -\sum_{i,j} J_{ij}\delta(\sigma_i, \sigma_j) - \sum_i h_i(\sigma_i),$$

where J_{ij} is the adjacency matrix of the interaction network for the unclassified proteins, $\delta(i, j)$ is the discrete delta function that equals 1 if $\sigma_i = \sigma_j$ and 0 otherwise, and $h_i(\sigma_i)$ is the number of interacting partners of protein i with known function σ_i. The two terms in the scoring function globally optimize the functional coherence of neighboring un-annotated proteins, as well as between neighboring un-annotated and annotated proteins, respectively. Karaoz *et al.* proposed a similar approach that in turn could handle each function separately (Karaoz *et al.*, 2004). Their method first assigns a state s_i for each annotated protein i to be 1 if the protein has a specific function and -1 otherwise, and then it minimizes an 'energy' function E by assigning 1 or -1 states to un-annotated proteins:

$$E = -\frac{1}{2} \sum_{i=1}^{n} \sum_{j=1, j \neq i}^{n} w_{ij} s_i s_j,$$

where n is the number of proteins in the network, w_{ij} is the edge weight, and s_i is the state assigned to protein i. This function is optimized by a local search.

11.2.4   Challenges and future directions of topology analysis

Despite the capability of novel high-throughput experimental techniques to generate a huge amount of data for network assembly, the protein and regulatory networks of major model organisms are still largely incomplete, as well as error-prone (Yu *et al.*, 2008; von Mering *et al.*, 2002). This poses a major challenge for network analysis, including topology analysis. Several approaches are designed to address this issue, such as using a relatively small network of high confidence for analysis; that is, trading off accuracy with coverage (Missiuro *et al.*, 2009), or adding, removing, or replacing interactions in current large-scale networks to mimic false interactions (Wuchty *et al.*, 2003). A fundamental solution to this challenge would be the development of more advanced experimental techniques. While a tremendous effort of expert curations on the generated data might be necessary, integrative computational approaches may contribute to improve the data quality in the networks as well.

A related issue is that most results of topology analysis are based on static conglomerate networks, that is, assembled networks that contain all available data of certain organisms regardless of the contexts in which the data are collected. However, recent studies on dynamic networks suggest that network statistics, motifs, and identified important nodes based on static networks may be different from those of context-dependent networks (Luscombe *et al.*, 2004; Han *et al.*, 2004). This challenge requires future studies on topology analysis to not only use more complete and accurate data, but also take contexts into account; that is, to understand and be aware of the temporal, spatial, and environmental conditions under which a specific interaction takes place.

In addition, currently most network statistics/measures do not consider edge weights, which may have various meanings, such as the reliability/confidence of an interaction, or physical distance of two proteins. Because the data in the network usually contains false interactions, when calculating network statistics/measures, taking into account edge weights such as the edge confidence might result in more meaningful statistics that better characterize the network.

11.3   Network motif analysis

11.3.1   Motif analysis: concept and method

A network motif is an interconnected substructure of several nodes in a network that occurs significantly more frequently than random expectations (Milo *et al.*, 2002). Recently, substantial efforts have been made to identify and characterize motifs in different types of networks, including transcriptional regulatory networks (Shen-Orr *et al.*, 2002; Milo *et al.*, 2002; Dobrin *et al.*, 2004; Yeger-Lotem *et al.*, 2004; Mangan and Alon, 2003; Lee *et al.*, 2002), ecological food webs (Kashtan *et al.*, 2004; Milo *et al.*, 2002), and social networks (Kashtan *et al.*, 2004; Milo *et al.*, 2002).

Although the size and complexity of these networks varies (from tens of nodes in social networks to millions in the World Wide Web), the algorithms designed to identify network motifs are largely similar. A motif identification algorithm generally has the following two steps. First, a network is scanned for all possible n-nodes patterns that are connected substructures of n nodes, and the frequency of each pattern recurring in the network is recorded. Second, the frequency of each pattern is compared with that in randomized networks where edges are rewired and node degrees (both out-degrees and in-degrees in directed networks) are preserved in order to assess the statistical significance. A pattern is considered a motif if the frequency of the pattern in the network of interest is more than n (n is an integer usually larger than two) standard deviations greater than their mean number of appearances in the randomized networks (Shen-Orr *et al.*, 2002; Milo *et al.*, 2002).

Using this algorithm, the most frequently identified three-node motifs are the 'feed-forward loop' motif in transcriptional regulatory networks and neuron networks, the 'three chain' motif in food webs, the 'three-node feedback loop' motif in electronic circuits, and the 'feedback with two mutual dyads' motif in the World Wide Web (Shen-Orr *et al.*, 2002; Milo *et al.*, 2002; Lee *et al.*, 2002). The most frequently identified four-node motifs are 'bi-fan' in transcriptional regulatory networks, neuron networks and electronic circuits, 'bi-parallel' in neuron networks, food webs and electronic circuits, and 'four-node feedback loop' in electronic circuits. Due to the fact that distinct motifs are found overrepresented in different networks, motifs may be considered as the basic building blocks of networks (Shen-Orr *et al.*, 2002).

11.3.2   Applications of motif analysis

For molecular networks, the motif identification and analysis has been applied on homogeneous networks in which all the edges are of the same type (Kashtan *et al.*, 2004; Dobrin *et al.*, 2004; Lee *et al.*, 2002; Shen-Orr *et al.*, 2002; Milo *et al.*, 2002), and integrated networks where edges are of different types (Zhang *et al.*, 2005; Mangan and Alon, 2003). Studies have also been conducted to compare the motif changes between static conglomerate networks and dynamic networks that correspond to specific contexts (Luscombe *et al.*, 2004).

11.3.2.1   Motif analysis in homogeneous networks

The motif analysis on homogeneous molecular networks is mainly focused on transcriptional regulatory networks. Shen-Orr *et al.* systematically identified network motifs in the well-characterized transcriptional regulatory network of *Escherichia coli* (Shen-Orr *et al.*, 2002). The authors found three types of highly significantly recurring motifs in the network: namely 'feed-forward loop' (FFL), 'single input module' (SIM), and 'dense overlapping regulons' (DOR), with each motif having a specific regulatory pattern in controlling gene expression (Figure 11.3 (a)). An FFL is a three-node motif, in which a TF regulates the expression of another TF, and both TFs regulate a common TG. In the current *E. coli* transcriptional regulatory network, FFLs characterize 40 such regulation patterns. In a SIM, a number of TGs are controlled by the same TF. These TGs are likely to be co-expressed, and their protein products are likely to interact. Sixty-eight SIMs are currently available in the *E. coli* regulatory network, among which SIMs with more TGs occur less frequently. A DOR is a layer of interactions between a set of TFs and their TGs, with a many-to-many regulatory relationship between TFs and TGs.

These three frequently occurring motifs may have specific functions in the process of controlling gene expression. For example, an FFL may help resist a transient external stimulus in the cell. In an FFL, the first TF can either directly activate the TG or indirectly activate it through a second TF. For the former, when the activation signal of the first TF is transient, the second TF is activated to regulate the TG; therefore the TG is immediately inactivated when the first TF shuts down. A SIM motif occurs in the system when a group of genes work simultaneously to perform a certain function or when the group of genes is involved in the same metabolic pathway. In the DOR motifs, the genes in the same motif share the common biological functions and genes seem to be clustered into different transcriptional groups by these motifs.

11.3.2.2   Motif analysis on integrated networks

The motif detection method has also been applied on integrated networks, which integrate multiple types of molecular data. Yeger-Lotem *et al.* identified composite motifs in an integrated network combining PPIs and transcriptional regulations in *Saccharomyces cerevisiae* (Yeger-Lotem *et al.*, 2004). The identified motifs include a 'two-protein mixed-feedback loop' (TML) motif, five types of three-protein motifs (namely 'protein clique,' 'co-regulated interacting proteins,' 'feed-forward loop,' 'co-pointing,' and 'mixed-feedback loop'; Figure 11.3 (b)), and several four-protein motifs that are composed of the combinations of three-protein motifs. These motifs based on integrated networks consist of at least two types of edges, and thus are more complex than motifs identified from homogenous networks.

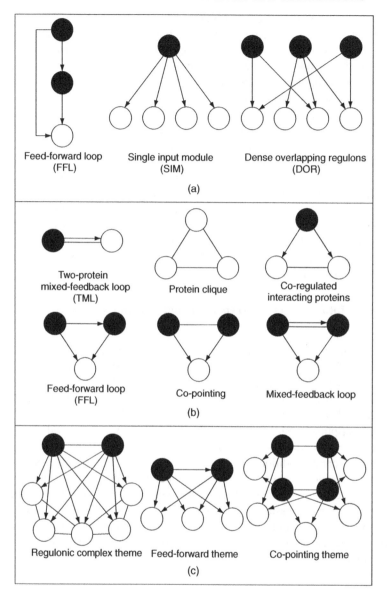

Figure 11.3 Frequently identified motifs and motif themes in molecular networks. A black node represents a TF, and a white node represents a TG or its protein product. A directed edge denotes the transcriptional regulatory interaction, while an undirected edge denotes a PPI, sequence homology or correlated expression. (a) Frequently identified motifs in transcriptional regulatory networks are FFLs, SIMs, and DORs (Shen-Orr et al., 2002). (b) Two- and three-node motifs identified in an integrated network of PPIs and transcriptional regulations (Yeger-Lotem et al., 2004). (c) Three representative motif themes extracted from an integrated network (Zhang et al., 2005).

Zhang *et al.* reported three- and four-node motifs in an integrated *S. cerevisiae* network of multiple interaction types, including PPIs, genetic interactions, transcriptional regulations, sequence homology and expression correlation (Zhang *et al.*, 2005). In addition, the authors defined network themes as family-like patterns that encompass multiple motifs of the same type, and found several network themes, such as the 'feed-forward' theme, the 'co-pointing' theme, and the 'regulonic complex' theme (Figure 11.3 (c)). The 'feed-forward' theme or the 'co-pointing' theme each contains multiple FFLs or 'co-pointing' motifs, respectively. The 'regulonic complex' theme differs from the 'co-pointing' theme in that the regulated proteins in the motif theme frequently interact with one another. These network themes are related to specific biological processes. For example, in nucleus, histone is responsible for the packaging of the newly synthesized DNA into structural units called nucleosomes. The six components (Hhf1, Hfh2, Hht1, Hht2, Hta1 and Hta2) in the histone octamer are co-expressed and regulated by two TFs (Hir1 and Hir2), and together they form a 'regulonic complex' theme (Figure 11.3 (c)). These network themes may also represent fundamental building blocks of networks.

11.3.2.3   Motif changes between static and dynamic networks

The application of motif analysis introduced above was all based on static conglomerate networks, where all interactions are assembled into a network regardless of the contexts or conditions in which the interactions take place. For motifs identified in static conglomerate networks, the interactions within the same motif may actually correspond to different contexts, thereby creating errors. To address this issue, Luscombe *et al.* applied motif analysis on five dynamic yeast transcriptional regulatory networks corresponding to three endogenous and two exogenous conditions, and found that different motifs are enriched in networks of different conditions (Luscombe *et al.*, 2004). For example, SIMs occur much more frequently in the sub-networks under exogenous conditions than under endogenous conditions, while FFLs are more favored in endogenous conditions. This observation is possibly due to the different functional roles of SIMs and FFLs: exogenously enriched SIMs sense the exogenous stimuli and transmit the signal into the cell to induce a wide change of gene expressions, while endogenously favored FFLs are able to sense the persistent input endogenous signal (Luscombe *et al.*, 2004).

11.3.3   Challenges and future directions of motif analysis

Although experiments in living cells as well as computational models have helped to determine the functions of some motifs, only a small fraction of motifs have determined functions and many more motifs remain to be characterized. In addition, many computationally predicted motifs need further experimental verifications in different systems (Albert and Albert, 2004). Another limitation that hinders the study of network motifs is the incompleteness and errors in large-scale networks. As the networks become better characterized, more and more new motifs will be identified and functions of these motifs will be assigned. Currently, the investigations of network motifs at levels of signaling and regulation have just begun and much work needs to be done in the future.

11.4   Network modular analysis and applications

Scale-free molecular networks are composed of modular structures (Hartwell *et al.*, 1999). A functional module in a molecular network is commonly defined as a group of molecules and interactions that together constitute a functional unit and perform a specific function (Hartwell *et al.*, 1999; Eisenberg *et al.*, 2000). In a molecular network, functional modules are widely believed to correspond to densely connected sub-network structures that are relatively isolated from the rest of the network (Figure 11.4; Barabasi and Oltvai, 2004; Palla *et al.*, 2005; Girvan and Newman, 2002). Recently, researchers have extensively studied the extraction and properties of functional modules from medium- and large-scale molecular networks, including PPI networks, regulatory networks and metabolic networks, mainly through network clustering, or modular analysis (Guimera and Nunes Amaral, 2005; Chen and Yuan, 2006; Palla *et al.*, 2005; Rives and Galitski, 2003). Using modular analysis, the topological and functional properties of community-like sub-network structures have been revealed. Together with functional annotations and genomic data, the analysis of extracted modules facilitates the prediction of protein functions, and the discovery of novel biomarkers, as well as the identification of novel drug targets.

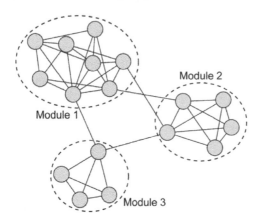

Figure 11.4   Modular structures in a PPI network. In the network, each node represents a protein and each edge represents a physical interaction. Three densely connected regions in the network are denoted as three modules.

In order to extract modular structures from medium- and large-scale molecular networks, for example PPI networks, various graph clustering approaches are adopted. These clustering methods can be categorized into four groups based on the underlying methodology: density-based, partition-based, centrality-based, and hierarchical clustering methods. In addition to the four main categories of clustering algorithms, a number of other clustering approaches exist, such as spectral clustering methods (Bu *et al.*, 2003) and the dynamic signal transduction system (STM) algorithm that simulates a signal transduction model (Hwang *et al.*, 2006). The input of the module extraction methods can be a homogeneous molecular network, or a network integrated with other genomic data such as gene expression. These network clustering algorithms may have different objectives, such as detecting functional units, or molecular complexes, or signaling pathways, or other defined sub-network structures (Sharan *et al.*, 2007).

11.4.1   Density-based clustering methods

A density-based clustering approach searches for densely connected sub-network structures, such as fully connected subgraphs, or cliques, in a network. The simplest and most straightforward density-based clustering method is the exhaustive clique enumeration, as introduced by Spirin and Mirny (Spirin and Mirny, 2003). The authors applied three different clustering methods, including a clique enumeration method, a superparamagnetic clustering method, and a Monte Carlo procedure optimization method on a PPI network from the MIPS database (Spirin and Mirny, 2003; Mewes *et al.*, 2002). Despite using an exhaustive search, the clique enumeration method is still efficient in PPI networks because PPI networks are very sparse. In addition, to enumerate cliques of n nodes, the method only needs to start from those cliques with $n - 1$ nodes, because any non-clique with $n - 1$ nodes cannot form a clique of n nodes (Spirin and Mirny, 2003).

The clique enumeration is overly stringent as it ignores many densely connected components other than cliques; some of these ignored components may well correspond to functional modules. To overcome this limitation, Palla *et al.* proposed a clustering method that extracts k-clique communities relaxed from cliques in multiple networks, including a social network, a word association network, and a small-scale PPI network (Palla *et al.*, 2005). Specifically, being less stringent than cliques, a k-clique community is a union of all k-cliques (complete subgraphs of size k) that share $k - 1$ nodes. k-clique community clusters allow overlaps, which makes the method appropriate for large-scale molecular networks because functional modules may have overlaps. This method was extended in further studies and implemented as the CFinder software (Zhang *et al.*, 2006a; Adamcsek *et al.*, 2006; http://angel.elte.hu/clustering).

Further relaxing the connectivity requirement of modules, Altaf-Ul-Amin *et al.* proposed another density-based clustering approach and applied their method on two large-scale PPI networks of *S. cerevisiae* and *E. coli* to extract protein complexes (Altaf-Ul-Amin *et al.*, 2006). Starting with single nodes, this algorithm iteratively adds one node to a cluster, and evaluates the connectivity between the newly added node and original nodes in the cluster, defined as the *cluster property* of the node, as well as the connectivity *density* of the newly formed cluster. Clusters grow node-by-node when both *cluster property* and *density* values are above certain thresholds.

Functional modules may frequently overlap with one another because they should bind together to accomplish certain biological processes, especially in signaling pathways (Zotenko *et al.*, 2006). Based on this idea, Zotenko *et al.* introduced a tree representation of complexes to mimic the transduction of signals (Zotenko *et al.*, 2006). Each node in the tree, which corresponds to a functional unit, is a clique of nodes in a modified *chordal* graph from the original network. According to graph theory, a *chord* is an edge that connects two non-connecting nodes of a cycle, and a *chordal* graph is a graph that does not contain chordless cycles of three or more edges. The authors applied this method to two signaling pathways and explained the functional roles of each module as a clique in the *chordal* graph.

In summary, a density-based clustering method captures densely connected, possibly overlapping, modular structures in networks. One limitation of density-based methods is the coverage. Because a density-based method does not partition the whole network, the modules extracted may constitute only a portion of the whole network, and other parts of the network, where none of their subcomponents meets the density requirement, are ignored.

11.4.2   Partition-based clustering methods

Partition-based clustering methods seek to explore a partition of a whole network by optimizing certain cost functions. For example, the Restricted Neighborhood Search Clustering (RNSC) algorithm partitions the node set of a network into clusters based on a cost function (King *et al.*, 2004). Starting with a random partition, the method iteratively reassigns the nodes on the border of one cluster to an adjacent cluster, and seeks to minimize the total cost of the partition by evaluating the cost function. The resulting clusters are then filtered based on additional parameters such as size, density and functional homogeneity. The authors applied RNSC on seven large-scale PPI networks of three different organisms, and extracted meaningful clusters corresponding to known protein complexes. One limitation of this method is that the number of clusters needs to be predefined while the true number is usually unknown.

Guimera and Nunes Amaral adopted a simulated annealing-based clustering method by optimizing a modularity score (Guimera and Nunes Amaral, 2005). Specifically, the modularity M of a network is defined as

$$M = \sum_{s=1}^{N} \left(\frac{l_s}{L} - \left(\frac{k_s}{2L} \right)^2 \right),$$

where N is the number of modules, L is the total number of edges in the network, l_s is the number of edges within module s, and k_s is the sum of the degrees of the nodes in modules (Guimera and Nunes Amaral, 2005). The meaning of the modularity score is that a good partition of a network should comprise as many within-module links and as few between-module links as possible, while the modularity should be zero when all nodes in a network are put in the same cluster. Using their method to extract modules from 12 complex metabolic networks, the authors classified nodes into universal roles based on their pattern of intra- and inter-module connectivities.

The advantage of partition-based methods, compared with density-based methods, is that all nodes in a network are assigned into certain clusters. One limitation is that optimizing a scoring function by exhaustive or heuristic search may be computationally complex. Partitioning the network may also lose the overlapping part of functional modules.

11.4.3   Centrality-based clustering methods

Another group of network clustering algorithms is based on centrality measures. For example, the edge betweenness centrality measures the number of the shortest paths that go through an edge, and edges with high betweenness scores denote bridge-like connectors for communication between two groups of nodes in a network (Girvan and Newman, 2002; Yu et al., 2007). Girvan and Newman proposed an edge-betweenness-based clustering method, which takes a whole network as the input, and iteratively removes an edge with the highest betweenness value (Girvan and Newman, 2002). Given a percentage of edge removal, the network is detached into a number of clusters. When using different percentages, the resulting clusters form a hierarchical structure. This algorithm was applied on several networks of known structures, and the results showed a high degree of agreement with known clusters (Girvan and Newman, 2002). An implementation of this algorithm using Java Universal Network/Graph Framework (JUNG) is available (http://jung.sourceforge.net/).

Dunn et al. applied the edge-betweenness clustering method and investigated functional modules in four large-scale PPI networks from high-throughput experiments (Girvan and Newman, 2002; Lehner and Fraser, 2004; Uetz et al., 2000; Ito et al., 2001; Dunn et al., 2005). In addition, they found that the extracted clusters were generally robust to false interactions in the PPI networks (Dunn et al., 2005). Chen and Yuan extended the edge-betweenness-based method to cluster a yeast interactome network with weighted edges, where the edge weight represented the dissimilarity between gene expressions of corresponding proteins (Chen and Yuan, 2006).

In a recent study, Missiuro et al. proposed a novel centrality measure called *information flow score* in interactome networks (Missiuro et al., 2009). The information flow score simulates the current flow on each node and edge in the circuit representation of an interactome network, and nodes and edges with high scores represent centralities of the network. A similar clustering method to the edge-betweenness clustering was designed to partition the network by iteratively removing an edge with the highest information flow centrality measure.

The centrality-based methods are computationally feasible for interactome networks and rather robust to errors and incompleteness in the networks.

Current centrality-based methods do not uncover overlapping modules. Although the resulting clusters form a hierarchy, the centrality-based methods differ from hierarchical clustering algorithms in that they do not use a pair-wise similarity/distance measure between node pairs.

11.4.4   Hierarchical clustering methods

An appropriate similarity/distance measure is the key for hierarchical clustering. When clustering a network, an intuitive distance measure is the pair-wise distance between node pairs. Rives and Galitski assumed that nodes within a module are likely to have the shortest path length profiles (Rives and Galitski, 2003). A hierarchical agglomerate average-linkage algorithm was used to cluster an all-pair-shortest-path matrix, where each value is $1/d^2$, and d is the shortest path distance between two nodes. The authors applied this method on small networks such as a yeast nuclear-protein network and a yeast regulatory network of filamentation, and extracted meaningful functional modules (Rives and Galitski, 2003).

A drawback of using the shortest path lengths directly as the distance measure is the problem of *ties-in-proximity* when distances between many node pairs are identical (Arnau *et al.*, 2005). To overcome this issue, Arnau *et al.* used a secondary similarity matrix converted from the direct shortest path length matrix. This matrix measures the probability that two nodes are clustered together using the direct distance matrix when *ties-in-proximity* occur. The authors applied this algorithm on several small- and medium-sized PPI networks, and extracted higher quality modules compared with those extracted using the direct distance matrix. The authors also provided an implementation of the method called UVCLUSTER.

Samanta and Liang used statistical significance of common interaction partners between node pairs compared with random networks as the similarity measure, and applied hierarchical clustering to cluster a yeast interactome from DIP into modules (Xenarios *et al.*, 2002; Samanta and Liang, 2003). The modules extracted were generally homogeneous in functions. In addition, 50% randomly generated interactions were added to the PPI network, and the algorithm was able to identify almost all of the original modules, indicating the robustness of the method against false interactions.

Brun *et al.* proposed a Czekanovski-Dice distance measure for hierarchical clustering based on the number of interacting partners shared by a node pair, and applied the algorithm to a medium-scale yeast PPI network to extract functional modules (Brun *et al.*, 2004). Based on the annotations of the modules, the authors assigned functions to previously uncharacterized yeast proteins.

Hierarchical clustering methods generate a hierarchy of clusters and are suitable for large-scale molecular networks. There are two limitations of hierarchical network clustering: First, different clustering parameters such as linkage methods and similarity cutoffs usually result in different output modules. Second, most hierarchical clustering methods cannot uncover overlapping modules in molecular networks.

11.4.5   Applications of modular analysis

Currently identified modules from large-scale networks, especially PPI networks, are often used to predict unknown protein functions by assigning the overrepresented functions of a module based on function enrichment analysis to its protein members. Another application of modular analysis is the identification of module-based prognostic biomarkers for diseases such as breast cancer.

11.4.5.1   Protein function prediction using modular approaches

A major application of network clustering and module identification is to assist protein function prediction. Despite obvious discrepancies among various module extraction methods, methods for predicting unknown protein functions based on identified modules and known functional annotations such as Gene Ontology (GO) or *Saccharomyces* Genome Database (SGD) annotations (Ashburner *et al.*, 2000; Cherry *et al.*, 1998) are largely the same. A commonly used straightforward method is to assign each function that is shared by the majority or proteins/genes in a module to every protein/gene in the same module. Each of the assigned functions is also considered as a function of the whole module. Alternatively, a hypergeometric enrichment p-value can be computed for each function:

$$p = \sum_{i=k}^{m} \frac{\binom{j}{i} \binom{n-j}{m-i}}{\binom{n}{m}},$$

where n is the number of nodes in the whole network, j is the number of nodes in the network annotated with the function and m is the module size. This statistical test is identical to the corresponding one-tailed Fisher's exact test. Given a threshold for p-value, the significantly overrepresented functions are then predicted for all proteins/genes in the module.

Modular approaches contribute greatly to protein function annotations. For example, Bu *et al.* assigned annotations to 76 uncharacterized proteins in yeast based on 48 quasi-clique-structured modules extracted by their spectral clustering method, with each module assigned at least one specific function from MIPS annotations (Bu *et al.*, 2003; Mewes *et al.*, 2002). Brun *et al.* proposed new annotations for 37 previously uncharacterized yeast proteins based on 126 modules extracted by their clustering method, and GO annotations from the Yeast Protein Database (YPD), and *Saccharomyces* Genome Database (SGD) (Brun *et al.*, 2004; Cherry *et al.*, 1998; Costanzo *et al.*, 2000). In another example, King *et al.* partitioned a yeast PPI network and annotated 81 proteins in yeast based on identified modules and SGD annotations (King *et al.*, 2004). With the continuous development of more advanced clustering methods for functional module extraction as well as further efforts on molecular data collection, uncharacterized protein function predictions will be accelerated to facilitate further experimental determinations of protein functions.

11.4.5.2   Biomarker discovery using modular analysis

Functional modules are functional units in molecular systems by definition (Hartwell *et al.*, 1999; Eisenberg *et al.*, 2000). Integrative analysis of modules from molecular networks with disease-related genes may help elucidate the underlying mechanisms and facilitate the biomarker discovery of a disease. For example, combining modular analysis on large-scale human PPI networks with genomic data, a recent study identified novel modular biomarkers for human breast cancer metastasis that better predict the disease outcome (Chuang *et al.*, 2007).

To predict breast cancer metastasis potential, Chuang *et al.* identified modular biomarkers that can distinguish the samples of patients who developed metastasis after surgery from those who did not (Chuang *et al.*, 2007; Auffray, 2007). The authors selected the sub-networks by first overlaying differentially expressed genes in breast cancer tumor samples onto an assembled human PPI network, followed by calculating an activity score for sub-networks/modules in all patient samples. They then assessed the predictive power of modules by computing Mutual Information between activity scores and sample metastasis potential (Auffray, 2007; Chuang *et al.*, 2007). The obtained modular biomarkers achieved a better predictive accuracy as well as higher reproducibility on different groups of breast cancer samples, compared with individual gene biomarkers from previous gene expression profiling studies (Chuang *et al.*, 2007; van 't Veer *et al.*, 2002; Wang *et al.*, 2005). Consistent with functional modules, the identified modular biomarkers are significantly enriched in common biological processes, which may provide further insights into the mechanisms of breast cancer metastasis.

11.4.6   Challenges and future directions of modular analysis

Although numerous clustering methods have been proposed to extract functional modules from molecular networks, the validation of the resulting modules is often insufficient. Functional enrichment analysis of members in modules is commonly conducted to provide validation for identified modules. However, sub-networks enriched in certain functions do not necessarily correspond to functional modules. Another limitation of the current modular analysis is that little systematic evaluation is done to compare various clustering algorithms. So far, only one quantitative evaluation of four clustering algorithms has been done using a test graph composed of 220 known protein complexes (Brohee and van Helden, 2006).

Another limitation of current modular analysis is the network itself. First, the incompleteness and errors in large-scale networks poses a great challenge to modular analysis. Although some studies used false interactions in the networks to show the robustness of their methods (as in Girvan and Newman, 2002; Arnau *et al.*, 2005; Chen and Yuan, 2006), the modules extracted from networks with false interactions were not exactly compatible with those from the original networks; that is, the members between corresponding modules were somewhat different. Second, the results of modular analysis based on dynamic networks may differ from those of static networks, because the same nodes may participate in different functional units under different conditions. For example, overlaying time-specific gene expression data onto yeast cell cycle protein interactions, de Lichtenberg *et al.* reported dynamic variations of protein complexes during the yeast cell cycle based on a time-dependent interaction network (de Lichtenberg *et al.*, 2005). To overcome this limitation, the networks on which the modular analysis is applied need to contain fewer false interactions and have a higher coverage, as well as to be better annotated about the contexts in which the interactions take place. This requires further efforts on both experimental techniques and computational methods.

In contrast with the traditional methodology for drug target discovery, where single molecules are often selected as potential drug targets, recently proposed network-based strategy, or network pharmacology, suggests that targeting a small group of molecules may be more effective against complex diseases (Hopkins, 2007; Hopkins, 2008). Although currently identified potential multi-gene drug targets, such as 'gang of four' in breast cancer metastasis (Eltarhouny *et al.*, 2008; Gupta *et al.*, 2007), do not necessarily correspond to members in the same functional modules, genes/proteins in the same modules may potentially play a role in multi-gene drug target discovery.

11.5   Network comparison

Molecular network comparison is another important aspect of understanding molecular systems by comparative research across various molecular networks, such as transcriptional regulatory networks, PPI networks, and signaling pathways.

11.5.1   Network comparison algorithms: from computer science to systems biology

Molecular network comparison is an extension from 'graph matching' algorithms in computer science (Conte *et al.*, 2004). Generally, a graph is defined as $G(V, E)$ where V is a set of nodes, E is a set of edges. The classic graph matching problem seeks a mapping function for nodes $f(a) = b$ between two graphs, a data graph $G_D(V_D, E_D)$ and a model graph $G_M(V_M, E_M)$. If such a mapping function exists, this mapping is called an isomorphism, or exact graph matching (Conte *et al.*, 2004). As we expect, for molecular networks such as PPI networks, exact graph matching is overly stringent and does not allow any mismatching or error in the data. In contrast, the more relaxed inexact graph matching uses a mapping function to maximize a matching score function (or equivalently, to minimize an error function), and seeks a best match instead of the perfect match between two graphs by optimization (Conte *et al.*, 2004). For current large-scale molecular networks that are error-prone (Yu *et al.*, 2008; von Mering *et al.*, 2002), the graph matching algorithms need to be both efficient and fault-tolerant (Sharan and Ideker, 2006). Therefore, inexact graph matching algorithms are dominant in matching molecular networks.

The algorithm complexity poses another challenge for extending and applying graph matching algorithms to large-scale molecular networks. Unlike the DNA sequence matching problem, which can be efficiently solved in polynomial time using dynamic programming or BLAST (Altschul *et al.*, 1997), matching graphs can be a very difficult problem, even NP-hard (Conte *et al.*, 2004), which means there is not yet any polynomial time algorithm to solve it. In theoretical computer science, finding the optimal mapping between two graphs is a classic problem called the 'Maximum Common Subgraph' (MCS) problem, which is unfortunately NP-hard. Due to the large scale of various molecular networks, applying optimal mapping algorithms is unpractical, thus more efficient algorithms are required. Accommodating the computational complexity and the precision of the result leads to *approximate inexact algorithms*, which can obtain a good enough local optimal matching result heuristically and probabilistically within an acceptable computing time. This category of algorithms is the most suitable for matching molecular networks.

11.5.2   Network comparison algorithms for molecular networks

In molecular network analysis, graph matching problems are more commonly referred to as network comparison. The above-mentioned *approximate inexact* graph matching algorithms have been extended and applied on molecular networks to extract biologically meaningful matches. Three of the most representative network comparison algorithms are briefly reviewed as follows.

11.5.2.1   PathBLAST

The progress made in sequence matching has inspired a network comparison algorithm called PathBLAST (Kelley *et al.*, 2003). PathBLAST looks for the maximally conserved interaction pathway across networks (Figure 11.5). The conserved pathways are formulated as high-scoring paths by a scoring function in the alignment graph:

$$S(P) = \sum_{v \in P} \log_{10} \frac{p(v)}{p_{random}} + \sum_{e \in P} \log_{10} \frac{q(e)}{q_{random}},$$

where $S(P)$ is a log probability score, which is a combination of sequence similarity and the likelihood that a PPI edge is true. The graph matching problem models the matching process with 'gap matching' (allowing a null vertex) and 'mismatching' (allowing one fault vertex matching) to make it fault tolerant and get a larger subgraph which can in return bring more insights to the biological process. Using PathBLAST, the authors found that the PPI networks of distantly related species, for instance *S. cerevisiae* and *Helicobacter pylori*, actually share a

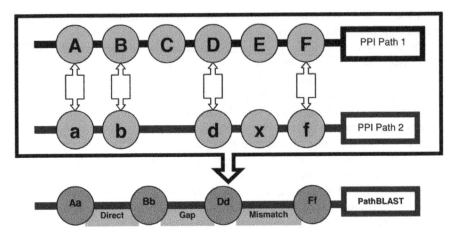

Figure 11.5   Illustration of PathBLAST. Each circle denotes a protein. For the two PPI paths (PPI Path 1 and 2, above) taken from two different PPI networks, PathBLAST will find the optimal matching path, which is the bottom 'PathBLAST' path, allowing three types of matching (direct, gap, and mismatch) in the Path-BLAST (Kelley et al., 2003).

large complement of evolutionarily conserved pathways, and that many pathways appear to have been duplicated and specialized within yeast (Kelley *et al.*, 2003).

11.5.2.2   Ogata's linear path algorithm

The work by Ogata and colleagues is a classic heuristic search algorithm (Ogata *et al.*, 2000). The method searches for correspondences between the reactions of specific metabolic pathways and the genomic locations of the genes encoding the enzymes catalyzing those reactions. Their network alignment graph combines the genome ordering information, represented as a network of genes arranged in a linear (or circular) path, with a network of successive enzymes in metabolic pathways. Single-linkage clustering was applied to this graph to identify pathways for which the enzymes are clustered along the genome (Ogata *et al.*, 2000).

11.5.2.3   IsoRank, IsoRank-Nibble

The previous two algorithms simplify the graph matching as sequence searching without considering the global topological structure of a network. Recent successful research in large-scale internet link analysis has also promoted the trend towards PageRank-like algorithms: IsoRank (Singh *et al.*, 2008) and IsoRank-Nibble (Liao *et al.*, 2009).

IsoRank constructs a scoring function to map all nodes between two networks, which is similar to PageRank's stochastic Google matrix (Singh *et al.*, 2008). Based on this function, an iterative procedure is performed by multiplying the matrix with the matching score vector. The algorithm converges within less than 100 iterations to a global optimum of the node mapping, which turns out to be the eigenvector of a linear equation system. After obtaining the mapping between two networks, a greedy matching is finally performed for constructing a mapping across multiple networks. IsoRank-Nibble improves the final matching stage in IsoRank by utilizing spectral clustering and achieves a higher accuracy in validations compared with IsoRank (Liao *et al.*, 2009).

11.5.3   Applications of molecular network comparison

There are three major applications for molecular network comparisons: network alignment, network integration, and network query; underlying all three of them is a graph matching algorithm. In these applications, both pair-wise and multiple network comparisons can be studied.

11.5.3.1   Network alignment

Network alignment is the process of globally comparing two networks of the same type, identifying regions of similarity and dissimilarity (Kelley *et al.*, 2003). The biological significance of finding conserved paths or conserved clusters is that they are likely to represent functional modules throughout the biological evolutionary process. Network alignment can be considered as the most important application of network comparisons, because it denotes the matching procedures that can connect nodes between different networks.

As an example of network alignment, PathBLAST identified five regions that were conserved across the protein networks of *S. cerevisiae* and *H. pylori* (Kelley *et al.*, 2003). This comparison was later extended to detect conserved protein clusters rather than paths, employing a likelihood-based scoring scheme that weighs the denseness of a given sub-network versus the chance of observing such topology at random (Kelley *et al.*, 2003). PathBLAST has recently been extended to perform a three-way prediction of thousands of new protein functions for yeast, worm, and fruit fly, with an estimated success rate of 58–63% across multiple networks (Sharan *et al.*, 2005).

Similarly, using IsoRank, a global alignment of PPI networks of five species (yeast, worm, fruit fly, mouse, and human) has been computed (Singh *et al.*, 2008). In addition, incorporating PPI data with sequence BLAST scores from ortholog predictions, IsoRank improves the protein alignment results between the yeast and fly networks over those from the existing sequence-based alignment approaches (Singh *et al.*, 2008).

11.5.3.2   Network integration

One type of network represents one aspect of a molecular system. Thus, integrating multiple types of networks may provide a more comprehensive and multi-view insight for the molecular system (Sharan and Ideker, 2006).

A fundamental problem is to identify in the merged network functional modules that are supported by interactions of multiple types.

An example of network integration of heterogeneous molecular networks is shown in Kelley and Ideker (2005). Integrating PPIs and genetic interactions (synthetic lethal) in yeast, interrelations between these two networks have been studied. Two structures in the integrated network are found: pairs of sub-networks of PPIs interconnected to each other by a dense pattern of genetic interactions; and clusters enriched for both physical and genetic interactions. The first structure was found to be more prevalent, suggesting that genetic interactions tend to bridge genes operating in two pathways with redundant or complementary functions, rather than occurring between protein subunits within a single pathway.

11.5.3.3 Network querying

Network querying allows researchers to submit a subgraph as a query, and the algorithm will search against a larger whole network and find the best matching parts. A typical Web server is also set up for querying networks across species to identify evolutionarily conserved patterns which will lead to revealing more biologically significant discoveries (Kelley et al., 2003).

Two tools for network querying are the PathBLAST Web server (Kelley et al., 2003) and MetaPathwayHunter (Pinter et al., 2005). PathBLAST can also be utilized to identify all matches to the query in the network under study (Pinter et al., 2005). As in the network comparison, the treatment here is only in queries that take the form of a linear path of interacting proteins. The MetaPathwayHunter algorithm is devised for metabolic networks (Pinter et al., 2005). It finds and reports all approximate occurrences of the query in the collection, ranked by similarity and statistical significance.

11.5.4 Challenges and future directions of network comparison

Research on molecular network comparison is still a young field that dates back only about one decade (Sharan and Ideker, 2006). The challenges and future directions in the current research of molecular network comparison are in three aspects. First, the scale of molecular networks poses a computational challenge. Due to the size of molecular networks that scale to about 10^5, more efficient algorithms are generally required. For example, IsoRank requires about $O(N^4)$ which is still too complex to be scalable for genome-scale networks. Time-sensitive applications like network querying need efficient algorithms. Second, more efforts are expected for the comparison of multiple networks. Currently most algorithms are based on pair-wise network comparison. In most algorithmic settings, each network is modeled as a graph, and thus how to combine the knowledge of each graph together towards a more comprehensive understanding of the biological evidence remains a challenge. Third, further development of novel strategies and theoretical frameworks is in needed to filter, interpret, and organize interaction data into models of cellular function (Sharan and Ideker, 2006).

11.6   Network analysis software and tools

To facilitate the study of molecular networks, especially large-scale networks, a number of software applications for network analysis and visualization, or network tools, have been developed. In this section, we comparatively summarize the main functionalities of several network tools commonly used by the research community of systems biology, with a focus on their relative advantages with respect to visualization and each of the four network analyses (topology, motif, modular analyses and network comparison) discussed in the previous sections (Table 11.2).

11.7   Summary

Molecular networks form the foundation of contemporary systems biology. High-throughput experiments as well as computational predictions have generated a huge amount of molecular data representable by networks, posing challenges for the extraction of biologically meaningful and useful information. Algorithms and methodologies originating from mathematics and computer science have been adapted and extended to analyze these molecular networks, with respect to topology, motifs, modules and network comparison, as discussed above. The extracted information and learned knowledge are then applied on specific biological/medical problems of interest, such as the identification of prognostic biomarkers of breast cancer metastasis (Chuang et al., 2007), target selection for structural genomics (Huang et al., 2008), and network-based drug target discovery (network pharmacology) (Hopkins, 2007; Hopkins, 2008), to facilitate the understanding of the underlying mechanisms of biological systems and benefit the health care of human beings.

Three challenges universally exist in different types of network analyses due to the data used to construct various networks. First, current molecular networks are still largely incomplete and error-prone (Yu et al., 2008; von Mering et al., 2002). This challenge requires the development of more advanced experimental techniques to generate large amounts of data of high accuracy, as well as the advance of computational methods with a decent learning ability or predictive power. In addition, expert curations can always help to refine the resulting networks. Second, network analysis results of previous studies are mainly based on static conglomerate networks by combining all available interactions. However, recent studies on network dynamics show that the analysis results are generally different between static conglomerate networks and dynamic networks of specific contexts with respect to network statistics, important nodes and network motifs (Luscombe et al., 2004; Han et al., 2004; Zhang et al., 2006b). Therefore, in the future it is important to understand which interaction is under which context in a conglomerate network (Xia et al., 2004). Finally, although

Table 11.2   Summary and comparison of network tools.

Network tool	Pajek (Batagelj and Mrvar, 2008)	Cytoscape (Shannon et al., 2003)	tYNA (Yip et al., 2006)	JUNG (O'Madadhain et al., 2003)
Main purpose	Visualization	Visualization and analysis	Visualization and analysis	Visualization and analysis
Visualization (layout and scalability)	Multiple layouts: e.g., 3D layout; good scalability (scalable to networks of thousands of nodes)	Various layouts: e.g., organic, circular, yfiles; scalable to genome-scale networks	Single layout; more layouts with SVG viewer; good scalability	Various layouts; good scalability
Topology analysis	N/A	By plug-ins: e.g., tYNA plug-in	Direct; major network statistics[a] as well as eccentricity and betweenness included	Major network statistics as well as eccentricity and betweenness included
Motif analysis	N/A	By plug-ins: e.g., Netmatch plug-in	Chain, cycle, FFL motifs identifiable	N/A
Modular analysis	Networks can be decomposed into clusters	By plug-ins: e.g., ClusterViz plug-in that uses multiple clustering methods	N/A	Multiple clustering algorithms implemented
Network comparison	N/A	Basic logic operations[b]; by plug-ins	Basic logic operations; filtering based on network statistics	N/A
System and compatibility	Standalone[c]; Linux/Win	Standalone; Linux/Mac/Win	Web-server	Java package
URL	http://vlado. fmf.uni-lj.si/ pub/networks/ pajek/	www. cytoscape.org/	http://tyna. gersteinlab. org/tyna/	http://jung. sourceforge.net/

[a]Major network statistics include at least the calculations of degree, clustering coefficient, and shortest path length.
[b]Basic logic operations include the union, intersection, and difference of two or more networks.
[c]For standalone software, the operating system compatibility for Windows, MacOS, or Linux is denoted as Win, Mac, or Linux, respectively.

Table 11.2   Summary and comparison of network tools (continued).

Network tool	N-Browse (Kao and Gunsalus, 2008)	Osprey (Breitkreutz et al., 2003)	NeAT (Brohee et al., 2008)	PINA (Wu et al., 2009)	VisANT (Hu et al., 2009)
Main purpose	Visualization	Visualization	Visualization and analysis	Visualization and analysis	Visualization and analysis
Visualization (layout and scalability)	Multiple layouts: e.g., interactive dynamic spring layout; moderate scalability	Multiple layouts; scalable to genome-scale networks	Several layouts; scalable to genome-scale networks	Multiple layouts; good scalability	Multiple layouts; scalable to genome-scale networks
Topology analysis	N/A	N/A	Direct; major network statistics included	Direct; major network statistics included; important nodes of centrality identifiable	Direct; major network statistics included; hubs and bottlenecks identifiable
Motif analysis	N/A	N/A	N/A	N/A	FFL motifs identifiable
Modular analysis	N/A	N/A	Multiple clustering algorithms implemented: e.g., MCL (Enright et al., 2002) and RNSC (King et al., 2004)	N/A	Network Module Enrichment Analysis implemented
Network comparison	Basic logic operations	N/A	Comparisons of large-scale networks with statistical significance	N/A	N/A
System and compatibility	Web-server	Standalone; Linux/Mac/Win	Web-server	Web-server	Web-server and standalone
URL	www.gnetbrowse.org/	http://biodata.mshri.on.ca/osprey/	http://rsat.ulb.ac.be/rsat/index_neat.html	http://csbi.ltdk.helsinki.fi/pina/	http://visant.bu.edu/

data integration has become one of the main themes of contemporary systems biology, how to represent, analyze and understand genome-scale integrated networks consisting of various types of available molecular data remains a challenge. To overcome this challenge is the ultimate goal of network and systems biology (Barabasi and Oltvai, 2004). However, the current edge representation is inconsistent and inadequate to accurately convey all the information in various networks. Therefore, a standardized and well-defined edge ontology is urgently needed in order to achieve this ultimate goal (Lu *et al.*, 2007).

Network biology is a very young and promising field. Although the major advances in each type of network analysis and applications have just taken place in recent years, network biology has already changed the conventional ways of studying biological systems by shifting the focus from single molecules to a system-level view. The network approaches have just proven useful in various practical aspects such as target identification in structural genomics, prognostic biomarker discovery, and network pharmacology, and much work remains to be done in the near future.

11.8   Acknowledgement

We thank Raj Bhatnagar for reading the manuscript critically.

11.9   References

Adamcsek, B., Palla, G., Farkas, I.J., *et al.* (2006) CFinder: locating cliques and overlapping modules in biological networks. *Bioinformatics*, **22**, 1021–3.

Albert, I. and Albert, R. (2004) Conserved network motifs allow protein-protein interaction prediction. *Bioinformatics*, **20**, 3346–52.

Albert, R., Jeong, H., and Barabasi, A.L. (2000) Error and attack tolerance of complex networks. *Nature*, **406**, 378–82.

Altaf-Ul-Amin, M., Shinbo, Y., Mihara, K., *et al.* (2006) Development and implementation of an algorithm for detection of protein complexes in large interaction networks. *BMC Bioinformatics*, 7, 207.

Altschul, S.F., Madden, T.L., Schaffer, A.A., *et al.* (1997) Gapped BLAST and PSI-BLAST: a new generation of protein database search programs. *Nucleic Acids Res.*, **25**, 3389–402.

Arnau, V., Mars, S., and Marin, I. (2005) Iterative cluster analysis of protein interaction data. *Bioinformatics*, **21**, 364–78.

Ashburner, M., Ball, C.A., Blake, J.A., *et al.* (2000) Gene ontology: tool for the unification of biology. The Gene Ontology Consortium. *Nat. Genet.*, **25**, 25–9.

Auffray, C. (2007) Protein subnetwork markers improve prediction of cancer outcome. *Mol. Syst. Biol.*, **3**, 141.

Balazsi, G., Barabasi, A.L., and Oltvai, Z.N. (2005) Topological units of environmental signal processing in the transcriptional regulatory network of Escherichia coli. *Proc. Natl. Acad. Sci. U.S.A.*, **102**, 7841–6.

Barabasi, A.L. and Oltvai, Z.N. (2004) Network biology: understanding the cell's functional organization. *Nat. Rev. Genet.*, **5**, 101–13.

Batagelj, V. and Mrvar, A. (2008) Analysis of kinship relations with Pajek. *Soc Sci Comput Rev*, **26**, 224–46.

Breitkreutz, B.J., Stark, C., and Tyers, M. (2003) Osprey: a network visualization system. *Genome Biol.*, **4**, R22.

Brohee, S. and Van Helden, J. (2006) Evaluation of clustering algorithms for protein-protein interaction networks. *BMC Bioinformatics*, **7**, 488.

Brohee, S., Faust, K., Lima-Mendez, G., *et al.* (2008) Network analysis tools: from biological networks to clusters and pathways. *Nat Protoc*, **3**, 1616–29.

Brun, C., Herrmann, C., and Guenoche, A. (2004) Clustering proteins from interaction networks for the prediction of cellular functions. *BMC Bioinformatics*, **5**, 95.

Bu, D., Zhao, Y., Cai, L., *et al.* (2003) Topological structure analysis of the protein-protein interaction network in budding yeast. *Nucleic Acids Res.*, **31**, 2443–50.

Chen, J. and Yuan, B. (2006) Detecting functional modules in the yeast protein-protein interaction network. *Bioinformatics*, **22**, 2283–90.

Cherry, J.M., Adler, C., Ball, C., *et al.* (1998) SGD: Saccharomyces Genome Database. *Nucleic Acids Res.*, **26**, 73–9.

Chua, H.N., Sung, W.K., and Wong, L. (2006) Exploiting indirect neighbours and topological weight to predict protein function from protein-protein interactions. *Bioinformatics*, **22**, 1623–30.

Chuang, H.Y., Lee, E., Liu, Y.T., *et al.* (2007) Network-based classification of breast cancer metastasis. *Mol. Syst. Biol.*, **3**, 140.

Cohen, R. and Havlin, S. (2003) Scale-free networks are ultrasmall. *Phys. Rev. Lett.*, **90**, 058701.

Conte, D., Foggia, P., Sansone, C., and Vento, M. (2004) Thirty years of graph matching in pattern recognition. *Intern J Pattern Recognit Artif Intell*, **18**, 265–98.

Costanzo, M.C., Hogan, J.D., Cusick, M.E., *et al.* (2000) The yeast proteome database (YPD) and Caenorhabditis elegans proteome database (WormPD): comprehensive resources for the organization and comparison of model organism protein information. *Nucleic Acids Res.*, **28**, 73–6.

Dobrin, R., Beg, Q.K., Barabasi, A.L., and Oltvai, Z.N. (2004) Aggregation of topological motifs in the Escherichia coli transcriptional regulatory network. *BMC Bioinformatics*, **5**, 10.

Dunn, R., Dudbridge, F., and Sanderson, C.M. (2005) The use of edge-betweenness clustering to investigate biological function in protein interaction networks. *BMC Bioinformatics*, **6**, 39.

Eisenberg, D., Marcotte, E.M., Xenarios, I., and Yeates, T.O. (2000) Protein function in the post-genomic era. *Nature*, **405**, 823–6.

Eltarhouny, S.A., Elsawy, W.H., Radpour, R., *et al.* (2008) Genes controlling spread of breast cancer to lung "gang of 4". *Exp. Oncol.*, **30**, 91–5.

Enright, A.J., Van Dongen, S., and Ouzounis, C.A. (2002) An efficient algorithm for large-scale detection of protein families. *Nucleic Acids Res.*, **30**, 1575–84.

Giot, L., Bader, J.S., Brouwer, C., *et al.* (2003) A protein interaction map of Drosophila melanogaster. *Science*, **302**, 1727–36.

Girvan, M. and Newman, M.E. (2002) Community structure in social and biological networks. *Proc. Natl. Acad. Sci. U.S.A.*, **99**, 7821–6.

Guimera, R. and Nunes Amaral, L.A. (2005) Functional cartography of complex metabolic networks. *Nature*, **433**, 895–900.

Gupta, G.P., Nguyen, D.X., Chiang, A.C., *et al.* (2007) Mediators of vascular remodelling co-opted for sequential steps in lung metastasis. *Nature*, **446**, 765–70.

Han, J.D., Bertin, N., Hao, T., *et al.* (2004) Evidence for dynamically organized modularity in the yeast protein-protein interaction network. *Nature*, **430**, 88–93.

Hartwell, L.H., Hopfield, J.J., Leibler, S., and Murray, A.W. (1999) From molecular to modular cell biology. *Nature*, **402**, C47–52.

Hishigaki, H., Nakai, K., Ono, T., *et al.* (2001) Assessment of prediction accuracy of protein function from protein–protein interaction data. *Yeast*, **18**, 523–31.

Hopkins, A.L. (2007) Network pharmacology. *Nat. Biotechnol.*, **25**, 1110–1.

Hopkins, A.L. (2008) Network pharmacology: the next paradigm in drug discovery. *Nat. Chem. Biol.*, **4**, 682–90.

Hu, Z., Hung, J.H., Wang, Y., *et al.* (2009) VisANT 3.5: multi-scale network visualization, analysis and inference based on the gene ontology. *Nucleic Acids Res.*, **37**, W115–21.

Huang, Y.J., Hang, D., Lu, L.J., *et al.* (2008) Targeting the human cancer pathway protein interaction network by structural genomics. *Mol Cell Proteomics*, **7**, 2048–60.

Hwang, W., Cho, Y.R., Zhang, A., and Ramanathan, M. (2006) A novel functional module detection algorithm for protein-protein interaction networks. *Algorithms Mol Biol*, **1**, 24.

Ito, T., Chiba, T., Ozawa, R., *et al.* (2001) A comprehensive two-hybrid analysis to explore the yeast protein interactome. *Proc. Natl. Acad. Sci. U.S.A.*, **98**, 4569–74.

Jeong, H., Tombor, B., Albert, R., *et al.* (2000) The large-scale organization of metabolic networks. *Nature*, **407**, 651–4.

Jeong, H., Mason, S.P., Barabasi, A.L., and Oltvai, Z.N. (2001) Lethality and centrality in protein networks. *Nature*, **411**, 41–2.

Jonsson, P.F. and Bates, P.A. (2006) Global topological features of cancer proteins in the human interactome. *Bioinformatics*, **22**, 2291–7.

Kanehisa, M. and Goto, S. (2000) KEGG: Kyoto encyclopedia of genes and genomes. *Nucleic Acids Res.*, **28**, 27–30.

Kao, H.L. and Gunsalus, K.C. (2008) Browsing multidimensional molecular networks with the generic network browser (N-Browse). *Curr Protoc Bioinformatics*, Chapter 9, Unit 9.11.

Karaoz, U., Murali, T.M., Letovsky, S., *et al.* (2004) Whole-genome annotation by using evidence integration in functional-linkage networks. *Proc. Natl. Acad. Sci. U.S.A.*, **101**, 2888–93.

Kashtan, N., Itzkovitz, S., Milo, R., and Alon, U. (2004) Topological generalizations of network motifs. *Phys Rev E Stat Nonlin Soft Matter Phys*, **70**, 031909.

Kelley, R. and Ideker, T. (2005) Systematic interpretation of genetic interactions using protein networks. *Nat. Biotechnol.*, **23**, 561–6.

Kelley, B.P., Sharan, R., Karp, R.M., *et al.* (2003) Conserved pathways within bacteria and yeast as revealed by global protein network alignment. *Proc. Natl. Acad. Sci. U.S.A.*, **100**, 11394–9.

King, A.D., Przulj, N., and Jurisica, I. (2004) Protein complex prediction via cost-based clustering. *Bioinformatics*, **20**, 3013–20.

Lee, T.I., Rinaldi, N.J., Robert, F., *et al.* (2002) Transcriptional regulatory networks in *Saccharomyces cerevisiae*. *Science*, **298**, 799–804.

Lehner, B. and Fraser, A.G. (2004) A first-draft human protein-interaction map. *Genome Biol.*, **5**, R63.

Li, S., Armstrong, C.M., Bertin, N., *et al.* (2004) A map of the interactome network of the metazoan C. elegans. *Science*, **303**, 540–3.

Liao, C.S., Lu, K.H., Baym, M., *et al.* (2009) IsoRankN: spectral methods for global alignment of multiple protein networks. *Bioinformatics*, **25**, 1253–8.

de Lichtenberg, U., Jensen, L.J., Brunak, S., and Bork, P. (2005) Dynamic complex formation during the yeast cell cycle. *Science*, **307**, 724–7.

Lu, L.J., Sboner, A., Huang, Y.J., *et al.* (2007) Comparing classical pathways and modern networks: towards the development of an edge ontology. *Trends Biochem. Sci.*, **32**, 320–31.

Luscombe, N.M., Babu, M.M., Yu, H., *et al.* (2004) Genomic analysis of regulatory network dynamics reveals large topological changes. *Nature*, **431**, 308–12.

Mangan, S. and Alon, U. (2003) Structure and function of the feed-forward loop network motif. *Proc. Natl. Acad. Sci. U.S.A.*, **100**, 11980–5.

Maslov, S. and Sneppen, K. (2002) Specificity and stability in topology of protein networks. *Science*, **296**, 910–3.

von Mering, C., Krause, R., Snel, B., *et al.* (2002) Comparative assessment of large-scale data sets of protein-protein interactions. *Nature*, **417**, 399–403.

Mewes, H.W., Frishman, D., Guldener, U., *et al.* (2002) MIPS: a database for genomes and protein sequences. *Nucleic Acids Res.*, **30**, 31–4.

Milo, R., Shen-Orr, S., Itzkovitz, S., *et al.* (2002) Network motifs: simple building blocks of complex networks. *Science*, **298**, 824–7.

Mishra, G.R., Suresh, M., Kumaran, K., *et al.* (2006) Human protein reference database – 2006 update. *Nucleic Acids Res.*, **34**, D411–4.

Missiuro, P.V., Liu, K., Zou, L., *et al.* (2009) Information flow analysis of interactome networks. *PLoS Comput. Biol.*, **5**, e1000350.

Ogata, H., Fujibuchi, W., Goto, S., and Kanehisa, M. (2000) A heuristic graph comparison algorithm and its application to detect functionally related enzyme clusters. *Nucleic Acids Res.*, **28**, 4021–8.

O'Madadhain, J., Fisher, D., White, S., *et al.* (2003) The Jung (Java Universal Network/Graph) Framework. Technical Report UCI-ICS 03-17, University of California at Irvine, School of Information and Computer Sciences.

Palla, G., Derenyi, I., Farkas, I., and Vicsek, T. (2005) Uncovering the overlapping community structure of complex networks in nature and society. *Nature*, **435**, 814–18.

Pastor-Satorras, R., Smith, E., and Sole, R.V. (2003) Evolving protein interaction networks through gene duplication. *J. Theor. Biol.*, **222**, 199–210.

Pinter, R.Y., Rokhlenko, O., Yeger-Lotem, E., and Ziv-Ukelson, M. (2005) Alignment of metabolic pathways. *Bioinformatics*, **21**, 3401–8.

Ravasz, E. (2009) Detecting hierarchical modularity in biological networks. *Methods Mol. Biol.*, **541**, 145–60.

Ravasz, E., Somera, A.L., Mongru, D.A., *et al.* (2002) Hierarchical organization of modularity in metabolic networks. *Science*, **297**, 1551–5.

Rives, A.W. and Galitski, T. (2003) Modular organization of cellular networks. *Proc. Natl. Acad. Sci. U.S.A.*, **100**, 1128–33.

Rual, J.F., Venkatesan, K., Hao, T., *et al.* (2005) Towards a proteome-scale map of the human protein-protein interaction network. *Nature*, **437**, 1173–8.

Said, M.R., Begley, T.J., Oppenheim, A.V., *et al.* (2004) Global network analysis of phenotypic effects: protein networks and toxicity modulation in Saccharomyces cerevisiae. *Proc. Natl. Acad. Sci. U.S.A.*, **101**, 18006–11.

Samanta, M.P. and Liang, S. (2003) Predicting protein functions from redundancies in large-scale protein interaction networks. *Proc. Natl. Acad. Sci. U.S.A.*, **100**, 12579–83.

Schwikowski, B., Uetz, P., and Fields, S. (2000) A network of protein-protein interactions in yeast. *Nat. Biotechnol.*, **18**, 1257–61.

Shannon, P., Markiel, A., Ozier, O., *et al.* (2003) Cytoscape: a software environment for integrated models of biomolecular interaction networks. *Genome Res.*, **13**, 2498–504.

Sharan, R. and Ideker, T. (2006) Modeling cellular machinery through biological network comparison. *Nat. Biotechnol.*, **24**, 427–33.

Sharan, R., Suthram, S., Kelley, R.M., *et al.* (2005) Conserved patterns of protein interaction in multiple species. *Proc. Natl. Acad. Sci. U.S.A.*, **102**, 1974–9.

Sharan, R., Ulitsky, I., and Shamir, R. (2007) Network-based prediction of protein function. *Mol. Syst. Biol.*, **3**, 88.

Shen-Orr, S.S., Milo, R., Mangan, S., and Alon, U. (2002) Network motifs in the transcriptional regulation network of Escherichia coli. *Nat. Genet.*, **31**, 64–8.

Singh, R., Xu, J.B., and Berger, B. (2008) Global alignment of multiple protein interaction networks with application to functional orthology detection. *Proc. Natl. Acad. Sci. U.S.A.*, **105**, 12763–8.

Spirin, V. and Mirny, L.A. (2003) Protein complexes and functional modules in molecular networks. *Proc. Natl. Acad. Sci. U.S.A.*, **100**, 12123–8.

Stelzl, U., Worm, U., Lalowski, M., *et al.* (2005) A human protein-protein interaction network: a resource for annotating the proteome. *Cell*, **122**, 957–68.

Tong, A.H., Lesage, G., Bader, G.D., *et al.* (2004) Global mapping of the yeast genetic interaction network. *Science*, **303**, 808–13.

Uetz, P., Giot, L., Cagney, G., *et al.* (2000) A comprehensive analysis of protein-protein interactions in Saccharomyces cerevisiae. *Nature*, **403**, 623–7.

Vazquez, A., Flammini, A., Maritan, A., and Vespignani, A. (2003) Global protein function prediction from protein-protein interaction networks. *Nat. Biotechnol.*, **21**, 697–700.

van 't Veer, L.J., Dai, H., Van De Vijver, M.J., *et al.* (2002) Gene expression profiling predicts clinical outcome of breast cancer. *Nature*, **415**, 530–6.

Wagner, A. and Fell, D.A. (2001) The small world inside large metabolic networks. *Proc. Biol. Sci.*, **268**, 1803–10.

Wagner, G.P., Pavlicev, M., and Cheverud, J.M. (2007) The road to modularity. *Nat. Rev. Genet.*, **8**, 921–31.

Wang, Y., Klijn, J.G., Zhang, Y., *et al.* (2005) Gene-expression profiles to predict distant metastasis of lymph-node-negative primary breast cancer. *Lancet*, **365**, 671–9.

Wu, J., Vallenius, T., Ovaska, K., *et al.* (2009) Integrated network analysis platform for protein-protein interactions. *Nat. Methods*, **6**, 75–7.

Wuchty, S., Oltvai, Z.N., and Barabasi, A.L. (2003) Evolutionary conservation of motif constituents in the yeast protein interaction network. *Nat. Genet.*, **35**, 176–9.

Xenarios, I., Salwinski, L., Duan, X.J., *et al.* (2002) DIP, the Database of Interacting Proteins: a research tool for studying cellular networks of protein interactions. *Nucleic Acids Res.*, **30**, 303–5.

Xia, Y., Yu, H., Jansen, R., *et al.* (2004) Analyzing cellular biochemistry in terms of molecular networks. *Annu. Rev. Biochem.*, **73**, 1051–87.

Yeger-Lotem, E., Sattath, S., Kashtan, N., *et al.* (2004) Network motifs in integrated cellular networks of transcription-regulation and protein-protein interaction. *Proc. Natl. Acad. Sci. U.S.A.*, **101**, 5934–9.

Yip, K.Y., Yu, H., Kim, P.M., *et al.* (2006) The tYNA platform for comparative interactomics: a web tool for managing, comparing and mining multiple networks. *Bioinformatics*, **22**, 2968–70.

Yook, S.H., Oltvai, Z.N., and Barabasi, A.L. (2004) Functional and topological characterization of protein interaction networks. *Proteomics*, **4**, 928–42.

Yu, H., Kim, P.M., Sprecher, E., *et al.* (2007) The importance of bottlenecks in protein networks: correlation with gene essentiality and expression dynamics. *PLoS Comput. Biol.*, **3**, e59.

Yu, H., Braun, P., Yildirim, M.A., *et al.* (2008) High-quality binary protein interaction map of the yeast interactome network. *Science*, **322**, 104–10.

Zhang, L.V., King, O.D., Wong, S.L., *et al.* (2005) Motifs, themes and thematic maps of an integrated Saccharomyces cerevisiae interaction network. *J. Biol.*, **4**, 6.

Zhang, S., Ning, X., and Zhang, X.S. (2006a) Identification of functional modules in a PPI network by clique percolation clustering. *Comput. Biol. Chem.*, **30**, 445–51.

Zhang, Z., Liu, C., Skogerbo, G., *et al.* (2006b) Dynamic changes in subgraph preference profiles of crucial transcription factors. *PLoS Comput. Biol.*, **2**, e47.

Zotenko, E., Guimaraes, K.S., Jothi, R., and Przytycka, T.M. (2006) Decomposition of overlapping protein complexes: a graph theoretical method for analyzing static and dynamic protein associations. *Algorithms Mol Biol*, **1**, 7.

12

Biological pathway analysis: an overview of Reactome and other integrative pathway knowledge bases

Robin A. Haw, Marc E. Gillespie, and Michael A. Caudy

12.1 Biological pathway analysis and pathway knowledge bases

A biological pathway is an ordered series of molecular events that results in a new molecular product, or a change in a cellular state or process. For example, metabolic and signaling pathway databases contain experimental data and information relating to genes, gene products, and small molecules. A major challenge for biologists, clinicians, and bioinformaticians is the integration of new experimental and computational results with previous knowledge about specific biological pathways.

Pathway Knowledge Bases (PKBs) are annotated collections of data assembled to achieve that integration. The knowledge required for such high-level conceptual integration comes from analysis by human curators and experts,

Knowledge-Based Bioinformatics: From Analysis to Interpretation   Edited by Gil Alterovitz and Marco Ramoni
© 2010 John Wiley & Sons, Ltd

working from primary research articles and reviews in the published literature. It is not yet possible to accurately and reliably perform such analyses by computational approaches alone, although emerging tools, such as Semantic Web technologies, may allow such computational analysis in the future. We therefore distinguish between biological 'databases' and 'knowledge bases' by the inclusion of human curation in knowledge bases, whereas large databases can be, and often are, created by purely computational methods.

Pathway knowledge bases, such as Reactome and the others reviewed here, are Web applications that provide custom interfaces to online data resources. These resources consist of data repositories and data analysis tools, as well as highly specialized data visualization tools. As a result, these integrative pathway knowledge bases are built using the same Web technologies that are used now to create custom 'Web portals' and 'data mashups' in 'Web 2.0' applications. In addition, these different pathway Web tools are now starting to provide further integration, by accessing each other's data, via specific 'Web services APIs' (Application Programming Interfaces). For example, the human-curated pathway data provided by Reactome is now used by multiple other pathway knowledge bases, including NCBI-BioSystems and NCI-PID, and plans are underway to integrate data and analysis results from other pathway tools into Reactome.

The authors of this review are all members of the Reactome project, and we therefore focus on Reactome. In particular, we provide a detailed use case for pathway data analysis with Reactome. We also give brief overviews of several closely related knowledge bases, pointing out some of the differences in the data models used, and the different types of information integrated in these various pathway knowledge bases.

12.2   Overview of high-throughput data capture technologies and data repositories

The four 'omes': genome, transcriptome, proteome, and metabolome reflect major areas of biological and clinical research that forms the framework of systems biology. Over the last decade, the rapid advances in genomic, proteomic, and metabolomic data capture technologies have changed the way that biologists, life scientists and clinicians study cellular processes by generating thousands or millions of individual data points representing gene sequences, mRNA or protein expression levels, or metabolite levels. Although this data can now be generated relatively easily and rapidly, there is a critical need for powerful tools to aid in the organization, interpretation, and analysis of these primary data sets, thus hindering the capacity for discovery.

Historically, the original DNA, RNA and protein sequences databases hosted by NCBI (Sayers et al., 2009), EBI (Brooksbank et al., 2010) and DDBJ (Sugawara et al., 2009) provided the first large data warehouses of biological information. These repositories provide the life science community with the 'molecules of life,' analogous to the pieces of a jigsaw puzzle, but in general they have not provided a means to demonstrate how the pieces fit together. The size of these databases quickly grew so large that they could no longer be distributed on traditional digital media, such as CDs and DVDs, let alone in the published literature. As a result, access to large databases has primarily been either through online applications, or by downloading the entire data set for local use.

DNA sequencing data: originally, the Sanger method (Sanger *et al.*, 1977) of capillary sequencing was the leading approach for DNA sequencing. Capillary sequencing machines could, with a high degree of accuracy, analyze about 600–800 bases of 384 DNA molecules simultaneously. Next-Generation Sequencing (NGS) has revolutionized the genome and metagenome sequencing research field. Roche 454, Illumina GA and ABI SOLiD platform designs and sequencing chemistries use different approaches, but they share the overall goal of massively parallel sequencing, generating megabases or gigabases of short-read sequence outputs. NGS has become considerably less expensive than the Sanger method, and to date has been successfully effective in whole human genome sequencing (Wheeler *et al.*, 2008), RNA sequencing (RNA-Seq) to study the mammalian transcriptome (Mortazavi *et al.*, 2008), and ChIP sequencing (ChIP-Seq) to identify binding sites of transcription factors and DNA-associated proteins (Johnson *et al.*, 2007).

Gene expression data: a DNA microarray is a plastic, glass or silicon chip (array) containing thousands or millions of microscopic spots of cDNA sequences or DNA oligonucleotides of a known DNA sequence on their surface. The ability of a probe nucleic acid molecule to hybridize to the microarray affixed sequence is used to determine the global expression patterns, under specified conditions (e.g., toxicant exposure, disease, development). DNA microarrays can be used to determine gene expression patterns (Schena *et al.*, 1995), copy-number variation (Pollack *et al.*, 1999), single nucleotide polymorphisms (Hacia *et al.*, 1999), identify sequences bound by protein (Ren *et al.*, 2000), detect alternative splicing variants (Hu *et al.*, 2001), and to analyze the sequence of mutant genomes (Hacia, 1999). MIAME-compliant public repositories to store and organize microarray data sets have accompanied the rapid growth in the number of data sets generated (Brazma *et al.*, 2001). The two most comprehensive microarray data repositories are the NCBI's Gene Expression Omnibus/GEO (Barrett *et al.*, 2009) and the EBI's ArrayExpress (Parkinson *et al.*, 2009). There are a number of smaller specialized microarray repositories such as the Stanford Microarray Database (Marinelli *et al.*, 2008).

Proteomics and protein interaction data: at its simplest, proteomics is the high-throughput identification and quantitative analysis of proteins. The majority of proteomic technologies focus on the separation of proteins (by molecular weight and pI), coupled with mass spectroscopy (MS) for protein or peptide identification and quantification (Shevchenko *et al.*, 1996). A number of MS-based technological approaches have been employed to study the proteome, such as MALDI-TOF-MS (Shevchenko *et al.*, 1996), SELDI (Wright *et al.*, 1999), and LC-MS (McCormack *et al.*, 1997). Mass spectrometry has played a significant role in the characterization of protein post-translational modifications (Ficarro *et al.*, 2002) and biomarker discovery (Zhao *et al.*, 2009). Several protein fragment databases facilitate the identification of peptides within MS or tandem MS profiles, such as Mascot (Perkins *et al.*, 1999) and XTandem (Craig and Beavis, 2004). An integrated public data repository, the PRIDE (PRoteomics IDEntifications) database (Vizcaino *et al.*, 2009), brings together protein databases, proteomics publications and information on post-translational modifications to support proteomic data analysis. Protein interaction capture technologies such as yeast two-hybrid (Uetz *et al.*, 2000), protein microarray (Zhu *et al.*, 2001), affinity chromatography followed by MS (Ho *et al.*, 2002), and phage-display (Crameri and Kodzius, 2001) have been successfully used to generate extensive interaction data sets. Currently, there are several large interaction data sets for yeast, bacteria, and fruit fly, stored in many interaction databases, including DIP (Salwinski *et al.*, 2004), BIND (Alfarano *et al.*, 2005), BioGRID (Breitkreutz *et al.*, 2008), IntAct (Aranda *et al.*, 2009), and MINT (Ceol *et al.*, 2010).

Metabolomics: this is the high-throughput identification and quantification of small molecule metabolites found in a biological sample isolated from a cell, organ or organism. The major experimental techniques employed in metabolomics are similar in nature to those found in analytical chemistry, such as chromatography, nuclear magnetic resonance (NMR), and mass spectrometry. A significant challenge with metabolomics is identifying and characterizing a large number of metabolites at the same time and using these metabolite signatures to identify disease biomarkers or to model metabolic processes. Metabolomics is used in drug discovery (Kel, 2006), toxicology (Griffin and Bollard, 2004), medicinal chemistry (Wishart, 2008), nutrigenomics (Gibney *et al.*, 2005), and functional genomics (Denkert *et al.*, 2006). There are a number of publically accessible metabolomics databases, examples of which are the Human Metabolome Database (Wishart *et al.*, 2009), Golm Metabolome Database (Kopka *et al.*, 2005), PubChem (Wang *et al.*, 2009), and Metlin (Smith *et al.*, 2005).

Data models and data formats for pathway knowledge bases: while the different pathway knowledge bases all share the ability to store pathway information, they each have their own data models and formats, and different query, visualization, and analysis tools, and also different data quality requirements. Some pathway knowledge bases focus on simple relationships between genes, proteins, and pathways, whereas others accommodate more elaborate data structures describing relationships between genes, proteins, small molecules, complexes, reactions, and pathways. It is important to note that expert's and curator's opinions of data models vary greatly, and this different interpretation of data that is seen in journal publications also propagates into the content of pathway knowledge bases.

12.3   Brief review of selected pathway knowledge bases

The number of pathway *knowledge bases* and *databases* continues to grow. There have been many efforts to capture biological pathways and provide human and machine-readable systems. As of November 2009, PathGuide (Bader *et al.*, 2006), a highly comprehensive list of pathway knowledge bases, lists over 300 interaction and pathway databases and knowledge bases, categorized according to the information they store: Protein–Protein Interactions, Metabolic Pathways, Signaling Pathways, Pathway Diagrams, Transcription Factors/Gene Regulatory Networks, Protein–Compound Interactions, Genetic Interaction Networks, and Protein Sequence Focused. Different projects have emerged to provide biological and pathway information at a higher level of abstraction. A common general goal of these projects is the support of quantitative computational modeling of biological pathways. Not only do resources need to be developed to store this information but also standard exchange formats and languages for distributing the data for analysis and use in visualization tools. A selection of these metabolic and signaling pathway knowledge bases are described in Table 12.1. Here we compare and contrast Reactome, KEGG, WikiPathways, NCI-PID, NCBI BioSystems, Science Signaling, and PharmGKB.

12.3.1   Reactome

Reactome (Matthews *et al.*, 2009; Vastrik *et al.*, 2007; Joshi-Tope *at al.*, 2005) is a free, open-source, expert-authored, manually curated, peer-reviewed and highly reliable knowledge base of human biological pathways and reactions. This database is used as an online textbook of biology and also to make discoveries about biological pathways, thus providing utility to both biologists and bioinformaticians. The focus of the Reactome knowledge base is human biological pathways. If a specific process has not been directly studied in humans, the species that the experiments have been performed in are directly curated, and the process is manually projected back into human.

Pathways are cross-referenced to NCBI Entrez Gene, Ensembl and UniProt databases, UCSC and HapMap Genome Browsers, KEGG Compound and ChEBI small molecule databases, PubMed, and GO. The core unit of the Reactome data model is the reaction. Entities participating in reactions are assigned to a specific pathway. Examples of biological pathways in Reactome include signaling, innate and acquired immune function, transcriptional regulation, translation, apoptosis, the influenza and HIV life cycles in infected host cells, and classical intermediary metabolism.

Each manually curated set of Reactome human reactions is used to computationally project reactions onto 22 evolutionarily divergent species for which high-quality whole-genome sequence data are available, and a comprehensive, high-quality set of protein predictions exist. These species include the laboratory

Table 12.1   Pathway knowledge bases and databases freely available for academic research.

Pathway database	Website URL	Pathway content	Curation	Peer reviewed?	Pathway visualization	BioPAX	SBML	Other data exchange formats	Web service API
Reactome	http://www.reactome.org	Metabolic, regulatory, signaling, and disease	Manual	Yes	Dynamic/static	BioPAX Level 3	SBML Level 2	PSI MITAB	Yes[a]
KEGG	http://www.kegg.jp	Metabolic, signaling, regulatory, disease, and drug	Manual	No	Dynamic	BioPAX Level 1	No	KGML	Yes
WikiPathways	http://www.wikipathways.org	Metabolic, signaling, regulatory, disease, and drug	Manual	No	Dynamic	In progress	No	KGML	Yes
NCI/Nature Pathway Interaction Database	http://pid.nci.nih.gov	Signaling and regulatory	Manual	Yes	Dynamic	BioPAX Level 2	No	PID XML	No[a]
NCBI BioSystems	http://www.ncbi.nlm.nih.gov/biosystems	Metabolic, signaling, regulatory, disease, and drug	Computational	No	Static	No	No	–	Yes

Name	URL	Pathway types	Curation		Dynamic/None	BioPAX	SBML	GPML	
BioCarta	http://www.biocarta.com	Metabolic and signaling	Manual	No	Dynamic	No	No	–	Yes
Pathway Commons	http://www.pathwaycommons.org	Metabolic, signaling, regulatory, disease, and drugs	Computational	No	None	BioPAX Level 2	No	–	Yes
HumanCyc	http://humancyc.org	Metabolic	Computational/manual	No	Dynamic	BioPAX Level 3	SBML Level 2	–	Yes[a]
PharmGKB	http://www.pharmgkb.org	Drug	Manual	Yes	Dynamic	BioPAX Level 3	No	–	Yes
BioModels	http://www.ebi.ac.uk/biomodels-main	Metabolic, regulatory, and signaling	Manual	Yes	None	BioPAX Level 2	SBML Level 2	–	Yes
Science Signaling	http://stke.sciencemag.org	Metabolic, regulatory, signaling, and disease	Manual	Yes	Dynamic	No	SBML Level 2	–	No
The Cancer Cell Map	http://cancer.cellmap.org	Signaling and regulatory pathways	Manual	No	None	BioPAX Level 2	No	GPML	Yes[a]

[a]Pathway data from this resource is accessible through the Pathway Commons' Web Service API.

mouse and rat, the nematode *Caenorhabditis elegans*, budding and fission yeasts, two plants and several bacteria. This projection of the manually curated human pathways onto model organisms provides a tool for the researcher to access orthologous biological pathways.

The Reactome data model is based fundamentally on the reaction. A reaction has input, and produces output that is different from the input by some quantifiable measure. The inputs and outputs of the reaction are entities such as protein, nucleic acid, organic molecules, chemical compounds, or complexes. The reaction itself can be catalyzed by an entity, and the molecular process that describes the catalyst's function is associated with the catalyst. Every reaction and entity in Reactome is associated with a species, and is assigned to a specific compartment (or compartments if the entity is a complex). Every reaction is an assertion of a minimal biological step, and each of these steps is supported by experimental evidence, that in practice is represented by a link to the appropriate literature reference.

Reactions are arranged in ordered steps, supporting the assertion that a reaction must be preceded by a previous reaction before it can occur. In this way, ordered reactions or pathways can be assembled. Pathways can contain reactions, pathways, or both. Each pathway represents a broader biological concept, and is used to divide the sets of reactions into human-recognizable domains of biology.

In a Reactome reaction, the expert author uses their knowledge of the field and its history to create a computational review of the biology. The curator assists the expert in breaking down each step of a pathway into a reaction. Together they work through a sub-domain of biology to create a human- and computer-accessible description of the biology, linking all of the gene, protein, literature, and controlled vocabulary data together. Reactome connects an expert's knowledge of how the biological pathway works with numerous data sources.

12.3.2   KEGG

The Kyoto Encyclopedia of Genes and Genomes (KEGG) has grown from 1 online database to 19 integrated knowledge bases organizing phenotypic, genomic, chemical and systemic information (Kanehisa *et al.*, 2008). Thirteen of these databases are manually curated and six are computationally derived. KEGG knowledge bases contain information about genomes (GENOMES), genes and proteins (GENES), chemical substances (LIGAND), chemical compounds (COMPOUNDS), approved drugs in US and Japan (DRUGS), relationships between disease genes, pathways, and drugs (DISEASE), and ontologies representing knowledge of biological systems (BRITE), (Kanehisa *et al.*, 2008). Each entry in a KEGG database has a unique identifier that is used to directly link corresponding database entities. Users can browse the different KEGG resources or search using the DBGET feature that searches across all the KEGG databases (Goto *et al.*, 1997). The KEGG PATHWAY knowledge base is a collection of graphical diagrams (pathway maps) and associated text information (pathway entries) that encompasses metabolism, environmental and genetic information processing, drug development, and cellular processes. Metabolic processes comprise the largest and most complete group of pathways in KEGG. Like Reactome, KEGG PATHWAY is a manually curated pathway database that focuses on the curation of biological pathways from a number of organisms, including human, plant and bacteria. Human disease pathways are further categorized into four groups: infectious diseases, metabolic disorders, cancers, and neurodegenerative disorders. Currently, KEGG PATHWAY contains 97 354 pathways from over 1215 species, generated by electronic inference based upon orthology data from 342 canonical pathways. KEGG PATHWAY links internally to the other KEGG resources that contain over 16 000 compounds, 9000 drugs and 10 900 glycans, and it links externally to NCBI, UniProt, GO and other third-party databases. Each reference pathway is manually drawn and is used as a template to computationally generate an organism-specific pathway. The KEGG website provides a semi-dynamic visualization for the presentation and navigation of its pathway information. Recently, a new global map of metabolic pathways called KEGG Atlas was developed that allows users to explore the complete metabolome and to map genomic, proteomic or metabolomic experimental data. Tutorials are available to guide users on how to navigate through pathways and the other KEGG resources.

12.3.3 WikiPathways

WikiPathways (Pico et al., 2008) is a recent addition to the pathway knowledge base group and is unique in its approach to curation. Pathway curation is open and community based, similar in principle to Wikipedia. Registered users of WikiPathways can create new pathways, and edit existing pathways, using a simple and intuitive graphical Web interface based upon PathVisio (van Iersel et al., 2008). WikiPathways can represent biological pathways from multiple species and many disciplines, including metabolism, signaling, gene regulation and physiological functions. Species-specific annotation is provided and external link outs to gene and protein databases such as Ensembl genome databases and small-molecule databases such as HMDB. Users can browse and search pathways by keyword, database identifier, species, and pathway categories. Community curation utilizes researchers at all levels, from graduate student to experts, in order to create and edit pathways. Once a user logs in, a pathway can be edited using the PathVisio program. A version history keeps a record of all pathway record edits. Each pathway record is displayed as an editable wiki page that is composed of a pathway diagram, a description of the pathway, pathway components and literature citations. Curation tags alert the author and community if edits are made to the pathway. Tutorials and documentation are available to guide users through pathway curation and use of the WikiPathways resources.

12.3.4 NCI-Pathway Interaction Database

The NCI-Pathway Interaction Database (NCI-PID) is a collaborative project between the US National Cancer Institute (NCI) and Nature Publishing Group

(NPG). It is an open-access online resource (Schaefer *et al.*, 2009). The NCI-PID is a partially human-curated collection of information about known biomolecular interactions and key cellular processes assembled into signaling pathways. The focus of NCI-PID is on signaling pathways and other regulatory pathways related to cancer. There are three types of data in the NCI-PID: (1) NCI-Nature curated data, created by Nature Publishing Group editors and reviewed by experts in the field; (2) BioCarta data, imported programmatically, in BioPAX Level 2 format, without post-translational modifications, and without expert review; (3) human pathways from Reactome, which were curated by humans for entry into Reactome, and which are annotated with UniProt IDs and post-translational modifications when imported into NCI-PID. All NCI-Nature curated data are generated according to the following principles and guidelines: human data is the primary focus for this database, although interactions in other mammals that are inferred to occur in humans are sometimes included with appropriate evidence codes. Biologically relevant networks of well-documented interactions are synthesized into predefined pathways. Molecular interactions are identified and authorized for inclusion by their presence in primary peer-reviewed literature. Editors identify interactions that are physiologically relevant, and assign evidence codes to each interaction, which are then reviewed by experts in the field. Consistency of nomenclature is achieved by using the HUGO (Human Genome Organization) gene symbols, with UniProt names or aliases, and/or Entrez Gene names or aliases for molecules. Gene Ontology (GO) controlled vocabulary terms are used to annotate biomolecules and biological processes. PID offers a range of search features to facilitate pathway exploration. Users can browse the predefined set of pathways or create interaction network maps centered on a single molecule or cellular process of interest. In addition, a batch query tool allows users to upload long lists of molecules, such as those derived from microarray experiments, and either overlay these molecules onto predefined pathways or visualize the complete molecular connectivity map. Users can also download molecule lists, citation lists and complete database content in extensible markup language (XML) and Biological Pathways Exchange (BioPAX) Level 2 format.

12.3.5   NCBI-BioSystems

The NCBI-BioSystems Database (Geer *et al.*, 2010) is designed to function as a data integration service that provides access to data from KEGG, Human Reactome, and EcoCyc. It provides a central repository of data that includes gene, protein, or pathway data, along with related literature, and/or molecular, and chemical data present within the Entrez biological database system. It is designed to facilitate computational analysis of pathway and biosystems data from these various sources. Thus, the NCBI-BioSystems database functions as a 'clearinghouse' for these other databases by integrating their data into the existing NCBI Entrez databases, which include the Gene, Protein, PubMed and PubChem databases. It links back to the original data source, in case more information is needed, and/or the user desires further data analysis. Programmatic API access is available via the NCBI Entrez programming utilities. These include SOAP Web services and other APIs.

12.3.6 Science Signaling

Science Signaling (previously Signal Transduction Knowledge Environment; STKE) provides the Database of Cell Signaling through free subscription to academics and researchers (Gough and Ray, 2002). Like Reactome, Science Signaling is an expert-authored and peer-reviewed curated database of signaling pathways. The Database of Cell Signaling encompasses many areas of biology across several organisms. Users can browse and search over 140 Connection maps, dynamically generated pathway diagrams based upon GIF and SVG formats. These signal transduction pathways are categorized into two distinct types. Canonical Pathways reflect a generalized view of a given signaling pathway across all species. Specific Pathways represent signaling pathways that exists in a particular cell, tissue or organism. The Canonical pathway is the foundation on which the specific pathways are organized. For example, a user viewing the Filamentous Growth Pathway in Yeast (Specific Pathway) can navigate to the ERK1/ERK2 MAPK Pathway, its Canonical Pathway. Components of these pathways are similarly classified. Each component of a specific pathway has a 'parent' canonical component, which is referenced within a canonical pathway; users can navigate between these component types. The Database of Cell Signaling allows the pathway expert to express an opinion regarding the strength of the evidence supporting the relationships between the components within the pathway; that is, Is the relation Demonstrated, Strongly Implied, Implied, or Speculative. Science Signaling also provides a printed and online journal, publishing novel peer-reviewed articles and reviews in the areas of network and pathway biology and signaling transduction research. The website includes a number of useful personalization features, allowing users to organize pathway information and database searches into online folders associated with their free subscription accounts. Hyperlinks also exist between the journal articles and relevant information in the Database of Cell Signaling, providing a direct connection between the published literature and the pathway knowledge base. External links from pathways and components to nucleotide or protein sequence data in NCBI Entrez database and model organism databases are also provided.

12.3.7 PharmGKB

PharmGKB, the Pharmacogenomics Knowledge Base, collects, encodes, and disseminates knowledge describing the impact of human genetic variations on drug response. The data set contains primary genotype and phenotype data, annotation of gene variants and gene–drug–disease relationships supported by literature reviews and primary literature. Curators summarize important pharmacogenomic genes and drug metabolism and therapeutic pathways (Klein *et al.*, 2004). PharmGKB is a publicly available Internet research tool developed by Stanford

University with funding from the National Institutes of Health (NIH), and is part of the NIH Pharmacogenetics Research Network (PGRN), a nationwide collaborative research consortium. The PharmGKB database is a central repository for genetic, genomic, molecular, and cellular phenotype data, including clinical information describing pharmacogenomics research study participants. The data includes, but is not limited to, clinical and basic pharmacokinetic and pharmacogenomic research in the cardiovascular, pulmonary, cancer, pathways, metabolic, and transporter domains.

12.4   How does information get into pathway knowledge bases?

Data curation (or biocuration) in the life sciences has been around for at least a decade. Biocuration started with the advent of the gene sequence databases such as GenBank (Benson *et al.*, 2009). Curation focused on the association of functional annotations to a DNA sequence. This process required the life scientist to submit the relevant information as the sequence was being submitted to GenBank. As the requirements to store additional information (e.g., structure, function) have increased, the role of the curator has evolved. Biocuration is now part of the foundation driving the establishment and ongoing development of biological data repositories. The Gene Ontology (GO) project is an example of life scientists and curators working together to create a database of gene annotations regarding the molecular function, biological process and cellular locations of gene products (Barrell *et al.*, 2009).

The biocurator collects, manages, annotates, analyses, and reviews biological data. Creating a curated data record is a multistep process, entailing collection, selection, review, and archival preservation. Collection and data organization are important to minimize duplication. Data selection is critical to ensure the storage of the necessary information, and understand the approaches and tools used in the identification of relevant information. The process of assuring record quality, or validation, ensures the fidelity of each data record. The review and validation processes incorporate documented record editing, curator feedback, and a mechanism for updating public records in the future. Archiving and preservation guarantees that the data is carefully organized, stored, accessible and maintained for the future.

Pathway curation is the process of representing a set of related biological events or reactions in a given biological context. Ideally, interacting molecules, complexes, and reactions are annotated only to an extent that promotes clear and accurate identification of the interacting entities. The quality, quantity, and completeness of pathway data in knowledge bases vary significantly. Pathway knowledge bases have different data formats, curation standards and author tools supporting the data capture. Life scientists, clinicians and researchers have the knowledge needed to build the pathways. Curators and editors reshape the structure of the knowledge to associate it with entries in primary databases. An overview of the Reactome curation process is shown in Figure 12.1. Databases such as Reactome, NCI-PID, Science Signaling, WikiPathways and KEGG have curators or editors to assist with pathway curation. We will use Reactome, a mature manually curated pathway knowledge base, as a case study for mapping biological pathways.

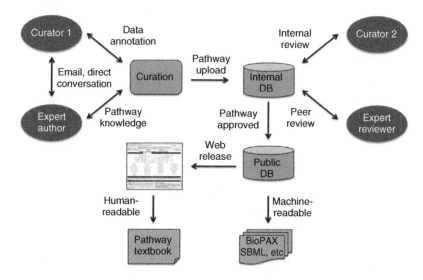

Figure 12.1   An overview of the Reactome curation process.

12.5   Introduction to data exchange languages

Searching for, assembling, integrating, and visualizing pathway data with other types of biological information and user data is a challenging and complex undertaking. A number of steps are needed to facilitate the interpretation of pathway and other biological data. Pathway data exchange languages should enable the use of formatted data without rewriting data for each software tool or different software environment and to guarantee the survival of pathway data beyond the life of the software used to create it. Pathway knowledge bases gather and exchange data in different formats: database dumps, flat file formats and data exchange languages such as BioPAX, SBML, or PSI MI and proprietary formats; for example, KGML and GPML. The BioPAX, SBML, and PSI MI exchange formats are described, and the adoption of these formats by pathway knowledge bases discussed.

12.5.1   SBML

Systems Biology Markup Language (SBML) was developed by the Systems Biology Workbench group and is supported by a relatively large user community, many third-party software tools, and several online data repositories. SBML is an XML format language for the exchange of computational models of biological pathways and processes (Hucka *et al.*, 2003). The current implementation, Level 2, focuses on the description of biochemical network and pathway models, such as intermediary metabolism, signal transduction pathways, and gene regulation. Each SBML model consists of a list of optional components, including compartments, species, species types, reactions, and events. Compartments describe the environment with defined boundaries in which the reaction takes place. There can be multiple compartments in an SBML model, and every entity should be assigned a compartment. The term *species* is used in SBML to describe the entities (protein, small molecule, etc.) that take part in the reactions, and the *species type* refers to the type of entity that can contribute in reactions. For each species it is possible to list the initial amount and concentration and if these values change in the course of the reaction. Reactions represent interactions between species or transformation, transport or binding reactions. Events define the changes that can occur in the model; for example, the event initiator, time limitations, and so on. The SBML model can contain mathematical functions, expressions, and units. Extra information and link-outs to external sources can be assigned to SBML note and annotation fields. SBML has become the *de facto* standard for pathway modeling, and work is in progress to draft SBML Level 3, a more modular structured language with a core set of features and optional packages. Some of the key additions to SBML Level 3 include layout and display information for pathway diagrams, hierarchical model composition, description of multi-component species, quantitative and steady-state models. An open-source library to help users read, write, and validate SBML files is available called libSBML (Bornstein *et al.*, 2008). Examples of software tool support for SBML are CellDesigner (Funahashi *et al.*, 2003) and Copasi (Hoops *et al.*, 2006).

12.5.2 BioPAX

Biological Pathway Exchange (BioPAX) is a collaborative effort among members of the BioPAX working group to develop a data exchange language to describe biological pathways data (Luciano, 2005). BioPAX is implemented in both an XML schema and OWL-DL (a standard for the representation of ontologies). At the time of this writing, BioPAX Level 3, Release 1.0 has just been approved. BioPAX Level 3 focuses on molecular and genetic interactions, metabolic pathways, signal transduction pathways, and gene relations. All objects in BioPAX are organized as a hierarchy of inherited classes. The BioPAX ontology consists of the root level Entity class and four subclasses: Pathway, Interaction, PhysicalEntity and Gene. Entity is a unique biological object or unit. Pathway is a series of interactions, often forming a network. Interaction refers to the relationship between entities, for example, protein–protein interaction. PhysicalEntity has five subclasses that describe the different objects that interact, for example, complex, DNA, RNA, protein, and small molecule. Finally, Gene is not a physical entity and is a simplification of the concept of a gene for genetic interactions. There are a number of subclasses for Interaction; each limits the possible role and the number of possible interactors, for example, control having catalysis and modulation. Specifying interacting molecules and interaction details in a pre-defined class hierarchy promotes efficient sharing of the information between databases. Additional data and hyperlinks to external sources can be assigned to BioPAX objects. A library to help users read and write BioPAX files is available called Paxtools. Software tools are also available to visualize and manipulate BioPAX files, such as Cytoscape (Shannon *et al.*, 2003).

12.5.3   PSI MI

The Proteomics Standards Initiative Molecular Interaction XML format (PSI MI) was defined by the Proteomics Standards Initiative (Hermjakob *et al.*, 2006). The aim is to create a standard data exchange format for molecular interaction and proteomic data representation to enable data comparison, validation, and exchange. A PSI MI record typically consists of one or more protein–protein interactions. The core element of the PSI MI 2.5 data model is the entrySet. Each entrySet consists of six entries. The source and availabilityList refers to the source of the interaction data, usually the original database and the copyright statement, respectively. Experimental parameters and publication details are referred to in the experimentList. A list of interactors and their references are defined by interactorList. The interactionList refers to the list of interactions. Finally, the attributeList describes the additional information that does not fall into one of the other entries. Each entry has further properties that capture additional interaction data. Furthermore, PSI MI makes extensive use of controlled vocabularies, which are viewed as an important part of encoding interaction data. PSI MI is widely supported by the interaction databases such as BioGrid, DIP, BIND, IntAct, and MINT. Even though PSI MI was not designed for the inclusion of pathway data, it is possible to represent pathway data as molecular interactions. The pathway is described by the interactorList and the interactionList. PSI MITAB format is part of the PSI MI 2.5 standard, and is used to describe binary interactions.

12.5.4   Comparison of data exchange formats for different pathway knowledge bases

Reactome supports the major pathway data exchange formats, and provides data in an open-source knowledge base. The data formats available from the Reactome Data Download page are: interaction, reaction, and pathway data in formats that include downloadable flat, MySQL, SBML Level 2, BioPAX Level 3 (and Level 2) and PSI MITAB. A SOAP-based Web Services API is also available to access the Reactome data programmatically. The Reactome Biomart tool assists with data mining and data analysis by allowing fast queries across Reactome data sets. For example, a user can create a query of the Reactome database for reactions that contain a given protein. Furthermore, a user can combine a query across the Reactome data set with a query from another database (e.g.,

ENSEMBL). Pre-computed queries are also available for common searches; for example, all genes in a given pathway.

The availability of Reactome data in the BioPAX format encourages disseminated curation. For example, NCI-PID is one of the major pathway knowledge bases to use Reactome's BioPAX export as a source of pathway data. PID itself distributes pathway and network data in BioPAX Level 2 or its native PID XML. WikiPathways uses GenMAPP Pathway Markup Language (GPML) to promote pathway data exchange. This representation is compatible with several third-party data analysis and visualization tools, including GenMAPP (Dahlquist *et al.*, 2002), PathVisio and Cytoscape. Science Signaling's Connections Maps database is available in SBML Level 2. There are terms and conditions of use but it is free for academic users. Some commercial companies such as Ariadne Genomics have integrated the Connections Map data set with their bioinformatics visualization and analysis tools.

KEGG PATHWAY data can be freely downloaded by academic researchers from the KEGG FTP site, or accessed programmatically via KEGG Web Services APIs (Moriya *et al.*, 2007). Pathway data is available in a licensed and proprietary XML-based KEGG Markup Language (KGML). There is also a KGML+ file that is a hybrid between an SVG file (pathway graphic) and a KGML file (text information). This provides a framework for automatic pathway diagram generation and computational analysis. KEGG data may be used for modeling and simulation purposes, but KEGG does not natively support SBML. There is a KEGGconverter tool that converts the KGML into SBML Version 2 (Moutselos *et al.*, 2009). KEGG data also can be converted to BioPAX Level 2 using the KGML-ED tool (Klukas and Schreiber, 2007).

BioPAX, PSI MI, and SBML are highly useful for the creation of various integrative pathway knowledge bases and databases that can be widely used to model, simulate, and visualize pathway and network data in various language formats. These are not rival standards but rather provide complementary approaches to describe related pathway and interaction information. Most pathway knowledge bases, like Reactome, provide qualitative mappings of functional relationships between pathway entities. These exchange formats allow the data present in various pathway knowledge bases to be used as a larger aggregate data source by the researcher.

12.6   Visualization tools

Typically, research articles, reviews, and textbooks present pathway diagrams as a simplified representation that doesn't convey enough detail about the pathway entities and relationships between them. The molecular mechanisms described in the diagram are more complex, while other pieces of information such as the transduction of the signal, the interactions between the entities are missing. With high-throughput data capture technologies it has becoming increasingly important to have tools available to integrate and visualize thousands of data points and entities in the context of the pathway diagrams. The challenge faced for data visualization is not just the increasing volume of data but also the increasing complexity of this information. To address this challenge, visualization tools are available to scientists to illustrate pathway diagrams. There is a shift from the reductionist approach (individual components and relationships) to scientific discovery in systems-level process analysis; exploring the entire systems simultaneously. The latter relies heavily upon visual representations of pathway entities and their relationships, behavior, and functionality.

The majority of pathway knowledge bases provide interactive visualization tools allowing users to interact with the pathway diagram, linking to other data about the molecules and relationships, such as Reactome, KEGG, PID, and Science Signaling. These same four pathway knowledge bases also provide pathway data in an XML format (i.e., BioPAX, SBML or proprietary format) that allows users the opportunity to generate interactive visualizations of pathway data using Cytoscape or CellDesigner.

Tools available through WikiPathways and Cytoscape allow users to manipulate pathway diagrams and visualize user-supplied data. Initiatives are currently underway to provide detailed descriptions of pathways in a standard, exchangeable, computational format that incorporates pathway and biological information. One such initiative that has been adopted by Reactome is the Systems Biology Graphical Notation; SBGN (Le Novere *et al.*, 2009). The aim of this standard is to define a unified framework for the diagrammatic representation of biological networks. The new entity-level pathway visualization tool for Reactome will allow a user to view 'textbook style' illustrations of pathways (Figure 12.2).

12.7 Use case: pathway analysis in Reactome using statistical analysis of high-throughput data sets

A typical user of the Reactome pathway analysis tools will have performed one of the types of data capture experiments described previously. This user will have generated a list (set) of genes that represent a measurable change that occurred within the experiment. The steps required to generate this list of genes is not within the scope of this review, but it is worth noting the general framework that would have occurred to arrive at this list.

The simplest conceptual experiments are those that compare gene or protein expression levels in the absence (control) or presence (experimental variable) of a compound, perhaps a toxicant, therapeutic agent, or signaling molecule. Whatever technique is used to collect the data, the initial experimental results are a large set of data that must first be analyzed for internal variability and preparation artifacts. Once internal controls are examined and integrated into a processed data set, a statistical comparison of the control and variable measures can be performed. Finally, the researcher generates a data set that, in this hypothetical case, would yield a list of identifiers (IDs) with expression patterns that are altered in the presence of an experimental variable. This list can contain expression levels and gene

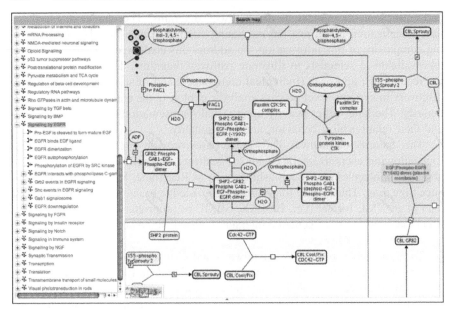

Figure 12.2   A screenshot of the Reactome's new entity-level viewer based upon Systems Biology Graphical Notation. This 'Google-maps-like' pathway visualization tool has a zoom in/out box available in the upper left and a 'birds-eye view' of a pathway provided in the lower left box. The entire pathway can be moved within the window by clicking and dragging it. There is a Reaction or pathway details section at the bottom of the page. Selecting from the event hierarchy to the left centers the map on that event and highlights the molecules and reactions. Scrolling over the Reaction names or the Molecule boxes will provide more information.

or protein identifiers, or can be more elaborate, with the experimental variables associated with a number of time points measured throughout the experiment.

The lists of identifiers generated by these methods often contain hundreds or thousands of distinct entities. Examining such a list manually, laboriously looking at identifiers and attempting to divine the biological significance of such a collection, is extremely difficult, or impossible. What is required for meaningful analysis is a method that analyses the set of distinct entities as a whole. Such a method would identify subsets of identifiers that share characteristics, including, but not limited to, function, biological mechanism, and logical biological pathway association. The primary data for making such assertions is contained within a number of different resources (described previously), but is not interlinked in a manner that allows for biological pathway analysis.

Pathway knowledge bases are precisely the resource required. The primary data used for construction of a pathway database has connected the relevant data resources with biological pathway knowledge. The data model used in the pathway resource, in this case Reactome, provides a biological pathway framework upon which experimental data may be analyzed, or 'mapped.' The core of this framework, the reaction, associates the protein, nucleic acid, organic molecules, chemical compounds, or complexes that are input and output with each of the primary data sets that describe these entities. Each reaction is associated with the experimental support for that reaction's occurrence and, if present, a catalyst entity and GO molecular function term.

The following is a description of pathway analysis in Reactome, and here we would like to note that pathway analysis is generally similar in the various pathway database resources described previously. The general concept of overlaying experimental data upon a curated framework is shared. This is an ever-growing opportunity for the researcher, and yet it poses challenges for the initial steps of analysis. Our description of these initial steps for pathway analysis should serve as a template for the researcher to access the high-quality, manually curated data set that is present within Reactome, as well as a general primer for access to other pathway databases.

The pathway tools are used to determine which reactions and/or pathways are statistically overrepresented in a set of genes as specified by a submitted list of identifiers. The user provides a given a list of identifiers, and the Skypainter tool will identify common events for these identifiers. Skypainter accepts a number of different types of identifiers directly, including: UniProt, Ref-Seq, Ensembl, OMIM, Entrez Gene, Affymetrix, GO, KEGG COMPOUND, and ChEBI. It should be noted that all purely numeric identifiers, such as from OMIM and Entrez Gene have to have the abbreviated database name and colon prepended to them, that is, MIM:602544, EntrezGene:55718.

If the user has a list of identifiers not included in this list, a number of online resources are available for interconversion between identifiers. A notable example of these resources is the DAVID Gene ID Conversion Tool (Huang da *et al.*, 2008)

The Reactome pathway tool picks out the total set of identifiers that participate in an event. This is the subset of identifiers (for the given species) that have been annotated in Reactome as participating in this event. This subset is then compared to the submitted list of experimental identifiers of which N identifiers participate in that given event (where N is the number of experimental identifiers that are in the original Reactome set). The pathway tool calculates the probability of observing at least N genes from an event (if the event is not overrepresented in the submitted list of genes) by performing the one-tailed version of Fisher's exact test. If a gene or protein identifier list is submitted to the pathway tool, the resulting report provides a p-value associated with each pathway containing identifier 'hits.' If the p-value is smaller than or equal to the significance level, that event is identified as overrepresented in the submitted list of genes.

Submitted lists can also contain identifiers followed (separated by space or tab) by a numeric value. A time series can be submitted by providing multiple values on the same line as each identifier, separated by a single space or tab. In this case, the pathway will provide a series of images representing each column of expression data, generating a movie of the time series.

The pathway tool provides a number of different views of the data: statistically overrepresented events in an ordered hierarchy of pathways and events, statistically overrepresented pathways as an ordered list, and a mapping from submitted identifiers to reactions. The current Reactome pathway tool, Skypainter, also provides graphical displays of reactions colored according to the number of genes or compounds (as specified by the submitted list of identifiers) participating in the given reaction, as well as links for downloading images. Future versions of the pathway tool are expected to replace the Skypainter functionality with a new pathway tool, as the view of the entire collection of biological reactions has become too large to interpret. Each of the outputs described are represented as a series of different panes.

It is easiest to understand the functionality of the tool by taking a 'hands-on approach.' All Reactome tools are available through the 'tools' menu item that can be accessed from any Reactome page. The current pathway tool for pathway analysis can be accessed by selecting 'Skypainter.' Once the user arrives at the input page, they will find a large text field where data can be directly typed, pasted, or uploaded from a local file. We can demonstrate the functionality of the pathway analysis tool by clicking on the word 'identifiers' in the directions above the text field.

By clicking on 'identifiers,' a demonstration set of data is loaded into the text field; see Figure 12.3. These identifiers are the UniProt identifiers that correspond to entries found in the OMIM Morbid Map (ftp://ftp.ncbi.nih.gov/repository/OMIM/morbidmap). OMIM is the Online Mendelian Inheritance in Man data set, a continuously updated catalog of human genes and genetic

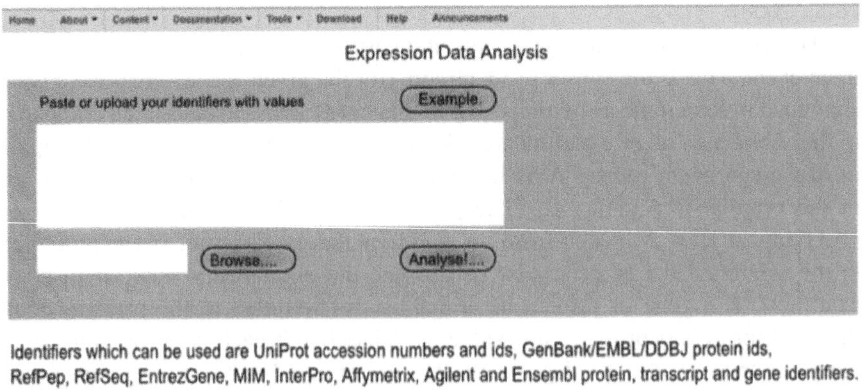

Figure 12.3 A text field is available on the tools pathway analysis page for data submission. The text field accepts typed or pasted text, and allows the user to upload a local file.

disorders (http://www.ncbi.nlm.nih.gov/omim/). Depending on the version of Reactome, clicking 'Paint!' or 'Analyze!' will send your data to the pathway tool.

The pathway tool analyzes the data and when complete sends the user to the results page. The results page provides a list of statistically overrepresented events as described previously; see Figure 12.4. Using Reactome version 30, the most overrepresented event is 'Formation of Fibrin Clot (Clotting Cascade),' with an unadjusted probability of 1.1×10^{-12} of finding the 19 submitted genes in this event (which contains a total of 29 genes) by chance. Each event is colored according to a heat map providing a visual cue as to which events contain the most significant overrepresentation.

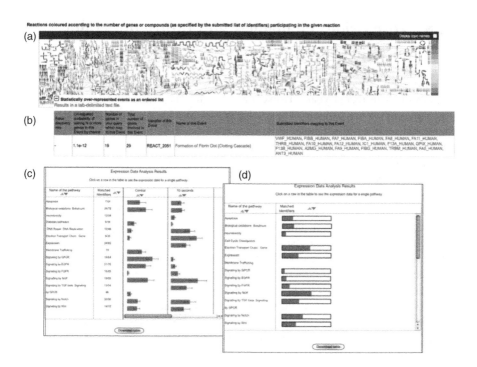

Figure 12.4   Examples of pathway analysis in Reactome – a view of the current Skypainter tool (a); the table of the worked OMIM example (b); future pathway analysis result tables for multiple (c) and single (d) time-point analysis. Online Mendelian Inheritance in Man, OMIM, is a Web resource created by the McKusick-Nathans Institute of Genetic Medicine, Johns Hopkins University (Baltimore, MD) and National Center for Biotechnology Information, National Library of Medicine (Bethesda, MD), URL: http://www.ncbi.nlm.nih.gov/omim/.

Scrolling down the page provides access to different representations of the data; future versions of the pathway tools will provide new functionality including basic ID mapping, and an enhanced table view of the data.

One of the unique opportunities that pathway tools provide is species comparison. Researchers working with a model organism can compare data generated within that model organism species to any other species listed in Reactome. The Reactome data set is meticulously annotated to ensure species fidelity within the logical steps of a pathway. A series of events identified within a data set will reference a single species only. Once a researcher has identified a pathway of interest, they can easily look through other species, identifying specific proteins and genes encoding the entities involved in the pathway of interest. This utility can be used as a framework for the logical extrapolation of results from a model system back to human.

12.8   Discussion: challenges and future directions of pathway knowledge bases

Pathway knowledge bases such as Reactome have made important contributions and advances in recent years. However, there are also major challenges remaining. Although most pathway knowledge bases provide qualitative mappings of functional relationships between pathway entities, there is a need to create computational and mathematical models of canonical pathways. There is also a need for much more quantitative data, such as reaction kinetics, entity stoichiometry, molecule concentrations, and other cell- or tissue-specific data.

Important pathway knowledge base functions, such as the prediction of drug/target associations and their relation to specific biological pathways, also have major challenges. In particular, there is the need for a greater exchange of chemical, physiological, pathological, reaction, and pathway data between different pathway resources. Pathway knowledge base projects must maintain and increase their commitment to collaboration and integrating biochemical, biological, biophysical and chemical information data exchange formats. Such collaboration will allow tremendous growth in computational pathway modeling.

The future directions of the pathway knowledge bases reviewed here are driven by the fact that they are all Web applications, built on existing Web technologies. As a result, the evolution of pathway knowledge bases has been closely related to the evolution of Web technologies, such as 'social networks' that allow individuals to interact in online groups or communities. One example of how social networks are influencing pathway tools is the appearance of 'community curation' efforts, such as WikiPathways. These allow end user biologists to contribute annotations as pathway authors, and thus increase the speed and accuracy of the curation process.

One of the most interesting and potentially important directions for pathway knowledge base development is the incorporation of ongoing developments of 'Semantic Web' technologies. At present, pathway knowledge bases, such as Reactome, rely on a manual curation approach that incorporates human knowledge via traditional literature review by curators and experts. This is necessary for the accurate association of genes and gene products identified through high-throughput or computational methods with specific biological pathways. However, in the future, Semantic Web technologies, such as RDF and OWL, may make it possible to accurately map new experimental or computational results to specific biological pathways without the need for human curation. For example, the use of RDF or other semantic technologies can allow precise RDF predicates or other semantic markup to be embedded easily by biologist authors directly into research articles as they are being written. If such articles are created, they can then later be analyzed computationally, so that new data can be mapped accurately onto specific biological pathways, in some cases without the need for a human curator. In principle, this also could allow automated formal reasoning processes to identify causal relationships between experimentally identified genes and molecules within specific biological pathways (Splendiani, 2008).

Given these new technologies, the power and utility of future pathway knowledge bases and other bioinformatics tools is likely to grow dramatically. Our ability to create such powerful new knowledge tools will be limited largely by our imagination.

12.9   References

Alfarano, C., Andrade, C.E., Anthony, K., *et al.* (2005) The Biomolecular Interaction Network Database and related tools 2005 update. *Nucleic Acids Res.*, **33**, D418–24.

Aranda, B., Achuthan, P., Alam-Faruque, Y., *et al.* (2009) The IntAct molecular interaction database in 2010. *Nucleic Acids Res.*

Bader, G.D., Cary, M.P., and Sander, C. (2006) Pathguide: a pathway resource list. *Nucleic Acids Res.*, **34**, D504–6.

Barrell, D., Dimmer, E., Huntley, R.P., *et al.* (2009) The GOA database in 2009 - an integrated Gene Ontology Annotation resource. *Nucleic Acids Res.*, **37**, D396–403.

Barrett, T., Troup, D.B., Wilhite, S.E., *et al.* (2009) NCBI GEO: archive for high-throughput functional genomic data. *Nucleic Acids Res.*, **37**, D885–90.

Benson, D.A., Karsch-Mizrachi, I., Lipman, D.J., *et al.* (2009) GenBank. *Nucleic Acids Res.*, **37**, D26–31.

Bornstein, B.J., Keating, S.M., Jouraku, A., and Hucka, M. (2008) LibSBML: an API library for SBML. *Bioinformatics*, **24**, 880–1.

Brazma, A., Hingamp, P., Quackenbush, J., *et al.* (2001) Minimum information about a microarray experiment (MIAME)-toward standards for microarray data. *Nat. Genet.*, **29**, 365–71.

Breitkreutz, B.J., Stark, C., Reguly, T., *et al.* (2008) The BioGRID Interaction Database: 2008 update. *Nucleic Acids Res.*, **36**, D637–40.

Brooksbank, C., Cameron, G., and Thornton, J. (2010) The European Bioinformatics Institute's data resources. *Nucleic Acids Res.*, **38**(Database issue), D17–25.

Ceol, A., Chatr Aryamontri, A., Licata, L., *et al.* (2010) MINT, the molecular interaction database: 2009 update. *Nucleic Acids Res.*, **38**(Database issue), D532–9.

Craig, R. and Beavis, R.C. (2004) TANDEM: matching proteins with tandem mass spectra. *Bioinformatics*, **20**, 1466–7.

Crameri, R. and Kodzius, R. (2001) The powerful combination of phage surface display of cDNA libraries and high throughput screening. *Comb. Chem. High Throughput Screen.*, **4**, 145–55.

Dahlquist, K.D., Salomonis, N., Vranizan, K., *et al.* (2002) GenMAPP, a new tool for viewing and analyzing microarray data on biological pathways. *Nat. Genet.*, **31**, 19–20.

Denkert, C., Budczies, J., Kind, T., *et al.* (2006) Mass spectrometry-based metabolic profiling reveals different metabolite patterns in invasive ovarian carcinomas and ovarian borderline tumors. *Cancer Res.*, **66**, 10795–804.

Ficarro, S.B., McCleland, M.L., Stukenberg, P.T., *et al.* (2002) Phosphoproteome analysis by mass spectrometry and its application to Saccharomyces cerevisiae. *Nat. Biotechnol.*, **20**, 301–5.

Funahashi, A., Tanimura, N., Morohashi, M., and Kitano, H. (2003) CellDesigner: a process diagram editor for gene-regulatory and biochemical networks. *BioSilico*, **1**, 159–62.

Geer, L.Y., Marchler-Bauer, A., Geer, R.C., *et al.* (2010) The NCBI BioSystems database. *Nucleic Acids Res.*, **38**(Database issue), D492–6.

Gibney, M.J., Walsh, M., Brennan, L., *et al.* (2005) Metabolomics in human nutrition: opportunities and challenges. *Am. J. Clin. Nutr.*, **82**, 497–503.

Goto, S., Bono, H., Ogata, H., *et al.* (1997) Organizing and computing metabolic pathway data in terms of binary relations. *Pac Symp Biocomput*, **1997**, 175–86.

Gough, N.R. and Ray, L.B. (2002) Mapping cellular signaling. *Sci. STKE*, **135**, 1632–3.

Griffin, J.L. and Bollard, M.E. (2004) Metabonomics: its potential as a tool in toxicology for safety assessment and data integration. *Curr. Drug Metab.*, **5**, 389–98.

Hacia, J.G. (1999) Resequencing and mutational analysis using oligonucleotide microarrays. *Nat. Genet.*, **21**, 42–7.

Hacia, J.G., Fan, J.B., Ryder, O., *et al.* (1999) Determination of ancestral alleles for human single-nucleotide polymorphisms using high-density oligonucleotide arrays. *Nat. Genet.*, **22**, 164–7.

Hermjakob, H. (2006) The HUPO proteomics standards initiative - overcoming the fragmentation of proteomics data. *Proteomics*, **6**(Suppl 2), 34–8.

Ho, Y., Gruhler, A., Heilbut, A., *et al.* (2002) Systematic identification of protein complexes in Saccharomyces cerevisiae by mass spectrometry. *Nature*, **415**, 180–3.

Hoops, S., Sahle, S., Gauges, R., *et al.* (2006) COPASI - a COmplex PAthway SImulator. *Bioinformatics*, **22**, 3067–74.

Hu, G.K., Madore, S.J., Moldover, B., *et al.* (2001) Predicting splice variant from DNA chip expression data. *Genome Res.*, **11**, 1237–45.

Huang da, W., Sherman, B.T., Stephens, R., *et al.* (2008) DAVID gene ID conversion tool. *Bioinformation*, **2**, 428–30.

Hucka, M., Finney, A., Sauro, H.M., *et al.* (2003) The systems biology markup language (SBML): a medium for representation and exchange of biochemical network models. *Bioinformatics*, **19**, 524–31.

van Iersel, M.P., Kelder, T., Pico, A.R., *et al.* (2008) Presenting and exploring biological pathways with PathVisio. *BMC Bioinformatics*, **9**, 399.

Johnson, D.S., Mortazavi, A., Myers, R.M., and Wold, B. (2007) Genome-wide mapping of in vivo protein-DNA interactions. *Science*, **316**, 1497–502.

Joshi-Tope, G., Gillespie, M., Vastrik, I., *et al.* (2005) Reactome: a knowledgebase of biological pathways. *Nucleic Acids Res.*, **33**, D428–32.

Kanehisa, M., Araki, M., Goto, S., *et al.* (2008) KEGG for linking genomes to life and the environment. *Nucleic Acids Res.*, **36**, D480–4.

Kell, D.B. (2006) Systems biology, metabolic modelling and metabolomics in drug discovery and development. *Drug Discov. Today*, **11**, 1085–92.

Klein, T.E. and Altman, R.B. (2004) PharmGKB: the pharmacogenetics and pharmacogenomics knowledge base. *Pharmacogenomics J.*, **4**, 1.

Klukas, C. and Schreiber, F. (2007) Dynamic exploration and editing of KEGG pathway diagrams. *Bioinformatics*, **23**, 344–50.

Kopka, J., Schauer, N., Krueger, S., *et al.* (2005) GMD@CSB.DB: the Golm Metabolome Database. *Bioinformatics*, **21**, 1635–8.

Le Novere, N., Hucka, M., Mi, H., *et al.* (2009) The Systems Biology Graphical Notation. *Nat. Biotechnol.*, **27**, 735–41.

Luciano, J.S. (2005) PAX of mind for pathway researchers. *Drug Discov. Today*, **10**, 937–42.

Marinelli, R.J., Montgomery, K., Liu, C.L., *et al.* (2008) The Stanford Tissue Microarray Database. *Nucleic Acids Res.*, **36**, D871–7.

Matthews, L., Gopinath, G., Gillespie, M., *et al.* (2009) Reactome knowledgebase of human biological pathways and processes. *Nucleic Acids Res.*, **37**, D619–22.

McCormack, A.L., Schieltz, D.M., Goode, B., *et al.* (1997) Direct analysis and identification of proteins in mixtures by LC/MS/MS and database searching at the low-femtomole level. *Anal. Chem.*, **69**, 767–76.

Moriya, Y., Itoh, M., Okuda, S., *et al.* (2007) KAAS: an automatic genome annotation and pathway reconstruction server. *Nucleic Acids Res.*, **35**, W182–5.

Mortazavi, A., Williams, B.A., McCue, K., *et al.* (2008) Mapping and quantifying mammalian transcriptomes by RNA-Seq. *Nat. Methods*, **5**, 621–8.

Moutselos, K., Kanaris, I., Chatziioannou, A., *et al.* (2009) KEGGconverter: a tool for the in-silico modelling of metabolic networks of the KEGG Pathways database. *BMC Bioinformatics*, **10**, 324.

Parkinson, H., Kapushesky, M., Kolesnikov, N., *et al.* (2009) ArrayExpress update - from an archive of functional genomics experiments to the atlas of gene expression. *Nucleic Acids Res.*, **37**, D868–72.

Perkins, D.N., Pappin, D.J., Creasy, D.M., and Cottrell, J.S. (1999) Probability-based protein identification by searching sequence databases using mass spectrometry data. *Electrophoresis*, **20**, 3551–67.

Pico, A.R., Kelder, T., van Iersel, M.P., *et al.* (2008) WikiPathways: pathway editing for the people. *PLoS Biol.*, **6**, e184.

Pollack, J.R., Perou, C.M., Alizadeh, A.A., *et al.* (1999) Genome-wide analysis of DNA copy-number changes using cDNA microarrays. *Nat. Genet.*, **23**, 41–6.

Ren, B., Robert, F., Wyrick, J.J., *et al.* (2000) Genome-wide location and function of DNA binding proteins. *Science*, **290**, 2306–9.

Salwinski, L., Miller, C.S., Smith, A.J., *et al.* (2004) The Database of Interacting Proteins: 2004 update. *Nucleic Acids Res.*, **32**, D449–51.

Sanger, F., Nicklen, S., and Coulson, A.R. (1977) DNA sequencing with chain-terminating inhibitors. *Proc. Natl. Acad. Sci. U.S.A.*, **74**, 5463–7.

Sayers, E.W., Barrett, T., Benson, D.A., *et al.* (2009) Database resources of the National Center for Biotechnology Information. *Nucleic Acids Res.*, **37**, D5–15.

Schaefer, C.F., Anthony, K., Krupa, S., *et al.* (2009) PID: the Pathway Interaction Database. *Nucleic Acids Res.*, **37**, D674–9.

Schena, M., Shalon, D., Davis, R.W., and Brown, P.O. (1995) Quantitative monitoring of gene expression patterns with a complementary DNA microarray. *Science*, **270**, 467–70.

Shannon, P., Markiel, A., Ozier, O., *et al.* (2003) Cytoscape: a software environment for integrated models of biomolecular interaction networks. *Genome Res.*, **13**, 2498–504.

Shevchenko, A., Jensen, O.N., Podtelejnikov, A.V., *et al.* (1996) Linking genome and proteome by mass spectrometry: large-scale identification of yeast proteins from two dimensional gels. *Proc. Natl. Acad. Sci. U.S.A.*, **93**, 14440–5.

Smith, C.A., O'Maille, G., Want, E.J., *et al.* (2005) METLIN: a metabolite mass spectral database. *Ther Drug Monit*, **27**, 747–51.

Splendiani, A. (2008) RDFScape: Semantic Web meets systems biology. *BMC Bioinformatics*, **9**(Suppl 4), S6.

Sugawara, H., Ikeo, K., Fukuchi, S., *et al.* (2009) DDBJ dealing with mass data produced by the second generation sequencer. *Nucleic Acids Res.*, **37**, D16–18.

Uetz, P., Giot, L., Cagney, G., *et al.* (2000) A comprehensive analysis of protein-protein interactions in Saccharomyces cerevisiae. *Nature*, **403**, 623–7.

Vastrik, I., D'Eustachio, P., Schmidt, E., *et al.* (2007) Reactome: a knowledge base of biologic pathways and processes. *Genome Biol.*, **8**, R39.

Vizcaino, J.A., Cote, R., Reisinger, F., *et al.* (2009) A guide to the Proteomics Identifications Database proteomics data repository. *Proteomics*, **9**, 4276–83.

Wang, Y., Xiao, J., Suzek, T.O., *et al.* (2009) PubChem: a public information system for analyzing bioactivities of small molecules. *Nucleic Acids Res.*, **37**, W623–33.

Wheeler, D.A., Srinivasan, M., Egholm, M., *et al.* (2008) The complete genome of an individual by massively parallel DNA sequencing. *Nature*, **452**, 872–6.

Wishart, D.S. (2008) Metabolomics: a complementary tool in renal transplantation. *Contrib Nephrol*, **160**, 76–87.

Wishart, D.S., Knox, C., Guo, A.C., *et al.* (2009) HMDB: a knowledgebase for the human metabolome. *Nucleic Acids Res.*, **37**, D603–10.

Wright Jr, G.W., Cazares, L.H., Leung, S.M., *et al.* (1999) Proteinchip(R) surface enhanced laser desorption/ionization (SELDI) mass spectrometry: a novel protein biochip technology for detection of prostate cancer biomarkers in complex protein mixtures. *Prostate Cancer Prostatic Dis.*, **2**, 264–76.

Zhao, Y., Lee, W.N., and Xiao, G.G. (2009) Quantitative proteomics and biomarker discovery in human cancer. *Expert Rev Proteomics*, **6**, 115–8.

Zhu, H., Bilgin, M., Bangham, R., *et al.* (2001) Global analysis of protein activities using proteome chips. *Science*, **293**, 2101–5.

13

Methods and challenges of identifying biomolecular relationships and networks associated with complex diseases/phenotypes, and their application to drug treatments

Mie Rizig

13.1 Complex traits: clinical phenomenology and molecular background

Complex traits represent many physiological functions of the human body including normal hemostatic functions such as pH control, blood glucose regulation, and temperature maintenance. The majority of innate and acquired behavioral characteristics in humans are also complex traits. These include memory, emotions, and learning. In addition, the majority of modern-day illnesses like hypertension, diabetes, obesity, cardiovascular diseases, immune-inflammatory, and neuropsychiatric disorders exhibit the multifaceted nature of complex traits. The umbrella of complex traits can be further extended to include the effects of drugs and

Knowledge-Based Bioinformatics: From Analysis to Interpretation   Edited by Gil Alterovitz and Marco Ramoni
© 2010 John Wiley & Sons, Ltd

chemicals on the body's organs and patients' responses to treatments. Thus, understanding complex traits is crucial to comprehending the status of health and diseases and in the exploration of management options.

Conventionally, a complex trait is defined as a phenotype whose features are regulated by multiple genetic and environmental factors. This is in contrast to monogenetic traits, which are directly controlled by variations in a single gene. Complex traits do not follow the rules of Mendelian inheritance because the relationships between their genetic variants and phenotypes are not linear, and their associated genes do not always interact additively.

Although the above definition captures most of the fundamental aspects of complex traits, it fails to reflect the intricacies of the clinical phenomenology and molecular backgrounds.

Clinically, complex traits cannot be defined using single qualities; they are usually described in terms of combinations of different heterogeneous phenotypes or symptoms. These phenotypes can involve multiple organs and/or tissues types. Some of the phenotypes or symptoms can be partially or entirely shared between two or more other complex traits.

With regard to their molecular origins, complex phenotypes do not arise solely from variations within DNA sequences of a particular gene or group of genes. They usually result from several interactive intracellular processes composed of a particular gene and/or genes and/or their products (e.g., RNA, proteins or metabolites). These intracellular machineries interact in different combinations and at different levels to form series of complex networks which are highly dynamic and fluid, and at the same time extremely resilient. In addition, they are highly specific to the affected cell or tissue types and they are ultra sensitive to internal and external environmental conditions.

A cardiovascular disease such as Coronary Artery Disease (CAD) provides an excellent example of the multifaceted nature of complex traits (Tegner *et al.*, 2007). CAD is a degenerative disease developing over years. It results from the gradual alteration of the composition of the arterial wall (both intra- and extracellularly) by stress from circulating blood cells and other plasma components. This gradual wear and tear eventually leads to the formation of atherosclerotic plaques. The rate of atherosclerosis development depends on environmental pressures and on the genetic makeup of the individual. Environmental pressures relevant to CAD include airborne pollutants (e.g., cigarette smoke), infections, food (specifically cholesterol), and behavioral factors (e.g., stress and exercise levels). The net effect of the environmental pressures is filtered through the individual's genetic makeup and is reflected in changes in blood flow and its constituents. Over years, environmental and lifestyle factors alter gene expression in organs. Specifically, changes in the expression of genes relating to energy metabolism and inflammation in the liver, fat, or skeletal muscle are believed to be particularly relevant for CAD. In turn, alterations in gene expression are reflected in the circulation, where metabolic and inflammatory markers synthesized in these organs can be detected. Thus, measurements of plasma constituents (e.g., cholesterol and triglycerides), blood glucose and insulin levels, and inflammatory markers such as C-reactive protein are the standard way to detect hypertriglyceridemia, hyper-cholesterolemia, insulin resistance, diabetes, states of inflammation and immune activation, and other CAD phenotypes. These, and other as yet unidentified constituents of blood and plasma, determine the rate of progression of atherosclerosis and severity of the cardiovascular injury (Tegner *et al.*, 2007).

Unraveling the molecular secrets of complex traits represents one of the greatest challenges facing geneticists in the twenty-first century. And as the discipline advances, it has become apparent that simply identifying the genetic variants underling these complex phenotypes is insufficient to understanding the whole picture. It is also important to estimate the relationship or the contribution of these gene(s) and their products to the phenotype under study. This needs to be done both qualitatively and quantitatively; it is also crucial to clearly delineate the nature of the interaction. These issues and the questions arising from them will be discussed and developed throughout this chapter.

13.2 Why it is challenging to infer relationships between genes and phenotypes in complex traits?

Classically, genetic studies (i.e., linkage and association studies) were used to find loci associated with clinical traits. Fine mapping (positional cloning) of these loci was then carried out in order to identify the causal genes and DNA variants responsible for susceptibility to the disease-associated traits (Botstein *and Risch*, 2003). This approach has worked particularly well for Mendelian phenotypes. It has also begun to show some promising results for complex traits – particularly with the use of high-throughput technologies capable of performing whole genome association studies.

However, researchers' initial optimism has stalled slightly. Most genetics studies of complex diseases have simply produced long lists of loci associated with the phenotypes. For example, interrogating the OMIM (*On Line Mendelian Inheritance in Man*) database (www.ncbi.nlm.nih.gov/omim/) regarding genes involved in the etiology of diabetes reveals that over 700 different loci have been associated with the disease. With regards to other complex phenotypes like obesity, the figure is nearly 500 candidate genes, and for schizophrenia or stroke the number is approaching 200 potential candidate genes (correct at October 2009). These lists continue to grow at a fast rate.

Unfortunately, the vast majority of these genetic studies have failed to provide the functional information needed to understand the role that the associated genetic variants play in causing phenotype/disease states. Further, from the results of genetic studies alone it is impossible to infer definitely that the gene most proximal to a particular DNA variant associated with the phenotype/disease is the actual gene causing the phenotype/disease. As a result, molecular biologists have redirected efforts to explore alternative approaches to identify functional relationships between complex traits and their claimed candidate genes.

The central dogma of molecular biology states that DNA provides the information needed to code for proteins – the active components of the cell. Messenger RNA (mRNA) is synthesized from a DNA template, resulting in the transfer of genetic information from the DNA molecule. The mRNA then codes for the protein. In this sense, studying the intermediate phenotypes or gene products (i.e., mRNA, proteins and their metabolites) of a particular gene and then combining this information with DNA variations data from genetic studies can provide a context within which it will be possible to understand a gene's function in relevance to the phenotype/disease.

This hybrid approach is not new; it was previously used successfully in the context of monogenic traits. But its application in the identification of functional relationships between complex traits and their candidate genes is not without complications. The intracellular regulatory mechanisms by which cells control their functions in complex traits are not direct. These mechanisms may involve an intricate cascade of intracellular networks modulated by signals from the extracellular environment. The mentioned networks include DNA variants, RNA molecules, proteins, and metabolites. The resultant interactions between these networks and the environment determine the multifaceted physiological/pathological processes underlying the phenotypes of complex traits/diseases.

Pinpointing and analyzing these multifaceted networks is technically demanding, especially the identification of topological relationships between various components of a system and the effect each component has upon the production of another.

An example of a simplified intracellular regulatory network of a particular complex phenotype is shown in Figure 13.1. In an attempt to simplify the picture, a three-dimensional illustration composed of coordinates (x, y, and z) has been used. Each coordinate shows different levels of interactions: (1) a horizontal dimension or x coordinate represents interactions occurring between molecules of a similar nature (e.g., DNA–DNA, RNA–RNA, and protein–protein); (2) a longitudinal dimension or y coordinate represents interactions occurring between molecules of different types (e.g., DNA–RNA, DNA–protein, and RNA–protein); and finally (3) the third dimension or z coordinate represents the environment within which the interactions take place. It is necessary to include a third dimension because the products of DNA expression (i.e., mRNA and protein) are highly dynamic and sensitive to changes in the environment. Connectivity between the components of this network – at both the horizontal and longitudinal dimensions – is highly specific to the phenotype and at the same time it is extremely flexible, responding to changes within the environment. Alterations in the environment (e.g., change of the pH or temperature or applying new medications) results in the rewiring of the network's connectivity and the development of series of new varied networks between the same molecules, which will lead to the modification of the phenotype.

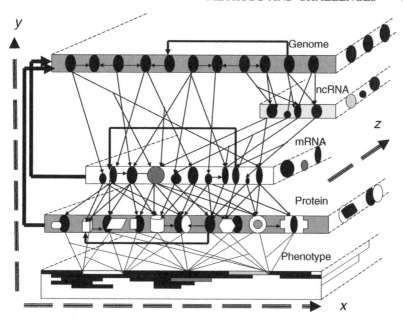

Figure 13.1   Schematic diagram of a simulated intracellular network in a complex phenotype. The top layer represents DNA in the genome. Changes in the DNA can occur with changes in transcription levels of mRNA, non-coding RNA (ncRNA) and protein. The x, y axes represent interactions between similar molecules (DNA–DNA, RNA–RNA, protein–protein), or dissimilar molecules (DNA–RNA, RNA–protein, DNA–protein) respectively. The quantity and the quality of these interactions determine the outcome of the phenotype or phenotypes. The z coordinate represents evolution of the connectivity of the network if environmental conditions change (see text).

Certain other genetic phenomena have amplified the difficulty of identifying and analyzing the intracellular networks associated with complex traits (Schork 1997). These are:

(1) Polygenicity: complex traits are caused by a number of DNA variants or mutations at different loci (which are likely to affect different physiological systems). All these variants must be present before a system is sufficiently challenged to produce the disease.

(2) Locus heterogeneity: the same complex trait is controlled in different pedigrees by different genetic loci. Variations or mutations in any of these loci confer disease susceptibility independently of each other, which means that a mutation on specific loci causing a specific disease in one population is not necessarily responsible for causing the same effect in a different population.

(3) Gene interaction or epistasis: the effects of one gene are influenced by one or more other genes. In this case, a mutation or genotype in a particular gene (i.e., epistatic or a modified gene) will confer susceptibility to a degree dictated by the presence of other mutations or genotypes in another gene or genes (known as hypostatic or modifier gene(s)).

(4) Pleiotropy: a single gene influences two or more phenotypic traits. Consequently, a new mutation in the gene may have an effect on one or all traits simultaneously. This can become a problem when selection of one trait favors one specific version of the gene (allele), while the selection on the other trait favors another allele. As a result, several possible relationships between these traits and the common controlling genetic locus can occur depending on whether the influencing gene causes the disease directly or through the modifications of other phenotypes. For further explanation see Figure 13.2.

(5) Environmental vulnerability: genes and their products (RNA, proteins, and metabolites) are highly modifiable by new environmental stimuli.

(6) *Cis*- and *trans*-regulation of gene expression: expression quantitative trait loci (eQTLs) mapping studies are valuable in connecting complex traits to their molecular causes. In these studies, gene expression levels are viewed as quantitative traits, and gene expression phenotypes are mapped to particular genomic loci by combining studies of variation in gene expression patterns with genome-wide genotyping (Gilad *et al.*, 2008). Substantial heritable variation in gene expression within and between populations was found to be related to variations within DNA regulatory elements (*cis*- and *trans*-regulatory elements) that control gene expression. A *cis*-regulatory element is a DNA sequence located on the same DNA strand or chromosome as the gene whose expression it affects. A *trans*-regulatory element is a DNA sequence associated with the regulation of a gene located outside the genomic region supporting the corresponding structural DNA region of the *trans*-regulatory element (i.e., a different DNA strand or different chromosome). DNA variations in *cis*-regulatory elements result in *cis*-acting or proximal expression quantitative trait loci (eQTLs). DNA variations in a *trans*-acting element result in *trans*-acting or distal eQTLs (Figure 13.3). The characterization of key drivers controlling eQTLs is particularly challenging when *trans*-acting element(s) or several combinations of *cis*- and *trans*-regulatory elements are involved (see Figure 13.3). For more information on *cis*- and *trans*-regulation of eQTLs, the interested reader is advised to refer to Gilad *et al.* (2008).

(7) Genetic circuits or feedback controlling mechanisms: these mechanisms are represented by complicated types of positive, negative, and feed-forward loops to ensure that the cells will have control over their gene and protein expression, even in the presence of noise from other cellular components (Figure 13.4). A positive feedback enables the generation of bi- or

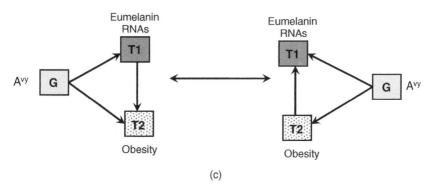

Figure 13.2   Diagrammatic representation of the possible relationships between two traits (T1 and T2) and a common controlling genetic locus (G) when feedback mechanisms are ignored. The graphs are depicted as Markov models showing simple linear relationships between the nodes. Assuming the two traits are linked to a common DNA locus and restricted by the fact that these variables are linearly related, there are three basic relationships one can infer among the traits:

(1) The first trait is a causal factor for the second trait with respect to the DNA locus. The example shown in Figure 13.2 (a) illustrates how the knockout of the leptin gene in the ob/ob mouse results in a complete lack of expression of the product (the first trait) which then leads to obesity (the second trait). Thus, leptin can be considered as a causal factor for obesity with regard to the ob/ob locus.

(2) The first trait is reactive to the second trait with respect to the DNA locus. The example shown in Figure 13.2 (b) illustrates how the knockout of the leptin receptor gene in the db/db mouse results in obesity (the first trait) because the animal's cells are not sensitive to leptin. Leptin levels increase (the second trait) because of the obese state of the animal, and so in this instance leptin expression would be reacting to the obesity phenotype instead of causing it with respect to the leptin receptor locus.

(3) The two traits are independently driven by the DNA locus. The example shown in Figure 13.2 (c) illustrates how eumelanin RNA levels and obesity phenotypes are induced by an allele acting independently and simultaneously on these two different traits in the Agouti A^{vy} mouse (Zhu et al. 2008).

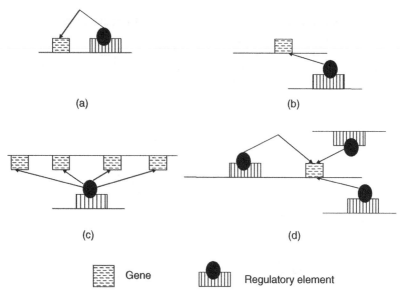

Figure 13.3   Diagram illustrating cis- *and* trans-*regulation of gene expression traits. The white dashed rectangles represent genes controlled by transcriptional units. The lined boxes with the circles represent transcriptional control units: (a) shows a* cis-*control unit acting on a gene. DNA variations in this control unit that affected the gene's expression would lead to* cis-*acting (proximal) eQTLs; (b) shows a* trans-*control unit regulating the indicated gene. DNA variations in this control unit affecting the gene's expression would lead to distal eQTLs; (c) shows a single* trans-*control unit regulating multiple genes. DNA variations in this single control unit could lead to a cluster of distal eQTLs (known as an eQTL hot spot); (d) shows a* cis-*control unit and multiple* trans-*control units regulating the indicated genes. DNA variations in these control units would lead to a complex eQTL signature for the gene.*

multi-stable responses depending on the duration and the intensity of the initial stimulus. Bi-stability is the existence of two stable states, and it is generated by an abrupt transition in the dynamics of a system that generates the new state (e.g., activation of a gene expression pattern leading to a permanent cell phenotype). Bi-stability is the proposed mechanism behind the storage of information and cellular decisions (e.g., cell phenotype). A negative feedback loop is an efficient mechanism for maintaining product levels in a tightly regulated range, and it is one of the most basic cellular mechanisms for controlling homeostasis. Feed-forward mechanisms provide a redundant mechanism for the transmission of molecular information by extending the duration of the signal and ensuring its arrival at the intended site. The study of these controlling mechanisms and their roles in regulating complex traits/phenotypes remains a challenge

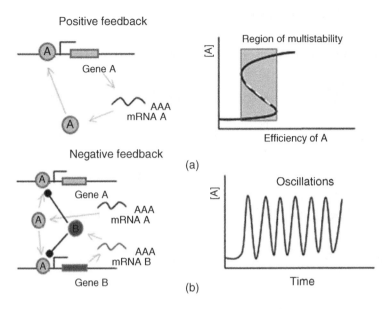

Positive feedback

Gene A

AAA
mRNA A

Region of multistability

[A]

Efficiency of A

(a)

Negative feedback

Gene A

AAA
mRNA A

AAA
mRNA B

Gene B

Oscillations

[A]

Time

(b)

Figure 13.4   Diagrammatic representation for genetic circuits controlling positive and negative feedbacks loops within biological networks. (a) In a positive feedback loop (left), the expression of a gene product is stimulated by its own expression, thus enabling the generation of bi-stable responses depending on the duration and intensity of the initial stimulus. (b) Another basic type of control is negative feedback loops. When levels of gene A increase, expression of gene B is induced, which represses expression of gene A (and its own expression). When levels of A fall below a threshold, its production is stimulated because of the absence of the repressor B. This control mechanism yields oscillations around a mean value, which, in turn, depends on other parameters of the system such as synthesis and degradation rates. (c) Feed forward loops (FFLs) consist of three genes: A, B, and C. Gene A is a regulator of B and C. An FFL is coherent if the sign (activation or suppression) of the path A–B–C is the same as the sign of the path A–C. If the signs do not match, the FFL is incoherent. FFLs can reject transient inputs and activate only after persistent stimulation (i.e., TrkA and calcium signaling). (d) The bi-fan motif, composed of two source nodes directly cross-regulating two target nodes, is able to act as signal sorter, a synchronizer, or a filter (i.e., glycine pathway, right). It also provides temporal regulation of signal propagation. (e) In the single-input-module motif (SIM), modulator A controls the expression of a group of species. This motif can generate an ordered expression program for each of the components under regulation of A (i.e., Rtg1 mitochondrial pathway in yeast, right). mRNA, messenger RNA; NMDAR, N-methyl D-aspartate receptor; PI3K, phosphatidylinositol 3 kinase; PKC, protein kinase C; IRS1, insulin receptor substrate 1; AC5, adenylate cyclase 5; SRC, steroid receptor coactivator; GLYCR, glycine receptor; BAS1, biogenic amine synthesis related family member 1; UPB11, ubiquitin specific peptidase 11; STO1, stomatin-1. Courtesy of Villoslada et al. (2009).

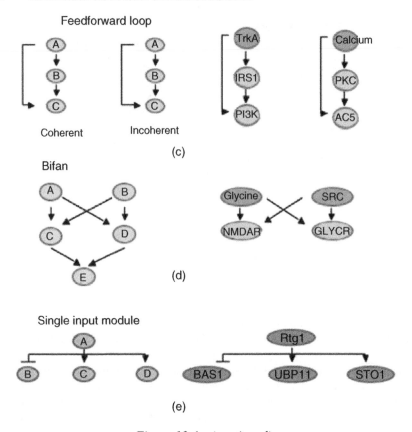

Figure 13.4   (continued)

(Villoslada *et al.*, 2009). Figure 13.4 contains diagrammatic representations for some of the basic genetic circuits responsible from controlling intracellular biological networks.

(8) Epigenetic phenomena such as DNA methylation and chromatin remodeling.

Clearly the complex phenomena outlined above have complicated the task of dissecting the genetic background of complex phenotypes. In order to overcome these difficulties, a systematic approach is needed to firstly map the genes associated with the phenotype and, secondly, to characterize the interactive networks between those genes and their intermediate phenotype (RNA–Protein), and finally, to determine the roles those networks have in causing the phenotype/disease or other co-morbidities (Figure 13.5). In this way, certain molecular targets can be selected specifically from these networks and used as diagnostic, prognostic, and therapeutic targets.

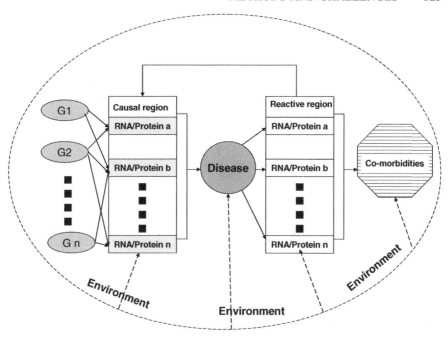

Figure 13.5   Model of how multiple genetic loci, gene expression or protein traits, and disease traits interact in complex phenotypes. G1, G2 and so on to Gn represent variations in DNA that lead to changes in RNA or protein activity. As a result of these changes, broader perturbations to the transcriptional network are induced that ultimately lead to common human diseases like obesity. The disease state can in turn lead to additional changes in the transcriptional network that may give rise to co-morbidities of the disease. In addition to the genetic causes of disease, environmental factors also play a critical role, as common human diseases are the result of complex interactions among multiple genetic loci and between the genetic loci and environmental factors.

In the next section, the different approaches used to identify functional relationships between key driving molecules of complex traits and their phenotypes will be reviewed. The applications and challenges to modern medicine of these approaches will also be explored.

13.3   Bottom-up or top-down: which approach is more useful in delineating complex traits key drivers?

Traditional candidate gene approaches such as positional cloning, transgenic or antisense knockdown, and chemical activation or inhibition of transcription were used to identify functional relationships between candidate genes/genes' products and their associated phenotypes. These approaches are called 'bottom-up' or 'local' since they start with a particular molecule or molecules and gradually build up the relationships between them and the final phenotype. Such methodologies were adopted simply because the technologies available at the time were limited.

The advantage of the bottom-up design is: it logically constructs simple hypotheses and then formulates ways of investigating them. The major drawback is: it generates fragmented information and is unable to construct models for whole biological systems. In essence, it failed to combine all the information it generated. Further, because it was hypothesis driven, it required *a priori* knowledge about each step of the system under study, and in reality such details were often lacking.

Measuring entire biological systems simultaneously without any prior hypothesis is crucial to understanding clinical and cellular responses in complex phenotypes. As explained above, the behavior of intracellular systems in complex traits depends heavily on the dynamics of the system. The problem is that if some parts are missing, the reconstituted system model may not behave like the real system at all. Thus, it is crucial to understand the architecture of global regulatory networks by looking at the overall pattern of the biological process and at the same time to find the single biological changes giving rise to a particular alteration in the phenotype.

With the development of high-throughput technologies and bioinformatics, a new holistic approach (also known as a 'top-down' approach) was increasingly adapted to dissect complex biological systems. The top-down approach is a reverse engineering method which looks at a snapshot of the entire biological system without any prior hypothesis. The system is then gradually dissected in an attempt to gain insight into the compositional subsystems.

This approach allows interactions of phenotype-relevant gene(s) and/or gene(s) products to be analyzed simultaneously. This can be done by moving gradually from the whole body or organ level to the intercellular level and then to individual cell models (Katagiri, 2003). At the whole-body and intercellular level, this approach can be used to identify 'principal networks' consisting of many (but not all) the key interactions. The principal networks of disease at the whole-body level will demonstrate how sub-phenotypes common to several complex disorders (e.g., inflammation, immunity, metabolism, cell proliferation, translation) can be integrated into a complex disease setting. In the next step, key aspects of the principal networks, relevant to the phenotype under study, can be investigated in relationship to disease development at the intercellular level using animal models or specific cell types. Finally, those interesting aspects of the principal network, isolated during disease development in animal models or specific cell types, can be further tested using environmental perturbations with compounds or metabolites or genetic perturbations, such as small interfering RNA (siRNA), gene deletion or expression studies (Katagiri, 2003).

The difficulties of the top-down approach are that it requires repeated rounds of manipulations of the biological system in order to assess its response to such different conditions. These responses are best captured using techniques including whole-genome measurement tools. Such processes generate an incredibly large amount of data. To make sense of this information, researchers need to be *au fait* with extensive statistical analytical algorithms, comprehensive data mining strategies, and computational modeling approaches (Tegner *et al.*, 2007). The accuracy of each one of these steps is crucial for the quality of the data and to guide the next round of experimental measurements in an efficient way. If the data generated fails to come up to standard or the system is not robust enough, the old adage of 'garbage-in-garbage-out' will rear its ugly head.

In contrast to the bottom-up approach, which employs simple, direct, and intuitive logic algorithms, the top-down approach demands more statistical and bioinformatical skills because complex relationships and dynamics must be derived from large data sets. This type of analysis requires sophisticated statistical packages and scientists who are truly interdisciplinary (i.e., biologists with broad mathematical backgrounds and mathematicians with an extensive knowledge of biology).

Which approach is more useful in delineating complex traits key drivers? The answer to this question is: Ideally both top-down and bottom-up approaches should be used in combination to ensure all bases are covered. The top-down approach distils the system into smaller parts. The bottom-up approach reconstitutes the elementary steps into larger parts. If the results of these approaches meet in the middle, and if they are consistent, an investigator can be confident he/she is on the right track. In essence, information from the reductionist approach acts as a constraint in the building of a holistic model (Katagiri, 2003).

13.4   High-throughput technologies and their applications in complex traits genetics

The release of the human genome sequencing project in 2001 constituted a landmark in the evolution of molecular technologies. The vast amount of accumulated information resulting from this discovery allowed the initiation of large-scale studies using high-throughput technologies.

Initial studies focused largely on understanding patterns and frequencies of structural genetic variants within DNA sequences, such as Single Nucleotide Polymorphisms (SNPs), and their contribution to human biology. This was followed by an explosion of new techniques addressing functional genomics and other DNA products including mRNA transcripts, proteins and their metabolites.

Large numbers of whole Genome-Wide Association Studies (GWAS) have been used successfully to determine disease susceptibility genes or mutations in large numbers of complex diseases, for example in diabetes, inflammatory bowel diseases, and neuropsychiatric disorders. Gene expression and protein studies have been successfully applied to distinguish and classify subtypes of diseases such as leukemia (Frankfurt *et al.*, 2007), lymphoma (Iqbal *et al.*, 2009), melanoma (Duncan, 2009) and breast cancers (Correa Geyer and Reis-Filho, 2009). In fields of infectious diseases and developmental genetics, high-throughput technologies have allowed the rapid identification and sub-typing of bacteria, viruses, and parasites (Bekal *et al.*, 2003; Klaassen *et al.*, 2004; Gall *et al.*, 2009). The technological developments have also revolutionized the fields of pharmacogenomics and drug discovery. It has now become feasible to examine DNA variants which control drugs' pharmacodynamics and to tailor patients' management based on their unique genetic makeup, to ensure maximum efficacy and reduce side effects (personalized medicine; Jain, 2009; John *et al.*, 2009).

Hybrid approaches including different combinations of expression experiments, genotyping, and protein high-throughput data (known as convergent functional genomics approaches) were developed to assist in relating expression or protein profiles of diseases or treatments to hot spots or loci previously implicated genetically in a particular disease. These combined approaches played a dual job in investigating the patho-physiological role of candidate genes as well as the identification of new drug targets using simultaneous and integrated models. Table 13.1 lists the currently available high-throughput technologies and their applications in relation to complex traits molecular biology.

13.5   Integrative systems biology: a comprehensive approach to mining high-throughput data

While high-throughput technologies have provided a sudden increase in the amount of data on the possible molecular mechanisms underlying complex traits of health and disease, interpretation of this information remains a dilemma. The accumulated facts and figures are usually devoid of context and cannot fully explain the functional role(s) such genes or their products play in the phenotypes. It has become increasingly obvious that a comprehensive approach of mining is necessary to extract meaningful information from these extensive lists of data.

Systems biology was introduced as a method of lateral thinking. It aims to provide a structured framework to mine high-throughput data by using an understanding of biological processes as integrated systems. In this approach, all parts of the biological system are studied simultaneously, taking into consideration their dynamic interactions within temporal, spatial, and physiological contexts. It is important to stress that systems biology approaches not only focus on the molecular parts (i.e., the genes, proteins, or metabolites) but they also take into account their interactive relationships within the system.

To achieve this goal, functional models are constructed using experimental data from as many diverse sources as possible, including genome, transcriptome, proteome, interactome, metabolome, phenome, epigenomic, metagenomics data, biochemical and kinetic experiments, and information from the literature. The identified functional modules are then linked and compared to available models in public repositories. Mathematical and computational analyses are then used to generate new functional models based on the initial set of data and the available information within the public databases. These combined models are then refined through many iterative cycles of computational simulation, prediction, and validation with other experimental results (Ng *et al.*, 2006a). Figure 13.6

Table 13.1   Methods of high-throughput technologies currently in use to investigate complex traits.

Application	Concept	Technique	
Structural genomics (genomics)	Identification of structural genetic variants through linkage and associations studies	Recombinational cloning High-throughput pyrosequencing Genome-wide association studies (SNPs arrays) aCGH (array comparative genomic hybridization) arrays DNA methylation arrays	
Functional genomics (transcriptomics)	Identification of changes within gene expression in phenotypes under study	cDNA/oligonucleotide microarrays Serial analysis of gene expression (SAGE) High-throughput RNA interference assays Total analysis of gene expression (TOGA)	
Proteomics	Identification of changes within protein structure and function including protein/protein and DNA/protein interactions	Protein identification and quantification	2D gel electrophoresis Mass spectrometry Isotope-coded affinity tag (ICAT) Isobaric tag for relative and absolute quantification (iTRAQ) Stable isotope labeling with amino acids in cell culture (SILAC)
		Protein/protein interactions	Yeast 2-hybrid screening system
		DNA/protein interactions	Chromatin immunoprecipitation Microarray technology (ChIP-chip) DNA adenine methyltransferase fusion proteins chips (DamID)

(continued overleaf)

Table 13.1   *(continued)*

Application	Concept	Technique
Metabolomics	Identification of metabolites related to phenotypes under study	Gas chromatography/mass spectrometry (GC-MS) Liquid chromatography/mass spectrometry (LC-MS) Nuclear magnetic resonance (NMR) Electrospray ionization/mass spectrometry
Cellular	Identification of specific cellular components within the cell	Tissue arrays Cellular arrays, for example pMHC cellular microarrays spotted with pMHC complexes, peptide and MHC class I or peptide-MHC class II

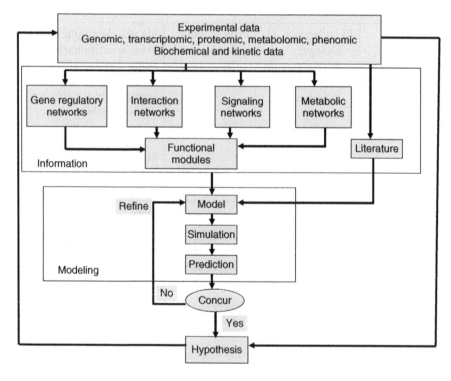

Figure 13.6   Integrative framework for systems biology.

illustrates the framework of mining high-throughput data in a context of integrative systems biology. This integrative approach is essential to understand the complexity of common human diseases like obesity and diabetes and to generate new hypotheses. Of course, once a model has been constructed, further study and experimentation is needed to more convincingly demonstrate understanding of the state of a given molecular network, the interactions, and, finally, how the networks change in response to different genetic and environmental contexts.

There are currently over 150 publicly available databases, visualization and simulations tools devoted to systems biology targets, including databases for gene expression, proteomic and protein interaction, metabolomic and metabonomic, and phenomic (Tables 13.2 and 13.3). Several consortia have been brought together in an attempt to achieve high standards of data integration and unified languages for information exchange (Table 13.4). Numerous current projects have adopted a systems biology approach to aid in disease exploration and drug discovery (Table 13.5). For an extensive review on the resources available for integrative systems biology, the interested reader can refer to Ng *et al.* (2006a) and Davidov *et al.* (2003).

13.6 Methods applying systems biology approach in the identification of functional relationships from gene expression data

Genes do not function in isolation; they always group together into clusters or so-called functional classes or modules. Each module controls a particular common regulatory mechanism or intracellular signaling process. The task of high-throughput data mining is to translate experimental data into functional modules and then utilize this information to infer intercellular regulatory mechanisms.

Several challenges arise with this task, firstly in defining, secondly in identifying and visualizing, thirdly in calculating statistically significant changes, and finally, inferring regulatory systems in these functional modules.

Three main approaches are used in mining functional genomics data in the context of intracellular processes (Cavalieri and De Filippo, 2005). These are methods:

(1) using gene expression values (intensities) to identify correlations between genes (also known as clustering);

(2) classifying expression data into groups with specific annotated functions, for example using biological terms derived with the aid of bioinformatics initiatives such as the Gene Ontology (GO);

(3) using results from functional genomics in combination with existing biological information to construct novel biological networks.

Table 13.2   Examples of currently available resources and databases for integrative systems biology.

Type	Database name	Web link
Gene expression databases	ArrayExpress	www.ebi.ac.uk/arrayexpress/
	ChipDB	http://staffa.wi.mit.edu/chipdb/public/
	ExpressDB	http://twod.med.harvard.edu/ExpressDB/
	Gene Expression Atlas	http://symatlas.gnf.org/SymAtlas/
	Gene Expression Omnibus	www.ncbi.nlm.nih.gov/geo/
	GermOnline	www.germonline.org/
	Human Gene Expression Index	www.biotechnologycenter.org/hio/
	M-CHiPS	www.dkfz-heidelberg.de/mchips/
	NASCArrays	http://affymetrix.arabidopsis.info/narrays/experimentbrowse.pl
	Oncomine	www.oncomine.org/main/index.jsp
	Stanford Microarray Database	http://genome-www5.stanford.edu/
	Yale Microarray Database	http://info.med.yale.edu/microarray/
	Yeast Microarray Global Viewer	www.transcriptome.ens.fr/ymgv/
	CIBEX	http://cibex.nig.ac.jp/index.jsp
Proteomic databases	ProteinProspector	http://prospector.ucsf.edu/
	2D-PAGE/DIFF	www.mpiib-berlin.mpg.de/2D-PAGE/
	dbPTM	http://dbptm.mbc.nctu.edu.tw/
	GELBANK	http://gelbank.anl.gov/
	X! Tandem	www.thegpm.org/TANDEM/
	ExPASy	Proteomics Server www.expasy.org/
	JHU ProteinDB2	http://proteomics.jhu.edu/dl/pathidb.php
	Rice Proteome Database	http://gene64.DNA.affrc.go.jp/RPD/main_en.html
	OPD	http://bioinformatics.icmb.utexas.edu/OPD/
	DynaProt 2D	www.wzw.tum.de/proteomik/lactis/
	PRIDE	www.ebi.ac.uk/pride/
	RESID	www.ebi.ac.uk/RESID/
	SWISS-2DPAGE	www.expasy.org/ch2d/
	HPA	www.proteinatlas.org/
	PepSeeker	http://nwsr.bms.umist.ac.uk/cgi-bin/pepseeker/pepseek.pl
	SWICZ	http://proteom.biomed.cas.cz/
	Peptide Atlas	www.peptideatlas.org/

Table 13.2   *(continued)*

Type	Database name	Web link
Metabolomic databases	Human Metabolome Project/Human Metabolite Database	www.metabolomics.ca/
	METLIN Metabolite Database	http://metlin.scripps.edu/
	Golm Metabolome Database	http://csbdb.mpimp-golm .mpg.de/csbdb/gmd/gmd.html
	The European Nutrigenomics Organisation	www.nugo.org/metabolomics/
	BRENDA	www.brenda.uni-koeln.de/
Phenotypic databases	DRSC	www.flyrnai.org/DRSC-TOO.html
	FlyBase	http://flybase.bio.indiana.edu/
	RNAiDB	http://rnai.org/
	OMIM	www.ncbi.nlm.nih.gov/entrez/query .fcgi?db=OMIM
	Phenomic DB	www.phenomicdb.de/
	PhenoBank	www.worm.mpi-cbg.de/phenobank2/ cgi-bin/MenuPage.py
	MGI – Mouse Genome Informatics	www.informatics.jax.org/
	SGD (yeast)	www.yeastgenome.org/
	PharmGKB	www.pharmgkb.org/
	YDPM	www-deletion.stanford.edu/YDPM/ YDPM_index.html
Gene regulatory databases	TRRD	wwwmgs.bionet.nsc.ru/mgs/gnw/trrd/
	JASPAR	http://mordor.cgb.ki.se/cgi-bin/ jaspar2005/jaspar_db.pl
	cisRED	www.cisred.org/
	TRED	http://rulai.cshl.edu/cgi-bin/ TRED/tred.cgi?process=home
	TRANSFAC Public database	www.gene-regulation.com/pub/ databases.html#transfac
Interaction databases	pSTIING	http://pstiing.licr.org/
	GeneNet	wwwmgs.bionet.nsc.ru/mgs/gnw/ genenet/
	STRING	http://string.embl.de/
	IntAct	www.ebi.ac.uk/intact/index.jsp
	MINT	http://mint.bio.uniroma2.it/mint/
	HiMAP	www.himap.org/
	BIND	www.bind.ca/Action
	DIP	http://dip.doe-mbi.ucla.edu/
	HPRD	www.hprd.org/
	MIPS/MPPI	http://mips.gsf.de/proj/ppi/
	Reactome	www.reactome.org/

(continued overleaf)

Table 13.2   *(continued)*

Type	Database name	Web link
Pathways databases	BioCyc	www.biocyc.org/
	MetaCyc	http://metacyc.org/
	PANTHER	www.pantherdb.org/
	KEGG	www.genome.jp/kegg/
	Biozon	http://biozon.org/
	BioCarta	www.biocarta.com/genes/index.asp
	GenMAPP	www.genmapp.org/
	STKE	http://stke.sciencemag.org/
	AfCS	www.signaling-gateway.org/
	SPAD	www.grt.kyushu-u.ac.jp/spad/
Gene ontology databases	iHOP	www.ihop-net.org/UniPub/iHOP/
	OBO – Open Biomedical Ontologies	http://obo.sourceforge.net/
	GO – Gene Ontology	www.geneontology.org/
	CL – Cell Ontology	http://obo.sourceforge.net/cgi-bin/ detail.cgi?cell
	SO – Sequence Ontology	http://obo.sourceforge.net/cgi-bin/detail.cgi?sequence
	PATO – Phenotype Ontology	http://obo.sourceforge.net/cgi bin/detail.cgi?attribute_and_value
	The OBI Consortium – Ontology for Biomedical Investigations	http://purl.obolibrary.org/obo/obi
	FMA – Foundational Model of Anatomy	http://sig.biostr.washington.edu/ projects/fm/index.html
	OBO_REL – Relation Ontology	http://obo.sourceforge.net/ relationship/

Each one of these approaches addresses one or more of the challenges mentioned previously with varying degrees of adequacy, as will be explained in the following sections.

13.6.1   Methods using quantitative expression data to identify correlations in expression between genes (clustering)

Clustering (also known as unsupervised learning) is a statistical approach identifying unknown classes of co-regulated genes using quantitative measures of expression from whole-genome expression profiles. Clustering research utilizes several different methods or statistical algorithms, for example hierarchal clustering, graphical Gaussian modeling, and self-organizing maps. All these methods assume that the data generated is based on the mixture modeling of underlying probability distributions. The important feature of mixture modeling is that

Table 13.3   Examples of available tools for visualizing and simulating
experimental models in the context of integrative systems biology.

Pathways and Networks visualization Tools	Osprey	http://biodata.mshri.on.ca/osprey/servlet/Index
	Cytoscape	www.cytoscape.org/
	NetBuilder	http://strc.herts.ac.uk/bio/maria/NetBuilder/index.html
	Pajek	http://vlado.fmf.uni-lj.si/pub/networks/pajek/default.htm
	VisANT	http://visant.bu.edu/
	BioLayout	www.biolayout.org/
	Jdesigner/SBW	http://sbw.kgi.edu/software/jdesigner.htm
	Genmapp	www.genmapp.org/
Model visualization tools	PaVESy	http://pavesy.mpimp-golm.mpg.de/PaVESy.htm
	Cell Illustrator	www.genomicobject.net/member3/index.html
	CellDesigner	www.celldesigner.org/
	E-Cell	www.e-cell.org/
	PNK 2e	http://page.mi.fu-berlin.de/~trieglaf/PNK2e/index.html
	Cellware	www.bii.a-star.edu.sg/achievements/applications/cellware/
	BioUML	www.biouml.org
	CellNetAnalyzer/FluxAnalyzer	www.mpi-magdeburg.mpg.de/projects/cna/cna.html
	Virtual Cell	www.nrcam.uchc.edu
	SigPath	http://icb.med.cornell.edu/services/sp-prod/sigpath/mainMenu.action
	CellML	www.cellml.org/
	MCell	www.mcell.psc.edu
Simulation tools	SOSlib	www.tbi.univie.ac.at/~raim/odeSolver/
	Jarnac	http://sbw.kgi.edu/software/jarnac.htm
	SpiM	http://research.microsoft.com/en-us/projects/spim/
	PLAS	www.dqb.fc.ul.pt/docentes/aferreira/plas.html
	Cellerator	www.cellerator.info/index.html
	Dizzy	http://magnet.systemsbiology.net/software/Dizzy/
	Trelis	http://sourceforge.net/projects/trelis
	Monod	http://monod.molsci.org/
	WinSCAMP	http://sbw.kgi.edu/software/winscamp.htm

(continued overleaf)

Table 13.3   *(continued)*

Biominer	www.zbi.uni-saarland.de/chair/ projects/BioMiner/index.shtml
BIONETGEN	http://vcell.org/bionetgen/
Moleculizer	www.molsci.org/~lok/moleculizer/ moleculizer-doc/index.html
Dynafit	www.biokin.com/dynafit
Kinsolver	http://lsdis.cs.uga.edu/~aleman/kinsolver/
BSTLab	http://bioinformatics.musc.edu/bstlab
Ingeneue	http://rusty.fhl.washington.edu/ingeneue/ index.html
StochSim	www.ebi.ac.uk/~lenov/stochsim.html
Simpathica	http://bioinformatics.nyu.edu/Projects/ Simpathica
BIOCHAM	http://contraintes.inria.fr/BIOCHAM/
libSBML	http://sbml.org/libsbml.html
SBMLToolbox	http://sbml.org/software/sbmltoolbox/
MathSBML	http://sbml.org/software/mathsbml/index .html
SimCell	http://wishart.biology.ualberta.ca/ SimCell/

Table 13.4   Examples of currently available consortiums regulating standards and formats for integrative systems biology data exchange and depositions.

Data type of standard	Consortium	Web link
Microarray	MAGE-ML	www.mged.org/Workgroups/MAGE/mage-ml.html
Microarray	MIAME	www.mged.org/Workgroups/MIAME/miame.html
Biochemical	MIRIAM	http://www.biomodels.net/miriam/
Interaction	PSI MI	www.psidev.info/index.php?q=node/31
Proteomic	MIAPE	www.psidev.info/index.php?q=node/91
Proteomic	HUP-ML	www1.biz.biglobe.ne.jp/~jhupo/index-e.htm
Pathway	BioPAX	www.biopax.org/
Network	SBGN	http://sbgn.org/

posterior probabilities of class membership (i.e., the probability of belonging to the class, given the observed gene expression data) are obtained.

A major limitation of clustering techniques is that they assume that functionally related genes have similar gene expression levels; this assumption may result in important genes – in particular regulators – being overlooked. It is vital to realize in this context that transcription factors can change in function without necessarily changing in expression. For example, they may change in configuration through binding to an activator or by proteolytic activation. Further, it

Table 13.5   Examples of organizations and projects that adopt systems biology approaches for disease studies and drug discovery efforts.

Project	Web link	Description
Beyond Genomics Technology platform	www.bg-medicine.com/	Facilitates analysis of clinically relevant samples and integrates data from the gene, protein, metabolite, and clinic for biomarker and target identification
Cellnomica	www.cellnomica.com	Conducts novel multi-cellular modeling in drug discovery and development
Cellzome	www.cellzome.com	Applies functional proteomics technology for therapeutic target discovery, validation, and drug development
Department of Energy's Genomes to Life initiative	http://genomicscience.energy.gov/	Plans to design and exploit new high-throughput strategies to obtain a blueprint of how living systems function
Eli Lilly Centre for Systems Biology	www.lilly.com	Focuses on integration of proteomic and genomic technologies to support drug discovery efforts
Entelos	www.entelos.com	Bio-simulation company that develops computer models of human disease using novel PhysioLab technology
Institute for Systems Biology Broad-based program	www.systemsbiology.org	Uses systems biology to investigate the complex interaction of biological elements that form hierarchical networks that define systems
Kitano Symbiotic Systems project	www.sbi.jp/symbio/symbio2/	The project aims to understand and design biological systems, thus creating a new paradigm in biology with a focus on model organisms including fruit
Physiome Sciences	http://nsr.bioeng.washington.edu/	Bio-simulation company that has created and develops integrated software platform for computer-based biological models applicable to drug discovery

(continued overleaf)

Table 13.5   *(continued)*

Project	Web link	Description
SurroMed	www.surromed.com	Develops and implements biological marker discovery platform to profile biochemical components in blood and other biological samples
BioSeek	www.bioseekinc.com	Uses systems biology approach to study primary human cell disease models

Source: Davidov *et al.* (2003).

must be noted that clustering does not directly address the physical or regulatory relationships between functionally related genes; rather that it is a simultaneous quantitative measure of the abundance of certain transcripts under particular experimental conditions (Cavalieri and De Filippo, 2005).

For a full review on techniques for clustering gene expression data and their merits and disadvantages, the interested reader could consult Kerr *et al.* (2008) and Zhao and Karypis (2005).

13.6.2   Methods integrating functional genomics into cellular functional classes

In this approach, patterns of expression variation are examined and classified into classes of genes with predefined intracellular functions, such as those involved in metabolism, cell-division control, apoptosis, membrane transport, sexual reproduction, and signaling. The goal of this approach is to integrate information obtained at the genomic level with biological information gathered over years of research in various disciplines including molecular genetics, biochemistry, and cell physiology.

Two internationally agreed methods are currently in use to annotate amalgamated expression data topologically into particular intracellular functional modules. These are: the Gene Ontology method and the biological pathways method.

13.6.2.1   Gene Ontology (GO)

The Gene Ontology (GO) database (www.geneontology.org/) is an international consortium providing a controlled vocabulary to annotate and analyze the functions of genes and the attributes of their products in any organism (Ashburner *et al.*, 2000). The current ontologies of the GO project are cellular components, biological processes, and molecular functions.

The GO classification topologically represents data as directed acyclic graphs which present the information in a hierarchical fashion (Figure 13.7). The hierarchy is able to encompass complex relationships such that a more specialized term

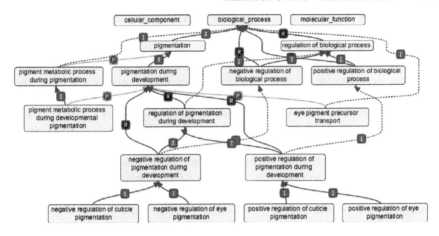

Figure 13.7   Screenshot from the ontology editing software OBO-Edit, showing a diagram of a part of a GO tree. The diagram illustrates the topological structure of GO terms. GO ontologies are described in terms of a graph, where each GO term is a node, and the relationships between the terms are arcs between the nodes. The relationships used in GO are directed – for example, a mitochondrion is an organelle, but an organelle is not a mitochondrion – and the graph is acyclic, meaning that cycles are not allowed in the graph. The ontologies resemble a hierarchy, as child terms are more specialized and parent terms are less specialized, but, unlike a hierarchy, a term may have more than one parent term. In the diagram, relations between the terms are represented by the arrows; the letter in the box midway along each arrow is the relationship type. Note that the terms get more specialized going down the graph, with the most general terms – the root nodes: cellular component, biological process and molecular function – at the top of the graph. Terms may have more than one parent, and they may be connected to parent terms via different relations.
Source: *GO ontology web documentation (www.geneontology.org/GO.ontology. structure.shtml).*

(child) can be related to more than one less-specialized term (parent). As a result of these different hierarchical levels, classes defined by the GO terms can be highly redundant. For example, the biological process term 'hexose biosynthetic process' has two parents, the hexose metabolic process and the monosaccharide biosynthetic process. This is because biosynthetic process is a type of metabolic process and hexose is a type of monosaccharide. When any gene involved in 'hexose biosynthetic process' is annotated to this term, it is automatically annotated to both hexose metabolic process and monosaccharide biosynthetic process.

A GO cellular component describes an anatomical structure within the cell (e.g., rough endoplasmic reticulum or nucleus) or a gene product group associated with such a structure (e.g., ribosome, proteasome or a protein dimer).

A biological process is a series of events accomplished by one or more ordered assemblies of molecular functions. Typically, GO biological processes cover a broad range of functions including those of cellular physiological processes or signal transduction. Examples of more specific terms are pyrimidine metabolic process or alpha-glucoside transport.

A GO molecular function describes activities, such as catalytic or binding activities, occurring at the molecular level. They do not specify where or when, or in what context, these actions take place. Molecular functions generally correspond to activities performed by individual gene products – although some activities are performed by assembled gene product complexes. Examples of broad functional terms are catalytic activity and transport activity; examples of narrower functional terms are adenylate cyclase activity or toll receptor binding.

It can be difficult to distinguish between a GO biological process and a GO molecular function, but the general rule is that a biological process must have more than one distinct step. It is important to note that a biological process is not equivalent to a pathway.

By way of a summary, one might talk about a gene product saying that it is associated with or located in one or more cellular components, it may be active in one or more biological processes during which it performs one or more molecular functions.

Even though GO terms are very useful in the classification of genes according to their functions, they are unable to connect or relate genes in the form of a pathway. At the moment, GO terms lack the dynamics or dependencies (i.e., causality, directionality, or the type of interactions between components of a given class) required to fully describe a pathway. All examples of GO ontologies provided in this section were taken from the GO consortium web site (http://wiki .geneontology.org/index.php)

13.6.2.2   Biological pathways

A biological pathway is a set of linked molecular components interacting with each other over time to generate a single biological effect. This interaction is commonly represented by a graph or diagram linking the interacting components. The graph includes the idea of directionality, where lines become arrows, and contain information about the organizing principles of the pathway. For example, the energy and metabolite flow where the product of a reaction is either the substrate or the enzyme that catalyzes a subsequent reaction.

A pathway can describe gene–protein interactions (e.g., genetic regulatory pathways) or protein–protein interactions (e.g., signaling transduction pathways) or protein–enzyme–metabolite (ligand) reactions (e.g., metabolic pathways).

The majority of pathways currently available are manually constructed using molecular information from published materials. However, some interactions can be created using prediction or kinetic modeling.

It is important to emphasize that the partitioning of interactions into pathways is somewhat arbitrary. For example, participants in one pathway can be involved in others; for example, malate dehydrogenase appears in six different metabolic pathways in some databases. In addition, the selection of the start and finish points of a pathway is flexible and dependent upon the investigator's priorities; such flexibility allows mapping of the genotype to the phenotype in a dynamic fashion.

The development of pathway databases is continuing apace with numerous resources currently available publicly (see Section 13.5). Examples include KEGG (www.genome.jp/kegg/), BioCyc (http://biocyc.org/), Reactome (www.reactome.org/), GenMAPP (www.genmapp.org/), and BioCarta (www.biocarta.com/). Pathguide, the pathway resources list (www.pathguide.org/) gives addresses of over 300 biological pathways resources. These databases provide advanced methods of visualizing and manipulating pathways, in addition to providing extensive repositories of well-established pathways from different species. Figure 13.8 provides an example, the Wnt signaling pathway graphical maps obtained from three different pathways databases.

A standard data exchange format for pathways has been identified by several projects including BioPAX (www.biopax.org), CellML (www.cellml.org/), PSI-MI (www.psidev.info/index.php?q=node/60) and SBML (http://sbml.org/Main_Page).

Numerous computer packages and databases are available either commercially or as an open source to help users map and visualize results obtained from gene expression experiments into metabolic or signal transduction pathways; examples include Rosetta Resolver (Rosetta Inpharmatics LLC), GeneSpring (Silicon Genetics, Agilent Technologies), Acuity (Axon instruments), GeneGO (GeneGo) and the Proteome Bioknowledge Library (Incyte). Freely available computer applications include GenMapp (www.genmapp.org/) and Cytoscape (www.cytoscape.org/).

These tools are not without limitations; for example, they fail to automatically indicate the statistical significance of the change of a pathway, making it hard to select the most interesting or important results (Cavalieri and De Filippo, 2005).

More recently, several software packages have started to address the issue of significance of the alteration in expression or enrichment analysis of a particular cellular pathway or GO process. Enrichment analyses calculate the significance of differentially expressed genes which are overrepresented in particular pathways, or functional classes. Programs addressing the issue of significance include Gene-Merge (http://genemerge.cbcb.umd.edu/), GOstat (http://gostat.wehi.edu.au/), Pathway Miner (www.biorag.org/index.php), and GoSurfer (http://bioinformatics.bioen.uiuc.edu/gosurfer/). These applications utilize the Fisher's exact test or hypergeometric distribution to calculate the probability that a particular pathway would contain as many, or more, affected genes as are actually observed; the null hypothesis being that the relative changes in gene expressions in the pathway are a random subset of those observed in the experiment as a whole. Several factors have to be considered by these applications in order to assess the significance of gene expression changes in a given pathway and to select the most appropriate statistical test. These factors are: firstly, the number of open reading frames (ORFs) having altered expression in each pathway; secondly, the total number of ORFs contained in the pathway; thirdly, the proportion of ORFs in the genome contained in a given pathway; and finally, correlation

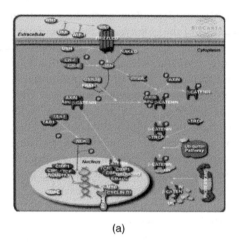

(a)

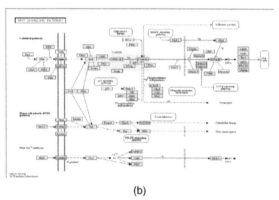

(b)

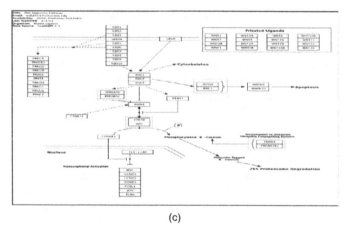

(c)

*Figure 13.8   Three graphical representations illustrating the different maps'
outputs for the Wnt signaling pathways obtained by searching (a) BioCarta,
(b) KEGG and (c) GenMAPP databases.*

of the pathways changed. The main limitation of these approaches is that the statistical packages inappropriately consider all the annotations as independent categories. As stated earlier, GO terms and biological pathways are highly interconnected systems with a great deal of redundancy. The fact that one gene can be contained in several GO categories or multiple pathways affects the use of hypergeometric statistics and/or the Fisher's exact test, thus making it difficult to correct for multiple hypotheses. Studies of different datasets have shown that the application of a standard Bonferroni correction is often too restrictive, and clearly discards a great deal of important biological information (Cavalieri and De Filippo, 2005).

In conclusion, both pathways and GO ontology methods have provided a big step forward in the understanding of the functional contribution of certain genes in complex phenotypes. Pathway analysis is more superior to GO ontology in providing information about the type of interactions between components taking into account directionality or causality. Both methodologies are restricted by the shortage of knowledge currently available about the functions of genes; and self-evidently they cannot provide information on genes of unknown functions. However, even in cases where genes are involved in well-characterized pathways, it is often not immediately obvious whether a particular gene is causing the phenotype via the identified pathway or whether the gene is involved in other pathways or more complex networks leading to the phenotype. For example, transforming growth factor, beta receptor II (*Tgfbr2*), a recently identified and validated obesity susceptibility gene, plays a central role in the well-studied transforming growth factor-beta (TGF-β) signaling pathway. In addition, *Tgfbr2* and other genes in this signaling pathway interact with hundreds of other genes (Figure 13.9), meaning that it is possible that perturbations in these and other genes may lead to diseases like obesity by controlling other pathways different from the TGF-β signaling pathway (Sieberts and Schadt, 2007). Therefore, considering genes in the context of single pathways does not necessarily provide a complete understanding of the role of a given gene in causing a specific disease.

13.6.3 Methods combining functional genomics results and existing biological information to construct novel biological networks

13.6.3.1 Biological networks: definition and topology

Historically, graph theory was first described in the city of Königsberg, located in the former Prussia. In the early eighteenth century, the citizens of Königsberg entertained themselves by wandering around the city's seven bridges to see whether it was possible to walk a route that crossed each bridge exactly once and then return to the starting point (Grigorov, 2005). In 1736, the Swiss mathematician Leonard Euler represented this problem as a cyclic graph and demonstrated that it was impossible to visit each of its edges only once and return to the

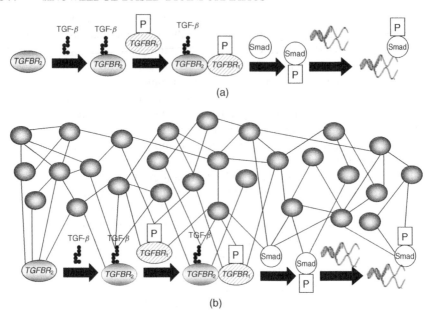

Figure 13.9   Tgfbr2 *as an example of a gene belonging to a simple linearly ordered pathway (in this case the TGF-β pathway); this gene interacts with other genes from other different pathways in order to perform its function. (a) The classic view of TGF-β signaling involves* Tgfbr2 *as a key component; (b) shows the same gene interacting with numerous genes in other pathways to form a network.* Tgfbr1, *Transforming growth factor-beta receptor type I;* Tgfbr2, *Transforming growth factor-beta receptor type II;* Smad, *Mothers against decapentaplegic homolog; P, Phosphorylation. (Sieberts and Schadt, 2007).*

beginning again (Euler, 1736). Later on, the development of graph theory 'wandered between randomness' (Erdos and Renyi, 1959) and order (Cayley, 1857). Both statistical and analytical methods were used to model network representations of natural systems (Grigorov, 2005). Since then, the words 'graph' and 'network' have been used interchangeably to describe the same concept. Lately, however, the term network has been adopted to describe the natural system itself, whereas graph has been employed to depict the mathematical object representing the topology of the system.

Network modeling has emerged as a popular method of exploring complex relationships within the context of biological systems. Interaction or association networks were developed using pair-wise relationships between genes or genes' products including protein interaction relationships (Han and Ju, 2003), and co-expression relationships (Zhang and Horvath, 2005; Ghazalpour *et al.*, 2006; Frankfurt *et al.*, 2007; Saris *et al.*, 2009), as well as other straightforward measures that may indicate association between any two components within the system.

The topology of biological networks is built of nodes and edges. Nodes are depicted either as circles or some other shape which are then connected to each other by lines representing the edges. Nodes in the networks typically represent genes or genes' products (e.g., mRNA, proteins or metabolites). The edges represent the relationships between any two components within the network. For example, an edge between two genes may indicate that changes in the activity of one gene lead to changes in the activity of the other gene, or, it may represent the idea that the two linked expression traits are correlated in a given population of interest. These edges can be 'directed' or 'undirected.' Directed edges have connections which are specific in their direction; that is, if an edge allows a connection between two nodes it can only occur in a particular way. To understand this, the edges can be thought of as roads between the nodes – but the roads are signed to be one way. For example, a gene may serve as the source of a direct regulatory edge to a target gene by producing an RNA or protein molecule that functions as a transcriptional activator or inhibitor of the target gene but not vice versa. Networks diagrams generated by directed edges without feedback edges between nodes are called directed acyclic graphs (or DAGs for short).

Edges can be changed in shape to represent the nature of the relationship between the nodes. For example, inductive relationships may be represented by arrowheads with an increase in the concentration of one node (where the node might represent a gene or protein) leading to an increase in the other. Inhibitory relationships are represented by interrupted dotted lines, with an increase in one leading to a decrease in the other. In other words, if the gene is an activator, then it is the source of a positive regulatory connection. If the gene is an inhibitor, then it is the source of a negative regulatory connection. Further, nodes can be color coded or changed in shape to represent changes in concentrations or status of the components (e.g., over or down expression, methylation or adenylation). Currently, quantitative information such as the strength of the interaction between the nodes within a network is usually reported as binary measurements rather than accurate measurements.

Other important topological features in network construction include degree, distance or shortest path length, clustering coefficient, and betweenness (Zhu et al., 2007a; see Figure 13.10). These terms are defined as follows:

(1) Degree: the number of links connected to one node is defined as its degree. In directed networks, the number of edges that end at the node is termed the 'in-degree,' and the number of edges that start from the node is termed the 'out-degree.' A node with high degree is better connected in the network and therefore may play a more important role in maintaining the network structure.

(2) Distance: the shortest path length between two nodes is defined as their distance. In an interaction network, the maximum distance between any two nodes is termed as the graph diameter. The average distance and diameter of a network measure the approximate distance between nodes in a network. A network with a small diameter is often termed as a 'small

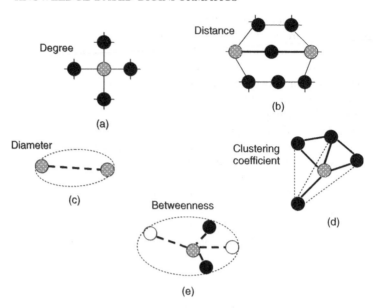

Figure 13.10   Graphical description of some commonly used topological terms in biological networks. (a) Degree measures the number of connections one node has. (b) Distance is the length of the shortest path between two nodes. (c) Diameter is the maximum distance between any two nodes in a network. (d) Clustering coefficient measures the percentage of existing links among the neighborhood of one node. (e) Betweenness is the fraction of those shortest paths between all pairs of nodes that pass through one vertex or link. All graphs are based on an undirected network. Courtesy of Zhu et al. (2007a).

world' network, in which any two nodes can be connected with relatively short paths.

(3) Clustering coefficient: the clustering coefficient of one node can be calculated as the number of links between the nodes within its neighborhood divided by the number of links that are possible between them. A high clustering coefficient for a network is another indicator of a small world.

(4) Betweenness: this is the fraction of the shortest paths between all pairs of nodes that pass through one vertex or link. Betweenness estimates the traffic load through one node or link assuming that the information flows over a network primarily following the shortest available paths.

Different parameters govern the behavior of intracellular networks over time. In most cases it is not clear how intracellular components at different levels in a cell interact. Nor is it easy to predict the complete state of these mechanisms at a given point in the future. For these reasons, dynamic modeling of a network and inference of the topology, that is, prediction of the 'wiring diagram' of the network, is a very challenging process.

Interestingly, experimentation with construction of networks in biological systems has demonstrated that biological networks exhibit scale-free and hierarchical connectivity structures, which provides an insight into how biological systems are ordered (Barabasi and Oltvai, 2004; Ghazalpour *et al.*, 2006; Lum *et al.*, 2006). The scale-free property (Figure 13.11) exhibited by these networks implies that most genes in a biological system are strongly connected to a small number of genes often referred to as hub nodes. These hub nodes work as the key drivers or regulators of that particular network. The hierarchical property

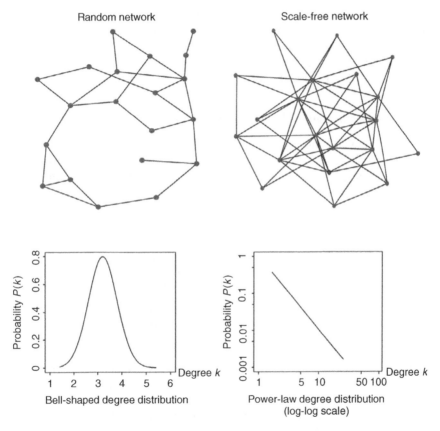

Figure 13.11   Topological comparison between a random network and a scale-free network. Degree distribution in random networks is bell-shaped. The scale-free network has more high-degree nodes and a power-law degree distribution, which leads to a straight line when plotting the total number of nodes with a particular degree versus that degree in log-log scales. Courtesy of Zhu et al. (2007a).

implies that biological networks are highly modular, with genes clustering into groups that are multiply interconnected with each other, but not as highly connected with genes in other groups. Both properties are extremely important in identifying key regulatory modules in complex phenotypes and in exploring new drug targets, as will be explained by examples in Section 13.8.

Biological networks mapping applies a variety of different algorithms; most of which use a combination of pattern prediction and systems biology to infer information. Methods using genomic information to infer regulatory networks rely on the identification of patterns using either the concept of partial correlation or conditional probabilities to indicate causal influence; these methods use DNA variants combined with other supplementary data of gene expression or proteins in the proposed networks.

The following section will discuss the types of biological networks available and their applications in molecular biology. The computation of these networks is beyond the focus of this chapter. For readers interested in mathematical methods and algorithms, the following references are recommended: Zhu *et al.* (2008); Klamt and von Kamp (2009); and Christensen *et al.* (2007).

13.6.3.2   Classification of biological networks

Biological networks are commonly classified according to the type of the molecules participating in them and the nature of their interactions. This method of classification divides biological networks into: (1) gene regulatory networks, (2) metabolic networks, (3) signal transduction or protein–protein interaction networks, and (4) transcriptional regulation networks (see Table 13.6).

Table 13.6   Classification of biological networks based on the participating molecules in the networks (nodes) and the nature of their interaction (edges).

Type of network	Nodes	Edges
Protein–protein network (signaling transduction networks)	Proteins	Interaction between proteins
Metabolic network	Metabolic products	A reaction transforming metabolite (A) into metabolite (B)
Transcriptional regulation network (protein–DNA)	Genes and proteins	A transcription factor (protein) regulates a gene
Gene regulatory networks (functional association network)	Genes	Expressions of gene (A) and gene (B) are correlated

From a functional viewpoint, biological networks can be classified into causal or association networks depending on whether the key nodes within the networks are causal or reactive factors in relationship to the phenotype.

13.6.3.3   Interaction (association) networks

Interaction or association networks are formed by considering pair-wise relationships between genes or genes' products (i.e., RNA, proteins, and metabolites). Examples of these networks include protein–protein interaction relationships (signaling transduction networks), co-expression relationships (gene expression networks), metabolic networks and other straightforward measures that may indicate association between two genes.

Co-expression networks analysis is one of the most popular methods currently available to explore functional modules and topological relationships within biological systems. Two approaches are used to build co-expression networks (Zhu et al., 2008). These approaches differ depending upon whether or not the strength of the interaction between two nodes within the network is considered. In the so-called weighted method, the level of significance of the strength of the interactions between nodes is set at a high threshold level of significance. And in the un-weighted approach, which is more liberal, no threshold of significance is applied.

In the weighted method, node–node relationships are encoded in a binary form. Two nodes in the network are connected by an edge if the significance level of the correlation measured meets some predetermined threshold (Davidson et al., 2003; Lum et al., 2006). Topological terms such as node degree, closeness, clustering coefficient, and betweenness (for explanations refer to Section 13.6.3.1) are used to structurally visualize the measures for strength or ties between the nodes within the network. Weighted networks are highly sensitive to the threshold being selected, and in most cases the selected threshold is somewhat arbitrary. In addition, the binary encoding can destroy information regarding the interaction strength between two genes. This may result in restricting the ability to identify higher-order relationships among the genes in the network.

In contrast, the un-weighted network assigns a connection weight to all pairs of genes by applying a more liberal threshold whose parameters are determined based on a biologically dependent scale-free topology (Zhang and Horvath, 2005).

Weighted gene co-expression networks analysis (WGCNA) was used successfully to identify functional modules in several complex diseases such as obesity and neurological disorders (Fuller et al., 2007; Miller et al., 2008; Saris et al., 2009). This type of analysis maintains the continuous nature of gene–gene interaction at the transcriptional level and is robust to parameter selection. However, building these networks is more computationally demanding as all pairs of nodes are simultaneously considered, so that as the number of nodes grows, the number of pairs to consider grows exponentially.

13.6.3.4   Inferring functional relationships from co-expression networks

Functional relationships can be inferred from co-expression networks using three types of integrative systems biology approaches. These approaches include:

(1) Identifying functional modules (sub-networks) of highly interconnected genes using methods such as clustering, GO ontology or pathways enrichment analyses.

(2) Applying prior structural knowledge by recruiting information from different systems biology resources to filter information and define nodes and edges within the network.

(3) Combining gene expression and genetic data to infer causal relationships.

In most experiments these methods are used concurrently or alternately, as will be illustrated by examples in Section 13.8, to recruit information and construct the desired networks according to the experimental needs.

Prior knowledge approaches, such as clustering, GO ontology and pathways enrichment analyses can be applied to co-expression networks to identify functional modules which are highly co-related to the diseases or the phenotype under study. Because of the scale-free and hierarchical nature of co-expression networks, those functional modules are likely to represent the hub or regulatory nodes of the networks.

In addition, it has been demonstrated that these types of modules are enriched for genes that associate with disease traits, and for genes that are linked to common genetic loci (Zhu *et al.*, 2008). In this way, one can identify those key groups of genes that are perturbed by genetic loci that lead to disease, and that therefore define the intermediate steps that actually define disease states (see examples in Section 13.8).

Interestingly, dividing co-expression networks into functional modules revealed a new concept called differential connectivity. Differential connectivity is defined as different patterns of connections between nodes with respect to different phenotypic groups within populations that have been profiled. For example, given tissue samples from two phenotypic groups of individuals (say, those who are responders to certain treatments versus those who are resistant), a gene is considered differentially connected between the two groups if the number of genes to which the gene is significantly correlated are significantly different (Figure 13.12). The most interesting observation related to this pattern of differential connectivity is that genes that are differentially connected between two groups are not necessarily differentially expressed between the two groups. Differential patterns of connectivity in gene expression data hold promising potential as a diagnostic and prognostic tool in the field of clinical medicine.

A structural prior approach can be used to add a functional context to the networks. Information of variable types, for instance transcription factors, promoter sequences, and protein–protein binding data, can be recruited from different databases listed in Section 13.5 and then added to the networks. The recruited

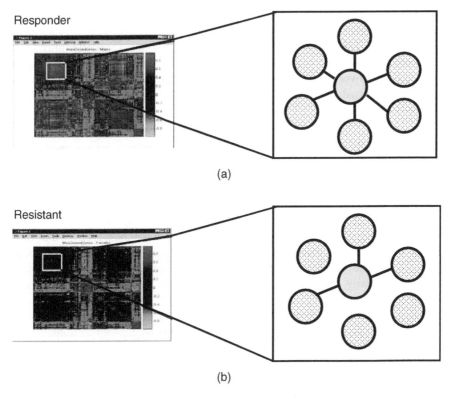

Responder

(a)

Resistant

(b)

Figure 13.12   Hypothetical example of differential connectivity patterns between expression profiles taken from a responding and a resistant patient to a particular drug. The right side of the figure illustrates topological diagrams of the gene expression areas enclosed within the white squares on the expression maps seen on the left: (a) shows a responder where the representative gene (node in the middle) is connected to six other genes (nodes at the periphery); (b) shows a resistant individual, where the connectivity of the representative gene is reduced by half.

information can be illustrated topologically through links between the nodes and these databases, or they can be added to the edges. This information can also be used as a filter to limit the number of possible edges.

Recently, more commercial programs and databases have provided the facility of visualizing networks within a context of functional classes. Examples of these packages are GeneXPress and CARRIE. GeneXPress (http://genexpress .stanford.edu/) is a new tool enabling users to combine cluster analysis with analysis of biological attributes. Following cluster analysis of microarray data, GeneXPress tests, firstly, for biological processes represented in a given cluster and, secondly, for *cis*-regulatory motifs shared by genes within a cluster. Once the analysis has been completed, the package will provide a p-value for the association, and integrate the pathway results with the structure of transcriptional regulatory networks. This is a substantial improvement compared to other tools currently available. Another recently developed program, CARRIE (http://cagt .bu.edu/page/CARRIE), uses promoter sequence and expression data to construct a regulatory network displayed in the form of an interactive graph. The network data can be visualized with the genes mapped on KEGG metabolic pathways (Cavalieri and De Filippo, 2005).

13.6.3.5   Combining gene expression and genetic data to infer causal relationships

The previously mentioned approaches are helpful for characterizations of the properties of biological systems, identification of highly connected (hub) nodes, and discovery of functional modules that aid in the description of sub-networks associated with disease. The major drawback they have, however, is that they cannot be used to link changes in functional data (expression and protein data) to their direct molecular causes (DNA variants); in other words, they have limited ability to identify direct causal associations among genes and between genes and phenotypes.

DNA variation is assumed to serve as a causal anchor for changes observed in any phenotype (given that variation in DNA leads to changes in transcription and other molecular trait activities). By integrating expression quantitative trait loci (eQTLs) data of a particular phenotype with sets of genes that are identified through genetic studies (linkage and association studies) it will be possible to infer causal association between genes and a phenotype.

One simple approach is to map the expression of *cis*-acting expression quantitative trait loci (eQTLs) in the genetic loci linked to the studied phenotype through linkage or association. From these expression data it is then possible to construct networks identifying relationships between genes. Projecting genetic information from association and linkage studies to the previously constructed gene networks will partition the network of genes associated with a disease into causal, reactive, and independent pieces with respect to a phenotype of interest, so that genes supported as causal for the phenotype can be identified.

These concepts have been applied to varying degrees by several groups to allow for the more general construction of gene networks by the integration of genetic and gene expression data (Cervino *et al.*, 2005; Ghazalpour *et al.*, 2006; Lum *et al.*, 2006; Fuller *et al.*, 2007; Sieberts and Schadt, 2007; Gilad *et al.*, 2008). The key to the success of this approach is the unambiguous flow of information, from changes in DNA to changes in RNA and protein function.

Probabilistic Bayesian networks are popular for inferring the correct structure of relationships both among genes and between genes and clinical phenotypes (Zhu *et al.*, 2007b). The relationships are usually depicted as Markov models and the genetic information is used to give a sense of directionality.

It is important to note that causal relationships used in this type of network are different from the standard causal interactions used in biochemistry and biology (i.e., two interacting molecules where one molecule physically or chemically interacts with another to increase or decrease its concentration or activity leading to further phenotypic changes).

The term causality in the probabilistic Bayesian causal networks is a statistical inference rather than direct chemical or physical interaction. Statistical associations between changes in DNA, changes in expression, and changes in phenotypes are examined for patterns of statistical dependency among these variables that support directionality between them. The directionality then provides the source of causal information (highlighting putative regulatory control as opposed to physical interaction).

13.7   Advantages of networks exploration in molecular biology and drug discovery

The modular architecture of biological networks that have been discovered has dramatically altered the perspectives of those working on the interactions and controls of biological entities. Probably the most important outcome of all these discoveries is the availability of new tools to visualize and prioritize experimental targets and leads.

Networks exploration has several valuable applications in the field of molecular biology and drug discovery.

With regard to molecular biology, networks will:

(1) Help in finding the key drivers of complex disease.

(2) Provide methods for data mining and organization.

(3) Provide methods of generating and modeling new hypotheses.

(4) Help in the identification of sub-networks associated with different disease subtypes. This will assist in the discovery of new biomarkers for the disease subtypes that can be used as diagnostic tools or used in classifying patients into treatment groups. In addition, it will help in exploring therapeutic targets that are specific to a given subtype.

(5) Provide a practical solution to deal with the problem of heterogeneity of complex diseases. For example, rather than looking at a complex phenotype as a single entity, the phenotype can be defined by different pathways associated with different molecular processes within subsets of networks. By using different mixtures of these sub-networks, the phenotype can then be defined molecularly in different populations as opposed to symptomatically.

(6) Provide computational models to predict biological responses. Such models can be used to make predictions that can be tested experimentally, as well as exploring questions that are not amenable to experimental enquiry. For example, assessing the impact of a novel drug or a genetic mutation.

With regard to drug discovery, networks will help to:

(1) Identify new drug targets. Genes will be selected for therapeutic intervention in a disease taking into account nodes in the network associated with other diseases, subtypes of the disease, and toxicity pathways, as well as other clinical phenotypes that may be adversely affected.

(2) Calculate the mechanism of action of a given compound. This application is particularly useful in preclinical studies. Network analysis can better predict possible side effects, for instance by assessing the impact of a compound on the gene networks of organs frequently involved in side effects, such as the liver, kidneys, skeletal muscle, and visceral fat deposits.

(3) Provide information on the genetic profiles and the current environmental pressures of individual patients. With this information, therapies could be tailored to the individual, and the individual network response to the treatment could be monitored.

13.8 Practical examples of applying systems biology approaches and network exploration in the identification of functional modules and disease-causing genes in complex phenotypes/diseases

Example 1

An integrative systems biology and networks exploration approach to study mechanisms of action of the antipsychotic drug clozapine in the mouse brain (Rizig *et al.*, 2009, unpublished data).

Background

Clozapine is an antipsychotic drug with superior clinical properties compared to other similar psychopharmacological agents. Particularly, superior efficacy has been proven, improving symptoms in patients suffering from so-called treatment-resistant schizophrenia. However, its widespread use is restricted by side effects such as agranulocytosis and diabetes. The mechanisms by which clozapine mediates its effects are not well understood. There may or may not be an overlap between pathways modulating clozapine's therapeutic and toxic effects.

Aims

- to identify the brain pathways and systems changed by clozapine.

- to link and compare the identified mechanisms with genes and pathways implicated previously in schizophrenia in an attempt to understand why clozapine is a superior drug to other antipsychotics.

- to understand the nature of interactions between the intracellular mechanisms responsible for both the beneficial and detrimental effects of clozapine.

Methods

Gene expression data were gathered from Affymetrix microarray experiments to compare the chronic gene expression profiles of clozapine and controls in the brains of mice after receiving oral preparations of the drug for up to 12 weeks. Pathways overrepresentation (enrichment) analysis was performed to identify signal transduction pathways significantly overrepresented in genes significantly altered by clozapine versus control. Mouse gene expression data were mapped to human signal transduction pathways via HUGO gene symbols using the pSTIING database http://pstiing.licr.org/ (Ng *et al.*, 2006b). Over 300 human signal transduction pathways from BioCarta, GenMAPP and KEGG have been integrated into pSTIING. Perl codes were written to extract this information and to compute the number of representations (gene hits) for each pathway. P-values for pathways overrepresented within sets of genes were then computed using the hypergeometric distribution implemented in the R-language.

Gene products associated with significantly overrepresented signaling pathways are collectively termed pathway-enriched gene sets. These pathway-enriched gene sets were projected onto human protein interaction and transcriptional regulatory information in the pSTIING database to generate functionally relevant networks (Figure 13.13).

These gene products were used as 'seeding nodes.' Using graph theory, gene products in the network were represented by nodes, with the 'seeding nodes' colored in light, mid or dark shades of gray. The network was generated by iteratively connecting up interacting nodes using solid edges (denoting physical interactions between gene products) or using dashed edges (representing transcriptional regulatory associations) following a heuristic that specifies the inclusion of 'seeding nodes' up to two degrees of separation. Network interaction or association 'partners' that were also differentially expressed were identified and grouped with the original pathway-enriched gene sets to further extend the network. The process was iterated until no new differentially regulated 'partners' were found, resulting in an extended network that was enriched for differentially expressed gene products triggered by clozapine treatment. A flowchart illustrating methods used for network construction is shown in Figure 13.14.

To put the information in a functional context, networks were fully integrated into the pSTIING database. Maps were displayed in dynamic graphical representations in both compressed and scalable vector graphics (SVG) formats. With a suitable SVG viewer such as Adobe SVG viewers, users can zoom in and out, search or move (pan) a network map around within the viewing screen. Nodes were coded to represent interacting components (i.e., dark gray nodes denote up-regulated gene products. mid-gray nodes denote down-regulated gene products and light-gray nodes denote interaction/association partners of induced gene products). Constructed regulatory networks comprise both transcriptional regulatory modules and molecular interaction cascades organized within signal transduction modules. The incorporation of transcriptional regulatory associations with molecular interaction information was essential in linking distinct signaling modules and in the creation of functional regulatory systems. Edges were used to represent interaction or transcriptional regulatory relationships between components (i.e., solid edges denote

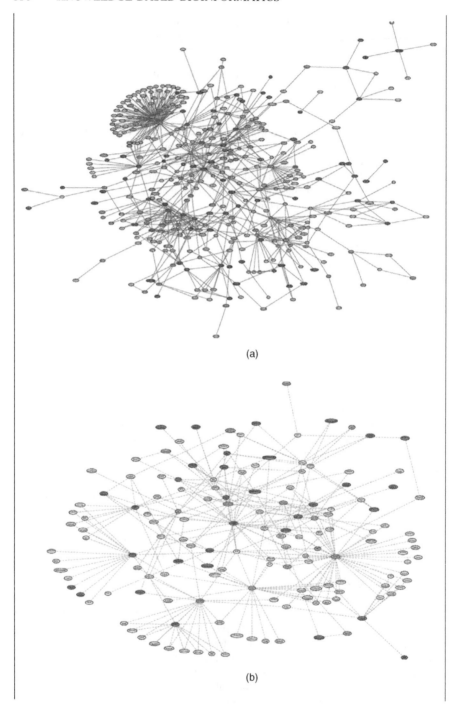

(a)

(b)

molecular interactions and dashed edges denote transcriptional associations). In signaling modules, these edges may be directed, denoting stimulatory or inhibitory interactions, or undirected, representing molecular interactions that are neither stimulatory nor inhibitory or where their stimulatory or inhibitory status is not known. The representation of a transcriptional association was abstracted to indicate the relationship between two proteins, one of which was a transcription factor that *trans*-activates the expression of the gene encoding the other protein. The networks are supported by all the functionalities available in pSTIING. For example, it is possible by clicking on a node to display only the query node in question and its immediate interaction partners. This feature is useful for simplifying network maps and to obtain more information about a particular node. The information can be displayed in a separate window in the form of protein name, synonyms, primary and secondary accession numbers, UniProt, gene symbol, Entrez Gene (Locus Link), RefSeq (nucleotide), OMIM, protein domains, homolog, tissue expression and localization, protein function, gene ontology, pathways, molecular interactions and interaction partners. Further, interaction maps can also be progressively expanded outwards to display the next interaction neighborhood or subsequent levels beyond, thus making it possible to 'grow' networks or extend signaling pathway modules in a desired direction beyond the original interaction map. It is also possible to query interaction information across species by using orthology groupings from Clusters of Orthologous Groups (COG/KOG) or UniRef 50/90 or Sequence similarity (BLASTP). Finally, pSTIING can provide the facility of linking other gene expression and proteomic experiments to interaction and regulatory networks via CLADIST, a clustering tool associated with pSTIING. This feature allows the contextual projection of co-expression patterns onto prior network information, facilitating the assembly of protein interaction and transcriptional regulatory networks into functional modules which facilitate direct comparisons between experiments. See Figure 13.15 for an illustration of some of these functions. For more information on the pSTIING database, the reader is advised to consult the paper by Ng *et al.* (2006b).

Results

Clozapine significantly changed gene expression in pathways related to several neuroprotective mechanisms. In addition, a significant number of genes implicated genetically or linked to pathways functionally related to schizophrenia were changed by clozapine. These included the glutamate receptor genes, retinoid receptor genes and microtubular associated proteins. Pathways involving carbohydrate metabolism and food intake regulation were

---

Figure 13.13   Signaling transduction and transcriptional networks influenced by clozapine in the mouse brain. Network was generated by projecting a set of differentially expressed gene products by clozapine versus control and associated with significantly overrepresented signaling pathways onto protein interaction and transcriptional regulatory information in the integrative pSTIING database. (a) Network is delineated to show just physical protein interactions alone. (b) Network showing transcriptional associations alone. Induced gene products are represented by the darkest nodes, while down-regulated gene products are denoted by mid-gray nodes. Solid lines represent physical interactions, while dashed lines denote transcriptional associations.

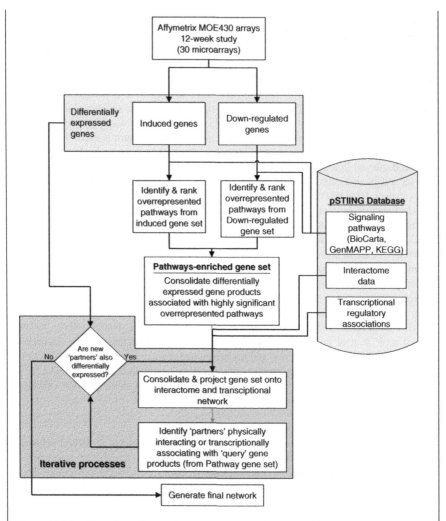

Figure 13.14   Flowchart illustrating methods used for network construction to identify functional classes from clozapine gene expression profile in the mouse brain (Rizig et al., 2009; unpublished data).

particularly affected by clozapine, which may help to explain the relationship between clozapine and diabetes. Graphical networks showed substantial interactions between pathways related to both the therapeutic and toxic effects of each drug. Modeling the interactions in the form of regulatory networks of transcription factors helped in the identification of 'cross points' or hubs. These points within the network may be investigated in the future to potentially modify drug responses.

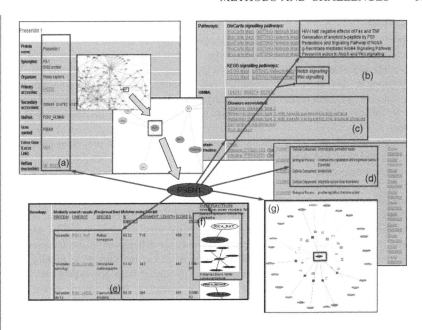

Figure 13.15   Illustration of functional information which can be recruited by the integration of the networks into the pSTIING database. Clicking an interaction node in the network, for example, provides more information about a particular protein in a separate window. In box (a) are the protein name, synonyms, primary and secondary accession numbers and protein function; these can be hyperlinked to several databases including UniProt, Swiss-Prot, Entrez Gene (Locus Link), RefSeq, OMIM, Pfam, InterPro, AmiGo, and iHop. Boxes (b), (c) and (d) provide expression, localization and homology information, pathways lists, and interaction partners. The viewer is provided with the facility to select from the list of interactors to refine the interaction map or to extend the network to the next level of protein connections. An interaction node provides interaction details and experimental evidence supporting that interaction, as shown in box (e) and (f). Pathways can be extended in a desired direction by displaying subsequent interaction neighborhood levels centered on interactors selected by users. It is possible to progressively expand outwards to display the next interaction neighborhood or subsequent level beyond, as shown in box (g). Thus it makes it possible to grow networks or extend signaling pathway modules in a desired direction.

Example 2

Integrating weighted gene co-expression network analysis (WGCNA) with genotyping data to identify weight-related functional modules and obesity driving genes in the mouse liver (Fuller *et al.*, 2007).

Background

Weighted gene co-expression network analysis (WGCNA) can be used to identify functionally relevant modules to a particular phenotype. Combining genetic markers and WGCNA can provide valuable information for prioritizing genes within a functional module. The investigators in this study have combined two different approaches of WGCNA (i.e., single-network and differential network analyses) with genetic markers (SNP data) in order to identify genes driving weight control in two genetically heterogeneous stains of mice.

Single-network analysis describes the structure and topological properties (pathways (modules) and their key drivers (e.g., hub genes)) of a single data set. In this approach, all samples, irrespective of their clinical trait, are used for the network and module construction. In contrast, differential network analysis aims to identify genes that are both differentially expressed and differentially connected between diverse data sets.

Since module genes tend to be highly connected in co-expression networks, screening for differentially connected genes is related to studying the preservation of modules between different networks. This approach can uncover differences in the modules and connectivity between phenotypically distinct groups, and at the same time highlight the conserved controlling modules. Combining both approaches with genotyping and QTLs data provides a systematic approach for linking the phenotype to its causing genes.

Aims

- To identify and characterize functionally interesting modules related to weight regulation in the mouse liver.

- To combine gene expression and genotyping data to identify expression quantitative trait loci (eQTLs) that perturb these modules.

- To report the genetic drivers for the weight-related identified modules.

Materials and methods

Gene expression data were collected from liver tissue of the female mice of two F2 crosses. The first F2 data set (BXH) is inter-cross between inbred strains C3H/HeJ and C57BL/6J. The second F2 data set (BXD) is inter-cross of two inbred strains C57BL/6J and DBA/2J. BXH mice are ApoE null (ApoE $-/-$) and thus hyperlipidemic, whereas BXD mice are wild type (ApoE $+/+$). BXH mice were fed a high-fat diet, and BXD mice were fed a high-fat, high-cholesterol atherogenic diet. Body weight and related clinical traits were measured in both sets of mice.

Two distinct network analysis approaches were used:

(1) Single-network analysis: in this approach data from all mice in each F2 inter-cross were used to identify trait-related modules and eQTLs as follows:

- A weighted gene co-expression network was constructed from genome-wide transcription data.

- Modules were identified using hierarchical clustering, and module centrality measures (intra-modular connectivity) were calculated. Connectivity of a particular gene (also known as degree) is defined as the sum of connection strengths with the other network genes. In co-expression networks, the connectivity measures how correlated a gene is with all other network genes.

- Network modules were then analyzed for biological significance.

- Genetic loci driving functionally relevant modules within the network were identified.

- Trait-related eQTLs are used to prioritize genes within functionally significant modules. Figure 13.16 illustrates the steps used in the single-network analysis.

(2) Differential network analysis: gene expression data from 30 mice at both extremes of the weight spectrum in the BXH dataset were selected. Two different networks were constructed: the first network using the 30 leanest mice and the second network using the 30 heaviest mice. The two networks were compared to identify the non-preserved modules, the differentially expressed genes, and the differentially connected genes.

Results

(1) Single network analysis

Hierarchical clustering of weighted gene co-expression networks constructed from the liver gene expression data of both strains revealed 12 functionally relevant modules. Modules with highest significant score were related to mouse weight and abdominal fat pad mass. The identified modules were highly conserved between BXH and BXD strains. A significant relationship was identified between the weight module expression and a single nucleotide polymorphism (SNP) marker on chromosome 19 (SNP19). This SNP was reported to be associated with weight gain in previous studies.

To determine the genes that mediate between this eQTL and body weight (clinical trait), the author ranked gene expression values based on their correlations with SNP19 and the weight trait. Genes were selected if they fulfilled all three criteria. First: their expression values were highly associated with the body weight; second: they were highly associated with a body weight-related eQTL (SNP19); and third: the genes have high intra-modular connectivity. Nine genes within the weight-related expression module fulfilled these criteria, including *Fsp27*, which encodes a pro-apoptotic protein, and *Gpld1* which encodes glycosyl phosphatidyl inositol-specific phospholipase D1. Searching the Mouse Genomics Informatics gene ontology database (www.informatics.jax.org/) and existing literature revealed that these genes have a potential functional relationship to body weight control.

(2) Differential networks analysis

To identify both differentially connected and differentially expressed genes between the lean and the obese mice, the author plotted the difference in connectivity between lean and obese mice versus the absolute value of the t-test statistic for each gene. This approach gave a visual demonstration of how differences in connectivity can be related to differences in expression between the two networks. Four sectors were identified as highly different between the two networks (i.e., lean and obese mice networks).The two sectors which contain most of the differentially connected genes (i.e., highly connected in Network 1 and poorly connected in Network 2) were analyzed for functional enrichment using the DAVID database (http://david.abcc.ncifcrf.gov/). These genes were found to be highly enriched for extracellular and cell–cell interaction compounds; notably 12 epidermal growth factor (EGF) or EGF-related factors.

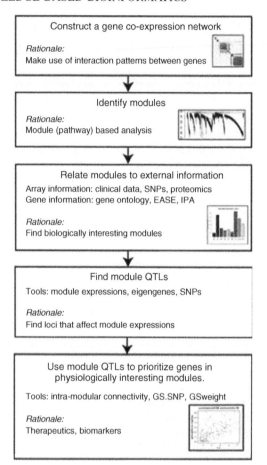

Figure 13.16   Overview of steps used by Fuller et al. (2007) for the weighted gene co-expression network analysis (single-network approach). (1) Genome-wide transcription data were used to build a weighted gene co-expression network. (2) Modules were identified and module centrality measures (intra-modular connectivity) were calculated. In co-expression networks, the intra-modular connectivity measures the strength of correlation between a gene and all other co-expressed network genes. (3) Network modules were analyzed for biological significance. (4) Genetic loci driving functionally relevant modules within the network were identified. (5) Trait-related mQTLs were used to prioritize genes within physiologically significant modules using information extracted from the strength of connection between network genes within the network (i.e., intra-modular connectivity) and gene significance (GS). Gene significance (GS) measures how correlated a gene expression is with a clinical trait. Mouse body weight could be used to define a physiologic trait-based gene significance measure. Likewise, SNPs can be used to define SNP-based gene significance measures. The higher this value, the more significant a gene is. Courtesy of Fuller et al. (2007).

The author concluded that: integrating weighted gene co-expression network analysis with genotype data has helped in identifying several functional modules in the mouse liver. These modules are roughly preserved between the BXD and BXH mice. Connected hub genes with the most significant intra-modular connectivity were found to have high correlation with weight control. Combining gene expression results of these hub genes with the genotyping data helped in identifying genes with potential causal relationships to the weight gain/loss phenotype. The differential network analysis has shown that genes that are differentially connected may or may not be differentially expressed. Changes in connectivity may correspond to large-scale 'rewiring' in response to environmental changes and physiologic perturbations. The list of module hub genes and the differential connectivity patterns have potential uses for diagnostic and therapeutic targets.

13.9 Challenges and future directions

Researchers in life and biomedical sciences have few options available for them to analyze the vast amount of data generated to elucidate how the cell's numerous fundamental components interact to give rise to complex phenotypes. Functional module identification and networks provide one of the few frameworks which systematically and simultaneously take account of all of the fundamental components. Statistical inferences from networks will be the limiting factor as the field of systems biology matures.

Understanding a particular network state that drives a particular disease (or other complex phenotypes that define living systems) will require not only knowledge of DNA and environmental variation and the changes these variation components induce in the network, but also information on the previous states of the network that led to the current state. While this more comprehensive reconstruction of biological networks is still outside the scope of what is presently doable, the types of approaches reviewed here represent solid first steps toward this ultimate goal.

Even though the number of networks that can be reconstructed from the fundamental components of living systems is truly daunting, as work progresses in this area, we will learn the rules that necessarily constrain the possible ranges of molecular interactions, and, as a result, we will begin to capture the more conserved network motifs that form the framework upon which all other interactions are based.

The complexity revealed by a systems biology-motivated approach to elucidating complex phenotypes such as multi-factorial illnesses or responses to treatment should be embraced, given the potential to develop a better understanding of the true diversity of diseases and the constellation of genes that need to be targeted to effectively treat these diseases.

As a result, physicians will have more options to effectively provide a diagnosis (identification of danger signs), prognosis (forecast of disease progression), and treatment plans (selection of an appropriate course among multiple alternatives) for their patients. At the same time, these options can be tailored to the specific genetic- and health-profile of each individual patient (personalized medicine) (Bolouri, 2008) (Ikediobi et al., 2009).

Coupling between molecular biology research tools and clinical IT systems in a systematic way is imperative for personalized medicine. Semantic ontology tools (for the association of meaning with data) and software for personal genomics are currently deficient. These tools are essential for physicians to interpret genetic, genomic, pharmacological, and medical data in order to make educated decisions about management options. We expect that the future years will witness great efforts from both academic and industrial fields to build integrative databases for phenotypes and personalized genomics in order to eliminate the gap between the clinical and the research arena.

13.10   References

Ashburner, M., Ball, C.A., Blake, J.A., et al. (2000) Gene ontology: tool for the unification of biology. The Gene Ontology Consortium. Nat. Genet., 25(1), 25–9.

Barabasi, A.L. and Oltvai, Z.N. (2004) Network biology: understanding the cell's functional organization. Nat. Rev. Genet., 5(2), 101–13.

Bekal, S., Brousseau, R., Masson, L., et al. (2003) Rapid identification of Escherichia coli pathotypes by virulence gene detection with DNA microarrays. J. Clin. Microbiol., 41(5), 2113–25.

Bolouri, H. (2008) Computational challenges of personal genomics. Curr. Genomics, 9(2), 80–7.

Botstein, D. and Risch, N. (2003) Discovering genotypes underlying human phenotypes: past successes for Mendelian disease, future approaches for complex disease. Nat. Genet., 33(Suppl), 228–37.

Cavalieri, D. and De Filippo, C. (2005) Bioinformatic methods for integrating whole-genome expression results into cellular networks. Drug Discov. Today 10(10), 727–34.

Cayley, A. (1857) On the theory of analytic forms called trees. Philos. Mag., 13, 1930.

Cervino, A.C., Li, G., Edwards, S., et al. (2005) Integrating QTL and high-density SNP analyses in mice to identify Insig2 as a susceptibility gene for plasma cholesterol levels. Genomics, 86(5), 505–17.

Christensen, C., Thakar, J., and Albert, R. (2007) Systems-level insights into cellular regulation: inferring, analysing, and modelling intracellular networks. IET Syst Biol, 1(2), 61–77.

Correa Geyer, F. and Reis-Filho, J.S. (2009) Microarray-based gene expression profiling as a clinical tool for breast cancer management: are we there yet? Int. J. Surg. Pathol., 17(4), 285–302.

Davidson, E.H., McClay, D.R., and Hood, L. (2003) Regulatory gene networks and the properties of the developmental process. Proc. Natl. Acad. Sci. U.S.A., 100(4), 1475–80.

Davidov, E.J., Holland, J.M, Marple, E.W., and Naylor, S. (2003) Advancing drug discovery through systems biology. DDT, 8(4), 175–83.

Duncan, L.M. (2009) The classification of cutaneous melanoma. Hematol. Oncol. Clin. North Am., 23(3), 501–13, ix.

Erdos, P. and Renyi, R. (1959) On random graphs. *Pub. Mathem.*, **6**, 290–7.

Euler, L. (1736) Solutio problematis ad geometriam situs pertinensis. *Commentarii Academiae Scientiarum Imperialis Petropolitanae*, **8**, 128–40.

Frankfurt, O., Licht, J.D., and Tallman, M.S. (2007) Molecular characterization of acute myeloid leukemia and its impact on treatment. *Curr Opin Oncol*, **19**(6), 635–49.

Fuller, T.F., Ghazalpour, A., Aten, J.E., *et al.* (2007) Weighted gene coexpression network analysis strategies applied to mouse weight. *Mamm. Genome*, **18**(6-7), 463–72.

Gall, A., Hoffmann, B., Harder, T., *et al.* (2009) Rapid haemagglutinin subtyping and pathotyping of avian influenza viruses by a DNA microarray. *J. Virol. Methods*, **160**(1-2), 200–5.

Ghazalpour, A., Doss, S., Zhang, B., *et al.* (2006) Integrating genetic and network analysis to characterize genes related to mouse weight. *PLoS Genet.*, **2**(8), e130.

Gilad, Y., Rifkin, S.A., and Pritchard, J.K. (2008) Revealing the architecture of gene regulation: the promise of eQTL studies. *Trends Genet.*, **24**(8), 408–15.

Grigorov, M.G. (2005) Global properties of biological networks. *Drug Discov. Today*, **10**(5), 365–72.

Han, K. and Ju, B.H. (2003) A fast layout algorithm for protein interaction networks. *Bioinformatics*, **19**(15), 1882–8.

Ikediobi, O.N., Shin, J., Nussbaum, R.L., *et al.* (2009) Addressing the challenges of the clinical application of pharmacogenetic testing. *Clin. Pharmacol. Ther.*, **86**(1), 28–31.

Iqbal, J., Liu, Z., Deffenbacher, K., and Chan, W.C. (2009) Gene expression profiling in lymphoma diagnosis and management. *Best Pract Res Clin Haematol*, **22**(2), 191–210.

Jain, K.K. (2009) Personalized clinical laboratory diagnostics. *Adv Clin Chem*, **47**, 95–119.

John, T., Liu, G., and Tsao, M.S. (2009) Overview of molecular testing in non-small-cell lung cancer: mutational analysis, gene copy number, protein expression and other biomarkers of EGFR for the prediction of response to tyrosine kinase inhibitors. *Oncogene*, **28**(Suppl 1), S14–23.

Katagiri, F. (2003) Attacking complex problems with the power of systems biology. *Plant Physiol.*, **132**(2), 417–9.

Kerr, G., Ruskin, H.J., Crane, M., and Doolan, P. (2008) Techniques for clustering gene expression data. *Comput. Biol. Med.*, **38**(3), 283–93.

Klaassen, C.H., Prinsen, C.F., de Valk, H.A., *et al.* (2004) DNA microarray format for detection and subtyping of human papillomavirus. *J. Clin. Microbiol.*, **42**(5), 2152–60.

Klamt, S. and von Kamp, A. (2009) Computing paths and cycles in biological interaction graphs. *BMC Bioinformatics*, **10**, 181.

Lum, P.Y., Chen, Y., Zhu, J., *et al.* (2006) Elucidating the murine brain transcriptional network in a segregating mouse population to identify core functional modules for obesity and diabetes. *J. Neurochem.*, **97**(Suppl 1), 50–62.

Miller, J.A., Oldham, M.C., and Geschwind, D.H. (2008) A systems level analysis of transcriptional changes in Alzheimer's disease and normal aging. *J. Neurosci.*, **28**(6), 1410–20.

Ng, A., Bursteinas, B., Gao, Q., *et al.* (2006a) Resources for integrative systems biology: from data through databases to networks and dynamic system models. *Brief. Bioinform*, **7**(4), 318–30.

Ng, A., Bursteinas, B., Gao, Q., *et al.* (2006b) pSTIING: a 'systems' approach towards integrating signalling pathways, interaction and transcriptional regulatory networks in inflammation and cancer. *Nucleic Acids Res.*, **34**(Database issue), D527–34.

Saris, C.G., Horvath, S., van Vught, P.W., *et al.* (2009) Weighted gene co-expression network analysis of the peripheral blood from amyotrophic lateral sclerosis patients. *BMC Genomics*, **10**, 405.

Schork, N.J. (1997) Genetics of complex disease: approaches, problems, and solutions. *Am. J. Respir. Crit. Care Med.*, **156**(4 Pt 2), S103–9.

Sieberts, S.K. and Schadt, E.E. (2007) Moving toward a system genetics view of disease. *Mamm. Genome*, **18**(6-7), 389–401.

Tegner, J., Skogsberg, J., and Bjorkegren, J. (2007) Thematic review series: systems biology approaches to metabolic and cardiovascular disorders. Multi-organ whole-genome measurements and reverse engineering to uncover gene networks underlying complex traits. *J. Lipid Res.*, **48**(2), 267–77.

Villoslada, P., Steinman, L., and Baranzini, S.E. (2009) Systems biology and its application to the understanding of neurological diseases. *Ann. Neurol.*, **65**(2), 124–39.

Zhang, B. and Horvath, S. (2005) A general framework for weighted gene co-expression network analysis. *Stat Appl Genet Mol Biol*, **4**, Article17.

Zhao, Y. and Karypis, G. (2005) Data clustering in life sciences. *Mol. Biotechnol.*, **31**(1), 55–80.

Zhu, J., Wiener, M.C., Zhang, C., *et al.* (2007b) Increasing the power to detect causal associations by combining genotypic and expression data in segregating populations. *PLoS Comput. Biol.*, **3**(4), e69.

Zhu, J., Zhang, B., and Schadt, E.E. (2008) A systems biology approach to drug discovery. *Adv. Genet.*, **60**, 603–35.

Zhu, X., Gerstein, M., and Snyder, M. (2007a) Getting connected: analysis and principles of biological networks. *Genes Dev.*, **21**(9), 1010–24.

Trends and conclusion

The successful sequencing of whole genomes has offered tremendous opportunities to understand various life processes/biological phenomena. Modern day research offers a remarkable set of challenges, such as computationally predicting the development of a single cell into an adult, synchronized functioning of the cells as a complete organism, reconstruction of an organism and many more. The answer to all these may not lie within the genome sequence itself but in the network of molecular interactions, thus emphasizing the importance of maintaining the database, not only of the building blocks but also their interactions for methodical functional analysis as demonstrated in the book.

Also there is a great potential for designing the drugs efficiently and cost effectively based on the knowledge of the target. Current bioinformatics has an essential role to play in interpreting genomic, transcriptomic, and proteomic data generated by high-throughput experimental technologies, as well as interpreting information gathered from research in the field of traditional biology and medicine. Sequence-based methods of analyzing individual genes or proteins, and annotation of the whole genome or transcriptome, may now lead towards personalized medicine.

This book discussed various aspects of knowledge-driven and data-analysis approaches for acquisition, maintenance, integration, and interpretation of biomedical data in Section 1 and 2. Section 3 covered various methods for exploring genomes of a wide variety of species. Section 4 illustrated biomolecular relationships and networks, their analysis and applications. The focus of this book has been on the integration and application of various methods for analysis and interpretation of the biomedical data, which can lead to break-through discovery for understanding systemic functional behavior of the cell and the organism.

To achieve a high level of coordination and management of data, participation of researchers working with molecular data and scientific literature is vital. This is apparent from the backgrounds of the contributing authors of each chapter in the book. To help in correspondence, email is included for each chapter. Also the correspondence to the editors can be addressed to Gil Alterovitz at gil_alterovitz@hms.harvard.edu (or ga@alum.mit.edu). To keep pace with the

Knowledge-Based Bioinformatics: From Analysis to Interpretation   Edited by Gil Alterovitz and Marco Ramoni
© 2010 John Wiley & Sons, Ltd

developments in this rapidly evolving field, readers can also participate in subject talks apart from reading papers. Best wishes to the readers in elucidating the structural and functional challenges.

Sincerely,
Gil Alterovitz, Ph.D.
Marco Ramoni, Ph.D.

Index

Printed and bound by CPI Group (UK) Ltd, Croydon, CR0 4YY

27/10/2024

14580207-0004